Consulting Editor

George A. Anastassiou
Department of Mathematical Sciences
University of Memphis

Radu Păltănea

Approximation Theory Using Positive Linear Operators

Birkhäuser
Boston • Basel • Berlin

Radu Păltănea
Transilvania University
Department of Mathematics
Braşov 500 036
Romania

AMS Subject Classifications: Primary: 41A36; Secondary: 41A35, 41A25, 41A44, 41A10, 41A28, 41A65, 41A46, 26B25, 32A70, 46A32, 47Axx, 47A67, 47A63, 47B38, 47A65, 39B62, 26A15, 26A51, 26B25, 26D15, 28Axx, 05A19, 05A20

Library of Congress Cataloging-in-Publication Data
Paltanea, Radu, 1953-
    Approximation theory using positive linear operators / Radu Paltanea.
        p. cm.
    Includes bibliographical references and index.
    ISBN-13:978-0-8176-4350-8   e-ISBN-13:978-1-4612-2058-9
    DOI: 10.1007/978-1-4612-2058-9

    1. Approximation theory. 2. Linear operators. I. Title.

QA221.P25 2004
511'.4–dc22   2004054852

ISBN-13:978-0-8176-4350-8   Printed on acid-free paper.

©2004 Birkhäuser Boston   *Birkhäuser* ℬ®

9 8 7 6 5 4 3 2 1   SPIN 10996522

Birkhäuser is a part of *Springer Science+Business Media*
www.birkhauser.com

Preface

We deal in this work with quantitative results in the pointwise approximation of functions by positive linear functionals and operators.

One of the main objectives is to obtain estimates for the degree of approximation in terms of various types of second order moduli of continuity. In the category of second order moduli we include both classical and newly introduced moduli. Particular attention is paid to optimizing the constants appearing in such estimates.

In the last decades, the study of linear positive operators with the aid of second order moduli was intensive, thanks to their refinements in characterization of the smoothness of functions. As promoters of this direction of research we mention Yu. Brudnyi, G. Freud, and J. Petree. Our approach is more akin to the approach taken by H. Gonska, who obtained the first general estimates for second order moduli with precise constants and with free parameters. Two new methods will be presented. The first one, based on decomposition of functionals and the use of moments, can be applied to diverse types of moduli and leads to simple estimates. The second method gives sufficient conditions for obtaining absolute optimal constants. The benefits of these more direct methods, compared with the known method based on K-functionals, consist in the improvement and even the optimization of the constants, and in the generalization of the framework.

Applications are given for the classical Bernstein operators and for two classes of Bernstein-type operators. So, for the Bernstein operators, we prove the analogous optimality result of Sikkema for the second order modulus. The first class of Bernstein-type operators on which we focus our attention is a class of certain generalized Brass operators that admit a construction based on discrete convolution. The second class consists of Durrmeyer integral operators with generalized weights. The Durrmeyer operators, introduced also by A. Lupaş, were intensively studied by M. Dierrennic and many other authors.

In the last part we consider the problem of approximation of vector functions by using certain generalizations of the positive and convex operators. The quantitative results are similar to those given in the scalar case.

Acknowledgments. The author is deepley indebted to Professor Heinz H. Gonska and to Professor George Anastassiou, for their support and for promoting the work. Also we are grateful to the Birkhäuser publishing house for accepting the publication of the book and also to Ann Kostant, Regina Gorenshteyn and Elizabeth Loew for their kind assistance.

Radu Păltănea

Contents

Approximation Theory
Using Positive
Linear Operators

1

Introduction

1.1 Operators and functionals. Moduli of continuity

The most constructive proofs of the Weierstrass theorem concerning the approximation of continuous functions on a compact interval by polynomials use some sequences of linear positive operators. So do the classical operators of Gauss–Weierstrass, Landau, Vallée-Poussin, Jackson, and Bernstein. We begin by constructing and studying a large class of such sequences of approximation operators.

We mention that the more usual tools of Analysis, like the Taylor polynomials, the Fourier series related to an inner product, or the Lagrange interpolations, are not appropriate to the problem of the uniform approximation of continuous functions.

An operator L defined on a linear space of functions, V, is called **linear** if

$$L(\alpha f + \beta g) = \alpha L(f) + \beta L(g), \text{ for all } f, g \in V, \ \alpha, \beta \in \mathbb{R}, \tag{1.1}$$

and is called **positive**, if

$$L(f) \geq 0, \text{ for all } f \in V, \ f \geq 0. \tag{1.2}$$

These properties are not necessary in order for a sequence of operators to give a uniform approximation to any continuous function. So, the sequence of the operators that assign to each continuous function its best polynomial approximation, of degree n, with respect to the sup-norm is not linear and is not positive. But these properties simplify study to a considerable degree. On the other hand, the value of the approximation may, though reduced somewhat in degree of goodness, consist of some other qualities. We mention for instance the preservation of some properties of the functions, the possibility of approximation of derivatives, or the possibility of characterization of certain classes of functions by the order of approximation that can be reached.

We specify now some notation that will be used in the whole paper. Particular notation will be introduced along the way.

Denote $\mathbb{N} := \{1, 2, \dots\}$. For $n \in \mathbb{N} \cup \{0\}$, Π_n is the space of algebraic polynomials of degree at most n. For any real number a, denote by $[a]$ the integer part of a, by $]a[$

the greatest integer that is less than a and by $\{a\}$ the fractional part of a. For any set $A \subset \mathbb{R}$ we consider $\sup A = \infty$ if A is not bounded above. If $b \le a$, we consider $(a, b] = \emptyset$ and $[a, b) = \emptyset$. The restriction of a function φ to a subset will also be denoted by φ.

We make the following conventions: $0^0 = 1$; $0 \cdot \infty = 0$; if $r > s$, then $\sum_{i=r}^{s} a_i$ is zero and $\prod_{i=r}^{s} a_i$ is the identity; if the integers n and k do not satisfy the condition $0 \le k \le n$, then $\binom{n}{k} = 0$.

Let I be an interval of the real axis. Consider the following spaces of real-valued functions on I:

$\mathcal{F}(I)$ — the space of all real-valued functions;

$\mathcal{F}_b(I)$ — the space of locally bounded functions ;

$B(I)$ — the space of bounded functions;

$C(I)$ — the space of continuous functions;

$C_c(I)$ — the spaces of continuous functions with compact support;

$\mathcal{D}(I)$ — the space of differentiable functions;   (1.3)

$C^k(I)$ — the space of k-continuously differentiable functions, $k \ge 0$;

$\mathcal{L}_\mu(I)$ — the space of μ -integrable functions, when μ is a positive measure.

In the case $I = [a, b]$ we write simply $\mathcal{F}[a, b]$ instead of $\mathcal{F}([a, b])$ and similarly for the other mentioned spaces. (For the spaces of vector-valued functions, analogous notation will be introduced in Chapter 6.)

For any function $f \in \mathcal{F}(I)$ denote by $\|f\|$ or $\|f\|_I$, the supremum

$$\|f\| := \sup_{x \in I} |f(x)|. \tag{1.4}$$

The supremum in (1.4), is the sup-norm of the space $B(I)$.

Denote by e_j, $j \in \mathbb{N} \cup \{0\}$, the monomial functions $e_j(x) = x^j$, $x \in I$. For any $A \subset I$ denote by $\chi_A \in \mathcal{F}(I)$ the characteristic function of the subset A.

If $f \in \mathcal{F}(I)$ and $x_1, \dots, x_k$ are distinct points of I, then denote by

$$[f; x_1, \dots, x_k] := \sum_{i=1}^{k} \left(\prod_{1 \le j \le k, \ j \ne i} (x_i - x_j)^{-1} \right) f(x_i), \tag{1.5}$$

the divided difference of the function f on the points $x_1, \dots, x_k$. Also, the finite difference of order $k \ge 1$ of the function f is defined by

$$\Delta_h^k f(x) := \sum_{j=0}^{k} \binom{k}{j} (-1)^{k-j} f(x + jh), \quad x \in I, \ x + kh \in I, \ h > 0. \tag{1.6}$$

Definition 1.1.1. [117] The function $f \in \mathcal{F}(I)$ is named **convex of order** $k \ge -1$, if we have $[f; x_1, \dots, x_{k+2}] \ge 0$, for any distinct points $x_1, \dots, x_{k+2}$ of I.

From this definition it follows that a function is convex of order -1 iff it is positive, it is convex of order 0 iff it is increasing and it is convex of order 1 iff it is usual convex.

Definition 1.1.2. A linear operator $L : V \to \mathcal{F}(I)$, where V is a linear subspace of $\mathcal{F}(I)$, is called **convex of order** k, $k \geq -1$, if for any function f that is convex of order k, it follows that the function $L(f)$ is convex of order k. Particularly, if L is convex of order -1, then L is **positive**.

For any linear operator $L : V \to \mathcal{F}(I)$, where V is a linear subspace of $\mathcal{F}(I)$, we write $L(f, x) := (L(f))(x)$, for $f \in V$, $x \in I$.

We mention an important particular case of linear positive operators. Let $\{\mu_x, \ x \in I\}$ be a family of positive Borel measures on I and for any $x \in I$, let $V_x := \mathcal{L}_{\mu_x}(I)$. Let $V := \bigcap_{x \in I} V_x$. We can consider a linear positive operator $L : V \to \mathcal{F}(I)$, defined by

$$L(f, x) := \int_I f(t)\, d\mu_x(t), \ f \in V, \ x \in I. \tag{1.7}$$

In the category of the operators of the form (1.7), two types are more usual: the discrete operators and the integral operators. The discrete operators are of the form

$$L(f, x) = \sum_{i=1}^{m} f(\xi_i)\Psi_i(x), \ f \in \mathcal{F}(I), \ x \in I, \tag{1.8}$$

where $\xi_i \in I$, and $\Psi_i \in \mathcal{F}(I)$, $\Psi_i \geq 0$, $1 \leq i \leq m$.

The integral operators are of the form

$$L(f, x) = \int_I f(t)K(t, x)\, dt, \ f \in C(I), \ x \in I, \tag{1.9}$$

where $K : I \times I \to \mathbb{R}$ is a positive continuous function.

From another point of view we can consider the following particular types of linear positive operators: 1) the operators that preserve the constant functions, i.e., $L(e_0) = e_0$; and 2) the operators that preserve the linear functions, i.e., $L(e_j) = e_j$, $j = 0, 1$, or equivalently $L(e_0) = e_0$ and $L(e_1 - xe_0, x) = 0$, for $x \in I$. This hierarchy cannot be continued, if $V = C(I)$, excepting the trivial case; see Corollary 1.1.2 below. Note that the study of linear positive operators can be simply reduced to the operators that preserve constants. Indeed, we can take $L_1 := \frac{1}{L(e_0)}L$. However, a simple reduction of a linear positive operator to an operator that preserves linear functions is not possible. The property of reproducing linear functions plays a crucial role in some problems of approximation.

$\star$

The study of approximation by linear positive operators is closely connected to the study of approximation by linear positive functionals. If V is a linear subspace of $\mathcal{F}(I)$, then a linear application $F : V \to \mathbb{R}$ is called positive if $F(f) \geq 0$ for any $f \in V$, $f \geq 0$.

Remark 1.1.1. Any linear positive operator $L : V \to \mathcal{F}(I)$ can be expressed as a collection of linear positive functionals $(F_x)_{x \in I}$. Indeed, for any fixed point $x \in I$ we obtain a linear positive functional $F_x : V \to \mathbb{R}$, defined by $F_x(f) := L(f, x)$, $f \in V$. Conversely, if we have a collection of linear positive functionals $(F_x)_{x \in I}$, $F_x : V_x \to \mathbb{R}$, where V_x are linear subspaces of $\mathcal{F}(I)$, then we can construct a linear positive operator $L : V \to \mathcal{F}(I)$, where $V = \bigcap_{x \in I} V_x$, by $L(f, x) := F_x(f)$, $f \in V$.

From Remark 1.1.1 it follows, particularly, that the study of operators of the form (1.7) is reduced to the study of functionals of the form

$$F(f) := \int_I f(t)\, d\mu(t), \ f \in \mathcal{L}_\mu(I), \tag{1.10}$$

where μ is a positive Borel measure. F is called the functional **induced by the measure** μ. We have $C(I) \subset \mathcal{L}_\mu(I)$, if I is a finite interval. The simplest functionals of this type are of the form

$$F(f) := \mu_1 f(a) + \mu_2 f(b), \quad f \in C(I), \tag{1.11}$$

where $a, b \in I, a \le b, \mu_1 \ge 0, \ \mu_2 \ge 0$.

We mention some basic properties of the linear positive functionals. Let V be a linear subspace of $\mathcal{F}(I)$ and let a linear positive functional $F : V \to \mathbb{R}$. We have

$$F(f) \le F(g), \text{ for } f, g \in V, \ f \le g, \tag{1.12}$$

$$|F(f)| \le F(|f|), \text{ for } f \in V, |f| \in V. \tag{1.13}$$

Moreover, if the functional F is induced by a positive Borel measure, then

$$|F(fg)| \le \sqrt{F(f^2)}\sqrt{F(g^2)}, \text{ for } f, g \in \mathcal{L}_\mu^2(I), \tag{1.14}$$

where $\mathcal{L}_\mu^2(I)$ is the space of square μ-integrable functions. For the linear positive operators one obtains similar properties as above.

Remark 1.1.2. Because the most important linear positive approximation operators are given on the space of continuous functions, we are led to consider linear positive functionals on this space. From the Riesz representation theorem, for any linear positive functional $F : C_c(I) \to \mathbb{R}$ there exists a positive regular Borel measure μ, such that we have the integral representation: $F(f) = \int_I f\, d\mu$, $f \in C_c(I)$. In other words, the functional F admits an extension to a functional of the form (1.10). But we can see that also any linear positive functional $F : C(I) \to \mathbb{R}$ admits an integral representation. This fact is immediate if I is compact. For arbitrary intervals, we can reason as follows. Let G be the restriction of the functional F on the subspace $C_c(I)$ and let μ be the positive Borel measure associated to μ. Then, formula (1.10) defines a continuation of the functional G to the space $\mathcal{L}_\mu(I)$. We show that for any $g \in C(I)$ we have the representation above. Indeed, let $g \in C(I)$, $g(t) \ge 0$, $(t \in I)$. Consider the sequence $(g_n)_n$, $g_n \in C_c(I)$ given by ; $g_n(t) := g(t)$, for $t \in [-n, n] \cap I$, $g_n(t) := (n + 1 - t)g(t)$, for $t \in (n, n + 1] \cap I$, $g_n(t) := (n + 1 + t)g(t)$, for

$t \in I \cap [-n - 1, -n)$, and $g_n(t) := 0$ for $t \in I \setminus [-n - 1, n + 1]$. Since $\int_I g_n \, d\mu = G(g_n) = F(g_n) \leq F(g)$, by applying the Beppo–Levi theorem we obtain that g is μ-integrable on I and $\int_I g \, d\mu \leq F(g)$. Let us suppose that $\alpha := F(g) - \int_I g \, d\mu > 0$. Then, if we consider the function $h \in C(I)$ given by $h(t) := |t| \cdot g(t)$, $(t \in I)$, and we define the sequence $(h_n)_n$ similar to $(g_n)_n$, we have $F(h) = F(h_n) + F(h - h_n) \geq F(h - h_n) \geq F(n(g - g_n)) \geq n \cdot \alpha$, for all natural numbers n. This implies $F(h) = +\infty$. Contradiction. Hence $F(g) = \int_I g \, d\mu$ for all $g \in C(I)$, $g(t) \geq 0$, $(t \in I)$. Considering the positive part and the negative part of a general function $g \in C(I)$, it follows that g is μ-integrable and $F(g)$ has the integral representation.

Note however that, since the domain of F is $C(I)$, the measure μ is compact supported. So, the fact that we consider an arbitrary interval I is not an effective extension: If the compact interval $J \subset I$ contains the support of μ, then there is a linear positive functional $F_1 : C(J) \to \mathbb{R}$, such that $F(g) = F_1(g|_J)$, $g \in C(I)$.

Denote by δ_x, the Dirac functional $\delta_x(f) = f(x)$, $f \in \mathcal{F}(I)$. An important property of the positive linear functionals induced by a positive Borel measures is given below.

Proposition 1.1.1. *Let a linear positive functional F be induced by a positive Borel measure μ. Suppose that there are $x \in I$ and $\sigma \in \mathcal{L}_\mu(I) \cup C(I)$ with the following properties:*

i) $\mu(\{x\}) > 0$.
ii) $\sigma(x) = 0$ and $\sigma(t) > 0$, for all $t \in I \setminus \{x\}$.
iii) $F(\sigma) = 0$.
Then

$$F(f) = F(e_0)\delta_x.$$

Proof. Let us show that $\mu(I \setminus \{x\}) = 0$. Denote $J_n := \{t \in I \mid 1/n \leq |t - x| \leq n\}$, $n \in \mathbb{N}$. We have

$$0 = F(\sigma) = \int_{I \setminus \{x\}} \sigma(t) \, d\mu(t) = \lim_{n \to \infty} \alpha_n, \quad \text{where } \alpha_n := \int_{J_n} \sigma(t) \, d\mu(t).$$

The sequence $(\alpha_n)_n$ is increasing and has the limit 0. But $\alpha_n \geq 0$. Hence $\alpha_n = 0$, $n \in \mathbb{N}$. Denote $\beta_n = \inf_{t \in J_n} \sigma(t)$. Since σ is continuous, we have $\beta_n > 0$, for all $n \in \mathbb{N}$. Also, $\alpha_n \geq \beta_n \mu(J_n)$. Therefore $\mu(J_n) = 0$, $n \in \mathbb{N}$ and then $\mu(I \setminus \{x\}) = \lim_{n \to \infty} \mu(J_n) = 0$. Consequently, if $f \in \mathcal{L}_\mu(I)$, we have $F(f) = f(x)\mu(\{x\}) = f(x)\mu(I) = f(x)F(e_0)$. $\qquad\square$

Corollary 1.1.1. *Let a functional F induced by a positive Borel measure μ and let $x \in I$. Suppose that one of the following two conditions holds:*

i) $\Pi_2 \subset \mathcal{L}_\mu(I)$ and $F(e_i) = e_i(x)$, $i = 0, 1, 2$ or,
ii) $\Pi_1 \subset \mathcal{L}_\mu(I)$, $F(e_j) = e_j(x)$, $j = 0, 1$ and x is an end point of I.
Then

$$F(f) = f(x), \quad \text{for all } f \in \mathcal{L}_\mu(I). \tag{1.15}$$

Proof. i) We can apply Proposition 1.1.1 with the choice $\sigma = (e_1 - xe_0)^2$.

ii) If, for instance, x is the left end of the interval I, then we have $F(|e_1 - xe_0|) = F(e_1 - xe_0) = 0$. Then we can apply Proposition 1.1.1 for $\sigma = |e_1 - xe_0|$.   □

By taking into account Remark 1.1.2, we have also

Corollary 1.1.2. *Let a positive linear functional $F : C(I) \to \mathbb{R}$. Suppose that one of the following two conditions holds:*

i) $F(e_i) = e_i(x)$, $i = 0, 1, 2$ or,

ii) $F(e_j) = e_j(x)$, $j = 0, 1$ and x is an end point of I.

Then

$$F(f) = f(x), \quad \text{for all } f \in C(I). \tag{1.16}$$

Remark 1.1.3. There are linear positive functionals which cannot be expressed in the form (1.10). We consider for instance $I = [0, 1]$, $V := \{f \in \mathcal{F}(I), \ |\exists \lim_{t \to 1} f(t) \in \mathbb{R}\}$ and $F : V \to \mathbb{R}$, $F(f) := \lim_{t \to 1} f(t)$, $f \in V$. We have $\Pi_1 \subset V$ and $F(e_0) = 1$, $F(e_1) = 1$. Suppose that there is a positive Borel measure μ on I, such that the functional F would be expressed by formula (1.10). Then, by Proposition 1.1.1, with $x = 1$, we must have $F(f) = f(1)$, for all $f \in V$. But this is not true.

Let $F : V \to \mathbb{R}$ be again a linear positive functional, where V is a linear subspace of $\mathcal{F}(I)$. Let also $x \in I$ be a fixed point. The numbers $F((e_1 - xe_0)^j)$, $j \in \mathbb{N} \cup \{0\}$, that are well defined if $\Pi_j \subset V$, are named the **moments** of the functional F, (with respect to x). We sometimes write

$$m_j := |F((e_1 - xe_0)^j)|, \ \ j = 0, 1, \dots . \tag{1.17}$$

★

The degree of approximation by positive linear functionals and operators depends on the smoothness properties of the functions. In the estimates of the degree of approximation, convenient tools for measuring the smoothness of functions are represented by the moduli of continuity of various types. We adopt here the following definition.

Definition 1.1.3. *Let $W \subset \mathcal{F}(I)$ be a linear subspace such that $\Pi_k \subset W$, $k \in \mathbb{N}$. A function $\Omega_k : W \times (0, \infty) \to [0, \infty) \cup \{\infty\}$ is called a **modulus of continuity** of order k on W, or shortly, a **modulus of order** k, if the following conditions are satisfied :*

$$\Omega_k(f, h_1) \leq \Omega_k(f, h_2), \quad f \in W, \quad 0 < h_1 < h_2, \tag{1.18}$$

$$\Omega_k(f + p, h) = \Omega_k(f, h), \quad f \in W, \quad p \in \Pi_{k-1}, h > 0 \tag{1.19}$$

$$\Omega_k(0, h) = 0, \quad h > 0. \tag{1.20}$$

*We say that the modulus Ω_k is **normalized** if there exists a constant $M > 0$, such that $\Omega_k(e_k, h) \leq Mh^k$, for all $h > 0$.*

Occasionally we extend certain moduli, for the value $h = 0$, by $\Omega_k(f, 0) = 0$, $f \in V$.

The moduli of continuity of order $k \geq 2$ are sometimes called, **moduli of smoothness**.

The usual modulus of continuity of order k is given by

$$\omega_k(f, h) = \sup\{|\Delta_\rho^k f(x)| \mid x, x + k\rho \in I, \ 0 < \rho \leq h\}, \tag{1.21}$$

where $k \in \mathbb{N}$, $f \in \mathcal{F}(I)$, $h > 0$. We allow here the supremum to be ∞.

Many other moduli can be constructed and are important in the process of estimating. In this work we shall use only moduli of order 1 and 2. Certain types of moduli of order two will be studied in Chapter 2. Extensions to classes of vector functions are given in Chapter 6.

1.2 Approximation of functions by sequences of positive linear operators

1.2.1 Basic theorems of convergence

The study of some particular approximation sequences of linear positive operators was extended at the beginning of the 1950s to general approximation sequences of such operators. The foundation of the theory of approximation by general sequences of linear positive operators was constructed by T. Popoviciu [118], H. Bohman [16] and P.P. Korovkin [56]. The theorem of T. Popoviciu is the following

Theorem 1.2.1. *Let a sequence of linear positive operators be of the form*

$$L_n(f, x) = \sum_{i=1}^{m_n} f(\xi_{n,i}) \Psi_{n,i}(x), \ f \in C[a, b], \ x \in [a, b], \tag{1.22}$$

where $\xi_{n,i} \in [a, b]$ and $\Psi_{n,i}$ are positive polynomials. Suppose that

$$L(e_0) = e_0 \tag{1.23}$$

and

$$\lim_{n \to \infty} L_n((e_1 - xe_0)^2, x) = 0, \ \text{uniformly related to } x \in [a, b]. \tag{1.24}$$

Then we have

$$\lim_{n \to \infty} L_n(f) = f, \ \text{uniformly on the interval } [a, b], \ \text{for any } f \in C[a, b]. \tag{1.25}$$

The theorems of H. Bohman [16] and P.P. Korovkin [56], see also [57], use in the hypothesis the convergence of the sequence $(L_n)_n$ on some "test" functions. For this, recall that the functions, $\varphi_0, \ldots, \varphi_m \in C[a, b]$ form a Chebychev system of order $m + 1$ on the interval $[a, b]$, if for any real numbers $\alpha_0, \ldots, \alpha_m$, the function $\varphi = \alpha_0\varphi_0 + \cdots + \alpha_m\varphi_m$ has at most m roots on $[a, b]$. The theorem of P.P. Korovkin can be formulated as follows:

Theorem 1.2.2. *Let a sequence of linear positive operators* $(L_n)_n$, $L_n : V \to \mathcal{F}[a, b]$ *where V is a linear subspace of $\mathcal{F}[a, b]$. Suppose that $\varphi_0, \varphi_1, \varphi_2 \in V \cap C[a, b]$ forms a Chebychev system on the interval $[a, b]$. If we have*

$$\lim_{n \to \infty} L_n(\varphi_j) = \varphi_j, \text{ uniformly for } j = 0, 1, 2, \tag{1.26}$$

then

$$\lim_{n \to \infty} L_n(f) = f, \text{ uniformly, for any } f \in V \cap C[a, b]. \tag{1.27}$$

The theorem of Bohman is the particular version of Theorem 1.2.2 when the operators L_n are of the form (1.22) and $\varphi_j = e_j$, $j = 0, 1, 2$.

Remark 1.2.1. Since $L_n((e_1 - xe_0)^2, x) = L_n(e_2, x) - 2xL_n(e_1, x) + x^2 L_n(e_0, x)$, relations (1.26), for the choice $\varphi_j = e_j$, $j = 0, 1, 2$, imply relation (1.24).

Remark 1.2.2. We can compare the conditions in the theorem of Korovkin with the following two simple sufficient conditions for approximation by linear and (only) continuous operators $(L_n)_n$, $L_n : C[a, b] \to C[a, b]$:

1) there is a dense subspace $Y \subset C[a, b]$ such that $\lim_{n \to \infty} \|L_n(f) - f\| = 0$, for any $f \in Y$ and

2) there is $M > 0$ such that $\|L_n\| \leq M$, $n \in \mathbb{N}$, where

$$\|L_n\| := \sup_{f \in C[a,b], \, \|f\| \leq 1} \|L_n(f)\|.$$

Note that the conditions in the theorem of Korovkin imply that the sequences of norms $(\|L_n\|)_n$ is bounded, because $\|L_n\| = \|L_n(e_0)\|$. It follows that the property of positivity enables us to replace the dense subspace Y by the subspace generated by three functions that form a Chebychev system on $[a, b]$.

The theory of approximation by linear positive operators was extended in many frameworks. The actual stages of the development of the theory in abstract spaces of functions can be formed in the monograph by F. Altomare and M. Campiti [5], see also T. Nishishiraho [75].

For the simultaneous approximation, i.e. the approximation of functions together with their derivatives by linear operators, a crucial property is the convexity of higher order of the operators, see Definition 1.1.2. The main result in this direction is a theorem of Sendov and Popov [124]. A simplified version of it is the following.

Theorem 1.2.3. *If $(L_n)_n$ is a sequence of linear positive operators, $L_n : C[a, b] \to C^p[a, b]$, $p \geq 1$, such that:*

i) L_n are convex of order k, for any $0 \leq k \leq p$, and

ii) $\lim_{n \to \infty} \|L_n(e_i) - e_i\|_{[a,b]} = 0$, for $i = 0, 1, 2$,

then for any $f \in C^p[a, b]$ and any subinterval $[c, d] \subset (a, b)$ we have

$$\lim_{n \to \infty} \|(L_n(f))^{(p)} - f^{(p)}\|_{[c,d]} = 0. \tag{1.28}$$

By a quantitative result of the Korovkin theorem with respect to a Chebychev system $\{\varphi_0, \varphi_1, \varphi_2\}$, we understand an estimate of the form

$$\|L_n(f) - f\| \le K(L_n, f), \tag{1.29}$$

where $K(L_n, f)$ is a quantity that tends to zero when the numbers $\|L_n(\varphi_i) - \varphi_i\|$, $i = 0, 1, 2$ tend to zero.

In the process of the (algebric) polynomial approximation of functions on a compact interval, the phenomenon of better approximation appears near the ends of the interval. This important fact was observed by Nikolski [74]. Also, in the case of a noncompact interval, the degree of approximation is not the same throughout the interval. For this reason, pointwise estimates of the form

$$|L_n(f, x) - f(x)| \le K(L_n, f, x) \tag{1.30}$$

are more appropriate to express the degree of approximation than the global estimate (1.29). The same argument is valid in the case of simultaneous approximation.

1.2.2 Estimates with the first order modulus

The simplest method of estimating the degree of approximation by positive linear functionals and operators is with the aid of the first order modulus of continuity given by:

$$\omega_1(f, h) := \sup\{|f(u) - f(v)|, \ u, v \in I, |u - v| \le h\}, \ f \in \mathcal{F}(I), \ h \ge 0. \tag{1.31}$$

We mention that estimates for the degree of approximation by linear positive operators with the first order modulus, or some modifications of it, are possible in the larger context when the domain of functions is a metric space. We refer to the papers M. Jiménez Pozo [50] and H. Gonska [38], [42]. However in the classical context of functions defined on an interval, the results are stronger.

Because the estimates for operators can be derived immediately from the estimates for functionals, we restrict ourselves to functionals.

We give such an estimate that uses a Chebychev system $\{\varphi_0, \varphi_1, \varphi_2\}$ on an interval $[a, b]$. Define the set of *polynomials*:

$$\mathcal{G} = \{a\varphi_0 + b\varphi_1 + c\varphi_2 \mid a, b, c \in \mathbb{R}\}.$$

First we define certain functions. The existence of them, stipulated in the following lemmas, follows from the basic properties of the Chebychev systems, see for instance [54]. Among these properties we mention that $\mathcal{G}$ is an interpolatory set of order 3 on $[a, b]$.

Lemma 1.2.1. *There is a unique polynomial $v \in \mathcal{G}$ such that $\min\limits_{t \in [a,b]} v(t) = v(a) = v(b) = 1$.*

Proof. It is immediate from the interpolatory property of $\mathcal{G}$. □

Lemma 1.2.2. *For any $x \in [a, b]$ there exists a unique polynomial $\sigma_x \in \mathcal{G}$, satisfying the following conditions:*

1) $\sigma_x(x) = 0$.

2) $\sigma_x(t) > 0$, for any $t \in [a, b] \setminus \{x\}$ and

3) if we write $\sigma_x = a\varphi_0 + b\varphi_1 + c\varphi_2$, then $\max\{|a|, |b|, |c|\} = 1$.

Morover the function $\Psi : [a, b] \to C[a, b]$, $\Psi(x) = \sigma_x$ is continuous.

Proof. For the existence of the functions σ_x see for instance [54]. Condition 3) assures the uniqueness. In order to prove the continuity, let x be fixed and let $x_n \to x$. From the Cesaro lemma we can choose from the sequence $(a_n, b_n, c_n)_n$, (which corresponds to $(x_n)_n$), a convergent subsequence, corresponding to a sequence of indices $(n_k)_k$. Then we obtain $\lim_{k \to \infty} \sigma_{x_{n_k}} = \sigma_x$, uniformly. Since the limit does not depend on the subsequence, we have $\lim_{n \to \infty} \sigma_{x_n} = \sigma_x$, uniformly. $\square$

Definition 1.2.1. *Define the function $\Phi : (0, b - a] \to \mathbb{R}$, by*

$$\Phi(h) = h \cdot \inf_{x \in [a,b]} \min_{t \in [a,b] \setminus (x-h, x+h)} \frac{\sigma_x(t)}{|t - x|}, \quad \text{for } h \in (0, b - a]. \tag{1.32}$$

Lemma 1.2.3. *The function Φ defined above has the following properties:*

i) $\Phi(h) > 0$, for $h \in (0, b - a]$ and Φ is increasing.

ii) Φ is continuous.

Proof. i) Since the function $x \mapsto \sigma_x$, is continuous, it then follows that for any $h \in (0, b - a]$, the function $x \to \min_{t \in [a,b] \setminus (x-h, x+h)} \frac{\sigma_x(t)}{|t-x|}$ is continuous. Then from the Weierstrass theorem we have

$$\Phi(h) = h \cdot \min_{x \in [a,b]} \min_{t \in [a,b] \setminus (x-h, x+h)} \frac{\sigma_x(t)}{|t - x|}. \tag{1.33}$$

Consequently $\Psi(h) > 0$. The monotonicity is immediate.

ii) The continuity of Ψ follows immediately, from relation (1.33) and the continuity of the application $x \mapsto \sigma_x$. $\square$

Theorem 1.2.4. *([110]) Let $F : V \to \mathcal{F}[a, b]$ be a positive linear functional, where V is a linear subspace of $\mathcal{F}[a, b]$. Let $\{\varphi_0, \varphi_1, \varphi_2\} \subset V \cap C[a, b]$ be a fixed Chebychev system on the interval $[a, b]$. For any $f \in V$, any $x \in [a, b]$ and any $0 < h \le b - a$, we have*

$$|F(f) - f(x)| \le |f(x)| \cdot |F(v) - v(x)|$$
$$+ \left(F(v) + \frac{F(\sigma_x)}{\Phi(h)} \right) \left(\omega_1(f, h) + |f(x)| \cdot \omega_1(v, h) \right). \tag{1.34}$$

Proof. Let F, f, x, h be fixed. First we have

$$|F(f) - f(x)| \le \left| F\left(\frac{f(x)}{v(x)} \cdot v \right) - f(x) \right| + \left| F(f) - F\left(\frac{f(x)}{v(x)} \cdot v \right) \right|$$
$$=: T_1 + T_2.$$

Using the linearity of F, we have

$$T_1 = \frac{|f(x)|}{v(x)} \cdot |F(v) - v(x)| \le |f(x)| \cdot |F(v) - v(x)|.$$

In order to estimate the term T_2, we take $t \in [a, b]$ arbitrarily. We have:

$$\left| f(t) - \frac{f(x)}{v(x)} \cdot v(t) \right| \le |f(t) - f(x)| + \left| f(x) \cdot \frac{v(x) - v(t)}{v(x)} \right|$$

$$\le |f(t) - f(x)| + |f(x)| \cdot |v(t) - v(x)|.$$

From the proprieties of the modulus ω_1 we have $|f(t) - f(x)| \le \left(1 + \frac{|t-x|}{h}\right) \omega_1(f, h)$ and $|v(t) - v(x)| \le \left(1 + \frac{|t-x|}{h}\right) \omega_1(v, h)$. It follows that

$$\left| f(t) - \frac{f(x)}{v(x)} \cdot v(t) \right| \le \left(1 + \frac{|t-x|}{h}\right) \left(\omega_1(f, h) + |f(x)| \cdot \omega_1(v, h) \right)$$

$$\le \left(v(t) + \frac{\sigma_x(t)}{\Phi(h)} \right) \left(\omega_1(f, h) + |f(x)| \cdot \omega_1(v, h) \right).$$

Define $\Psi(t) = \left(v(t) + \sigma_x(t)/\Phi(h) \right) \left(\omega_1(f, h) + |f(x)| \cdot \omega_1(v, h) \right)$, $t \in [a, b]$. Using the positivity of the functional F, we have

$$T_2 \le F\left(\left| f - \frac{f(x)}{v(x)} \cdot v \right| \right)$$

$$\le F(\Psi)$$

$$\le \left(F(v) + \frac{F(\sigma_x)}{\Phi(h)} \right) \left(\omega_1(f, h) + |f(x)| \cdot \omega_1(v, h) \right). \quad \square$$

Remark 1.2.3. From Proposition 1.1.1 it follows that if F is a functional induced by a positive Borel measure such that $F(\sigma_x) = 0$, for certain $x \in I$, then $F = \delta_x$. Consequently, if $F(\varphi_i) = \varphi_i(x)$, $(i = 0, 1, 2)$, then $F(\sigma_x) = \sigma_x(x) = 0$ and hence $F = \delta_x$. The implication $F(\varphi_i) = \varphi_i(x)$, $(i = 0, 1, 2) \Rightarrow F = \delta_x$ was observed first by Korovkin [57].

We derive now an estimate for the global approximation by linear positive operators.

Corollary 1.2.1. *Let $V \subset \mathcal{F}(I)$ be a linear subspace and let $\{\varphi_0, \varphi_1, \varphi_2\} \subset V$, be a Chebychev system. Let $L : V \to \mathcal{F}(I)$ be a linear positive operator. Define*

$$M := \frac{\displaystyle\sup_{x \in [a,b]} L(\sigma_x, x)}{\|\Phi\|}.$$

We have $M < \infty$. For all $x \in [a, b]$ define $h_x := \inf\{h \in (0, b - a], \; L(\sigma_x, x) = M\Phi(h)\}$ and set $h(L) = \displaystyle\sup_{x \in [a,b]} h_x$. Then for any $f \in C[a, b]$ we have:

$$\|L(f) - f\| \le \|f\| \cdot \|L(v) - v\|$$

$$+ (\|L(v)\| + M) \left(\omega_1(f, h(L)) + \|f\| \cdot \omega_1(v, h(L)) \right). \quad (1.35)$$

Proof. From the construction of the functions σ_x follows $|L(\sigma_x, x)| \leq \sum_{j=0}^{2} \|L(\varphi_j)\|$, for all $x \in [a, b]$. Hence $M < \infty$. If $L(\sigma_x, x) = 0$, then $h_x = 0$. If $M > 0$, then the existence of the numbers h_x follows from the continuity of the function Φ and the limit $\lim_{h \to 0} \Phi(h) = 0$. We apply Theorem 1.2.4, choosing $h = h_x$, for all $x \in [a, b]$. $\square$

Theorem 1.2.4 is a quantitative version of the Korovkin theorem, because we have:

Proposition 1.2.1. *Theorem 1.2.4 implies the theorem of Korovkin (Theorem 1.2.2).*

Proof. Let $(L_n)_n$ be a sequence of positive linear operators such that (1.26) holds, where $\{\varphi_0, \varphi_1, \varphi_2\}$ is a Chebychev system.

Let $x \in [a, b]$. From condition 3) in Lemma 1.2.2 it follows that,

$$\|L_n(\sigma_x) - \sigma_x\| \leq \sum_{j=0}^{2} \|L_n(\varphi_j) - \varphi_j\|.$$

We obtain $\lim_{n \to \infty} \|L_n(\nu) - \nu\| = 0$ and $\lim_{n \to \infty} \|L_n(\sigma_x) - \sigma_x\| = 0$ uniformly with regard to $x \in [a, b]$ from condition (1.26).

We shall apply Corollary 1.2.1, for $L = L_n$. It remains to show that $\lim_{n \to \infty} h(L_n) = 0$. If $M = 0$, then we have $h(L_n) = 0$. Consider now $M > 0$.

Note that for any $x \in [a, b]$, we have $L_n(\sigma_x, x) \leq \|L_n(\sigma_x) - \sigma_x\|$. Consequently,

$$\lim_{n \to \infty} \sup_{x \in [a,b]} L_n(\sigma_x, x) = 0.$$

Let $0 < \varepsilon \leq b - a$, arbitrarily chosen. Since $\Phi(\varepsilon) > 0$, there is $n_\varepsilon \in \mathbb{N}$, such that

$$L_n(\sigma_x, x) < M\Phi(\varepsilon), \quad \text{for all } x \in [a, b], \ n \in \mathbb{N}, \ n \geq n_\varepsilon.$$

It follows that $\Phi(h_x^n) < \Phi(\varepsilon)$. Since Φ is increasing, it follows that $h_x^n < \varepsilon$, for all $x \in [a, b]$. Hence $h(L_n) \leq \varepsilon$, for $n \geq n_\varepsilon$. $\square$

Particular Chebychev systems are the extended Chebychev systems, defined as follows. A set $\{\varphi_0, \varphi_1, \varphi_2\} \subset C^2[a, b]$ is an extended Chebychev system on the interval $[a, b]$, if the systems: $\{\varphi_0\}$, $\{\varphi_0, \varphi_1\}$ and $\{\varphi_0, \varphi_1, \varphi_2\}$ are Chebychev systems of orders 1, 2 and 3, respectively. Estimates in terms of an extended Chebychev system were obtained by Freud [33], Shisha and Mond [126] and Censor [21].

The simplest extended Chebychev system is the algebric system $\{e_0, e_1, e_2\}$. In this case, to the functions ν and σ_x, defined in Lemmas 1.2.1 and 1.2.2, correspond, respectively, the functions e_0 and $k(e_1 - xe_0)^2$, where $k = \min\left\{1, \frac{1}{2|x|}, \frac{1}{x^2}\right\}$. In this case, it is possible to obtain more precise estimates.

Remark 1.2.4. The moments $L(e_0, x)$, $|L(e_1 - xe_0, x)|$ and $L((e_1 - xe_0)^2, x)$ of the operators give a measure of the degree of approximation of the test functions e_i, $0 \leq i \leq 2$. Note the equivalence of the conditions $\lim_{n \to \infty} L(e_i) = e_i$, uniformly, for $i = 0, 1, 2$ with the conditions

$$\lim_{n \to \infty} L_n(e_1 - xe_0)^j = (e_1 - xe_0)^j, \quad \text{uniformly, with respect to } x, \text{ for } j = 0, 1, 2.$$

In the next theorem we discuss the posibility of estimating the functionals in terms of the moments m_0 and m_2, see (1.17), using the first order modulus.

Theorem 1.2.5. *Let $V \subset \mathcal{F}(I)$ be a subspace, such that $\Pi_2 \subset V$. If $F : V \to \mathbb{R}$, is a linear positive functional, then for any $f \in V$, $x \in I$, $r > 0$ and $h > 0$, we have*

$$|F(f) - f(x)| \leq |F(e_0) - 1| \cdot |f(x)|$$
$$+ \left(B_r F(e_0) + rh^{-2} F((e_1 - xe_0)^2) \right) \omega_1(f, h), \qquad (1.36)$$

where

$$B_r = \begin{cases} 1, & r \geq 1, \\ \max\{P_r(]1/2r[+1), \; P_r([1/2r])\}, & 0 < r < 1, \end{cases}$$

and P_r is the polynomial $P_r(u) = u + 1 - ru^2$, $(u \in \mathbb{R})$.

Conversely, suppose that there are constants $A, B, C \geq 0$, such that the inequality

$$|F(f) - f(x)| \leq A|f(x)| \cdot |F(e_0) - 1|$$
$$+ \left(B F(e_0) + Ch^{-2} F((e_1 - xe_0)^2) \right) \omega_1(f, h), \qquad (1.37)$$

holds (only) for all linear positive functionals of the form (1.11), any $f \in C(I)$, any $x \in I$, and any $h > 0$. Then we must have $A \geq 1$, $C > 0$, and if $C = r$, $r > 0$, then we must have $B \geq B_r$.

Proof. Consider $\omega_1(f, h) < \infty$. First we have

$$|F(f) - f(x)| \leq |f(x)| \cdot |F(e_0) - 1| + |F(f - f(x)e_0)|.$$

Let $t \in I$. There is $n \in \mathbb{N} \cup \{0\}$ and $q \in [0, 1)$, such that $|t - x| = (n + q)h$. Hence

$$|f(t) - f(x)| \leq (n + 1)\omega_1(f, h).$$

We have

$$\sup_{n \in \mathbb{N} \cup \{0\}, \; q \in [0,1)} [(n + 1) - r(n + q)^2] = \sup_{n \in \mathbb{N} \cup \{0\}} P_r(n) = B_r.$$

Consequently we have

$$|F(f - f(x)e_0)| \leq F(|f - f(x)e_0|)$$
$$\leq \left(B_r F(e_0) + rh^{-2} F((e_1 - xe_0)^2) \right) \omega_1(f, h).$$

Conversely, let us suppose that (1.37) holds, for any F, f, x and h as in the supposition.

If we take $I := [0, 1]$, $F := 0$, $f := e_0$, $x := 0$, $h := 1$, we obtain $A \geq 1$.

If we take $I := [0, 1]$, $F(g) := g(1)$, $(g \in C(I))$, $f := \frac{1}{\varepsilon} e_1$, $\varepsilon > 0$, $h := \varepsilon$ and $x := 0$, we obtain $\omega_1(f, h) = 1$, and $1/\varepsilon \leq B + (1/\varepsilon)^2 C$. Since $\varepsilon > 0$ can be taken arbitrarily, we obtain $C > 0$.

Finally, let $C = r > 0$ in (1.37). Choose, $n \in \mathbb{N} \cup \{0\}$, $q \in (0, 1)$, $I := [0, n + q]$, $F(g) := g(n + q)$, $(g \in C(I))$, $x := 0$, $h := 1$ and $f \in C(I)$ defined by:

$$f(x) := \begin{cases} \frac{1}{q}(t-k) + k, \ t \in [k, k+q], \ 0 \le k \le n, \\ k+1, \qquad\quad t \in [k+q, k+1], \ 0 \le k \le n-1. \end{cases}$$

We have $\omega_1(f, h) = 1$ and so consequently, relation (1.37) becomes $n + 1 \le B + r(n+q)^r$. It follows that $B \ge B_r$. $\qquad\qquad\square$

Corollary 1.2.2. *With the conditions in Theorem 1.2.5, we have*

$$|F(f) - f(x)| \le |F(e_0) - 1| \cdot |f(x)| + T_h \omega_1(f, h), \qquad (1.38)$$

where

$$T_h := \begin{cases} m_0 + h^{-1}\sqrt{m_0 m_2}, \ 0 < h < \sqrt{\frac{m_2}{m_0}}, \\ m_0 + h^{-2} m_2, \qquad h \ge \sqrt{\frac{m_2}{m_0}}. \end{cases}$$

Proof. We can majorize $B_r \le 1$, if $r \ge 1$ and $B_r \le 1 + \frac{1}{4r}$, if $0 < r < 1$. Then we take the best possible value for $r > 0$. Denote $U(r) := \left(1 + \frac{1}{4r}\right)m_0 + rh^{-2}m_2$. We have

$$\min_{r>0} U(r) = U\left(\frac{h}{2}\sqrt{\frac{m_0}{m_2}}\right) = m_0 + h^{-1}\sqrt{m_0 m_2}.$$

It follows that we can choose $r = \frac{h}{2}\sqrt{\frac{m_0}{m_2}}$, for $h < \sqrt{\frac{m_2}{m_0}}$ and $r = 1$ for $h \ge \sqrt{\frac{m_2}{m_0}}$. $\quad\square$

Remark 1.2.5. The estimate

$$\begin{aligned} |F(f) - f(x)| &\le |m_0 - 1| \cdot |f(x)| \\ &+ (m_0 + h^{-2}m_2)\, \omega_1(f, h), \ f \in C(I), \ x \in I, \ h > 0 \end{aligned} \qquad (1.39)$$

is a consequence of the general estimate (1.34). It was proved (for operators) by Mond [69].

The estimate

$$\begin{aligned} |F(f) - f(x)| &\le |m_0 - 1| \cdot |f(x)| \\ &+ (m_0 + h^{-1}\sqrt{m_0 m_2})\, \omega_1(f, h), \ f \in C(I), \ x \in I, \ h > 0, \end{aligned} \qquad (1.40)$$

was proved, (for operators), by Shisha and Mond [126] and in the case $m_0 = 1$, earlier by Mamedov [66].

2

Estimates with Second Order Moduli

2.1 A general approach

2.1.1 Introduction

In this chapter we continue the study of estimating the degree of an approximation using general linear positive operators by considering combinations of first and second order moduli, in terms of the moments of order 0, 1, and 2, see Remark 1.2.4. Estimates with such combinations of first and second order modulus, (and also with the absolute value of the function, which can be regarded as a modulus of order 0) are more refined then estimates using only the first modulus. A first observation is that, from estimates with the second order modules, one can derive estimates with the first order modulus. A second observation is the fact that such combinations decompose the error of approximation in three components, corresponding to three specific features of the functions that affect the error: amplitude, deviation from the linear functions, and deviation from the polynomials of degree 2. Roughly speaking, these moduli measure the deviation from the test functions of the algebraic Chebychev system.

By taking into account Remark 1.1.1, as in Section 1.2, it suffices to consider estimates for linear positive functionals. Especially we are interested in estimates of the form

$$|F(f) - f(x)| \leq A \cdot |F(e_0) - 1| \cdot |f(x)| + B|F(e_1 - xe_0)| \cdot h^{-1}\Omega_1(f, h)$$
$$+ (C\, F(e_0) + D\, h^{-2}F((e_1 - xe_0)^2))\Omega_2(f, h), \qquad (2.1)$$

where F is a linear positive functional on a linear subspace $V \subset \mathcal{F}(I)$, I being an arbitrary interval, such that $\Pi_2 \subset V$, Ω_1 and Ω_2 are moduli of continuity of order 1 and 2 on a subspace $W \subset \mathcal{F}(I)$, $f \in V \cap W$ and $h > 0$. This type of estimate has the following important properties: if $F(e_0) = 1$, then the first term drops and if $F(e_0) = 1$ and $F(e_1) = x$, then the first two terms drop. Recall that in Corollaries 1.1.1 and 1.1.2 there are given simple cases in which we have $F(f) - f(x) = 0$.

For testing the optimality of the constants we shall use the simple functionals of the form (1.11).

From estimates of the type (2.1) one can immediately derive estimates for linear positive operators $L : V \to \mathcal{F}(I)$, of the form

$$|L(f, x) - f(x)| \leq A \cdot |L(e_0, x) - 1| \cdot |f(x)|$$
$$+ B|L(e_1 - xe_0, x)| \cdot h^{-1}\Omega_1(f, h)$$
$$+ (C \, L(e_0, x) + D \, h^{-2}L((e_1 - xe_0)^2, x))\Omega_2(f, h), \quad (2.2)$$

when $f \in V \cap W$, $x \in I$ and $h > 0$.

G. Freud [33] obtained the first estimate, with unspecified constants, for general linear positive operators, with the usual higher order moduli. A basic method in estimates with higher order moduli is to use the K-modified functionals, introduced by Peteer [112]. This method was developed by many authors: P.L. Butzer and H. Berens [19], H. Johnen [51], H. Gonska [41], Z. Ditzian and V. Totik [30] and others. In this chapter we apply a new method consisting in decomposition of functionals. By comparing with the method of K-functionals, our method is more appropriate for obtaining estimates with good or even the best possible constants in front of the terms. In this mode one obtains an important diminution of the constants. Another advantage of this new method consists in the fact that it can be applied to a larger class of linear positive functionals, including the class of linear positive functionals defined on the space $C[a, b]$.

In this section we present a general result, given in [91], which will be applied for different choices of the first and second order moduli, in the next sections.

2.1.2 A general estimate for the degree of approximation by linear positive functionals

In the following theorem let D be an arbitrary set, such that $D = A \cup B \cup \{x\}$, $A \cap B = \emptyset$, $x \notin A$ and $x \notin B$. We do not exclude the cases $A = \emptyset$ or $B = \emptyset$. For any function $f \in \mathcal{F}(D)$ define the functions f_A, $f_B \in \mathcal{F}(D)$ by $f_A := f \cdot \chi_A$, $f_B := f \cdot \chi_B$.

Theorem 2.1.1. *Let V be a linear space of real-valued functions defined on D, with the property $f_A \in V$ and $f_B \in V$ for each $f \in V$. Let $F : V \to \mathbb{R}$ be a linear positive functional. Suppose that there are the functions $f, v, \eta, \theta \in V$ such that*

$$\theta(t) \geq 0, \ t \in D; \quad v(x) = 1; \quad \eta(x) = 0; \quad (2.3)$$

$$\eta(t) < 0, \ t \in A, \ (if \ A \neq \emptyset); \quad \eta(t) > 0, \ t \in B, \ (if \ B \neq \emptyset); \quad (2.4)$$

and

$$|\eta(t_2)[f(t_1) - f(x)v(t_1)] - \eta(t_1)[f(t_2) - f(x)v(t_2)]| \quad (2.5)$$
$$\leq \eta(t_2)\theta(t_1) - \eta(t_1)\theta(t_2), \ (t_1, t_2) \in A \times B, \ (if \ A \times B \neq \emptyset).$$

In the particular case where $F(\eta) = 0$, $A \neq \emptyset$ and $B \neq \emptyset$, we have

$$|F(f) - f(x)| \leq |f(x)| \cdot |F(v) - 1| + F(\theta). \quad (2.6)$$

In the general case, if we suppose moreover that there exists a number $k > 0$ such that

$$|f(t) - f(x)v(t)| \leq k|\eta(t)| + \theta(t), \quad \text{for all } t \in D, \tag{2.7}$$

then we have

$$|F(f) - f(x)| \leq |f(x)| \cdot |F(v) - 1| + k|F(\eta)| + F(\theta). \tag{2.8}$$

Proof. Step I. We can consider, for a choice, that $F(\eta) \geq 0$, that is $F(\eta_B) \geq -F(\eta_A)$. Otherwise we can consider the function $-\eta$ instead of η and interchange the sets A and B.

We have immediately

$$|F(f) - f(x)| \leq |F(f - f(x)v)| + |f(x)| \cdot |F(v) - 1|.$$

Step II. First we prove the theorem, both the particular case and the general case, when we have $F(\eta_A) < 0$ and $F(\eta_B) > 0$. This implies that $A \neq \emptyset$ and $B \neq \emptyset$. Consider the function $\varphi : D \times D \to \mathbb{R}$ defined by

$$\varphi(t_1, t_2) := \frac{\eta_B(t_2)}{F(\eta_B)} \cdot (f - f(x)v)_A(t_1)$$

$$+ \frac{\eta_A(t_1)}{F(\eta_A)} \cdot (f - f(x)v)_B(t_2), \quad (t_1, t_2) \in D \times D. \tag{2.9}$$

We have

$$F(f - f(x)v) = F_{t_1}(F_{t_2}(\varphi(t_1, t_2))), \tag{2.10}$$

where $F_{t_1}(F_{t_2}(\varphi(t_1, t_2)))$ denotes the value of the functional F applied to the function $t_1 \mapsto F_{t_2}(\varphi(t_1, t_2))$, and $F_{t_2}(\varphi(t_1, t_2))$ denotes the value of the functional F on the function $t_2 \mapsto \varphi(t_1, t_2)$. By taking into account the equality $\eta = \eta_A + \eta_B$ one obtains for all $(t_1, t_2) \in D \times D$:

$$\varphi(t_1, t_2) = \frac{\eta_B(t_2)}{F(\eta_B)} \cdot (f - f(x)v)_A(t_1) - \frac{\eta_A(t_1)}{F(\eta_B)} \cdot (f - f(x)v)_B(t_2)$$

$$+ \frac{F(\eta)}{F(\eta_A)F(\eta_B)} \cdot \eta_A(t_1)(f - f(x)v)_B(t_2).$$

First consider that $F(\eta) = 0$. We obtain from (2.5):

$$\varphi(t_1, t_2) \leq \frac{\eta_B(t_2)}{F(\eta_B)} \cdot \theta_A(t_1) - \frac{\eta_A(t_1)}{F(\eta_B)} \cdot \theta_B(t_2).$$

Consequently it follows that

$$F_{t_1}(F_{t_2}(\varphi(t_1, t_2))) \leq F(\theta_A) + F(\theta_B) \leq F(\theta).$$

In a similar mode we get

$$F_{t_1}(F_{t_2}(\varphi(t_1, t_2))) \geq -F(\theta).$$

Hence $|F(f - f(x)v)| \leq F(\theta)$ and then (2.6) follows.

Consider now that $F(\eta) \geq 0$ and suppose that relation (2.7) is satisfied. Using also relations (2.5) we obtain

$$\varphi(t_1, t_2) \leq \frac{\eta_B(t_2)}{F(\eta_B)} \cdot \theta_A(t_1) - \frac{\eta_A(t_1)}{F(\eta_B)} \cdot \theta_B(t_2)$$
$$+ \frac{F(\eta)}{F(\eta_A)F(\eta_B)} \cdot \eta_A(t_1)[\theta_B(t_2) + k\,\eta_B(t_2)].$$

Consequently it follows that

$$F_{t_1}(F_{t_2}(\varphi(t_1, t_2))) \leq F(\theta_A) + F(\theta_B) + kF(\eta) \leq F(\theta) + k\,|F(\eta)|.$$

In a similar mode we get

$$F_{t_1}(F_{t_2}(\varphi(t_1, t_2))) \geq -F(\theta) - k\,|F(\eta)|.$$

Hence $|F(f - f(x)v| \leq F(\theta) + k\,|F(\eta)|$ and then we get (2.8). Therefore the theorem is proved in the case $F(\eta_A) < 0$ and $F(\eta_B) > 0$.

Step III. Now we prove the theorem, both the particular case and the general case, if we have $A \neq \emptyset$ and $B \neq \emptyset$. Let us fix $a \in A$, $b \in B$ and for $\varepsilon > 0$ consider the linear positive functional $G_\varepsilon : V \to \mathbb{R}$ defined by

$$G_\varepsilon(g) := F(g) + \varepsilon\,(\eta(b)g(a) - \eta(a)g(b)),\ g \in V.$$

Then G_ε is linear and positive. We have $G_\varepsilon(\eta_A) = F(\eta_A) + \varepsilon\,\eta(b)\eta(a) < 0$ and $G_\varepsilon(\eta_B) = F(\eta_B) - \varepsilon\,\eta(b)\eta(a) > 0$. Also, if $F(\eta) = 0$, then $G_\varepsilon(\eta) = 0$. From Step II, it follows that if $F(\eta) = 0$, then

$$|G_\varepsilon(f) - f(x)| \leq |f(x)| \cdot |G_\varepsilon(v) - 1| + G_\varepsilon(\theta)$$

and in the general case, if relation (2.7) holds, then

$$|G_\varepsilon(f) - f(x)| \leq |f(x)| \cdot |G_\varepsilon(v) - 1| + k\,|G_\varepsilon(\eta)| + G_\varepsilon(\theta).$$

Since ε is arbitrary, by passing to limit $\varepsilon \to 0$ we obtain (2.6) and (2.8), respectively.

Step IV. It remains to prove the theorem in the case where at least one of the sets A or B is empty. In this case we assume relation (2.7) and we must prove (2.8). If $A = \emptyset$ it follows from (2.7) that

$$F(f - f(x)v) = F((f - f(x)v)_B) \leq F(\theta_B) + kF(\eta_B) \leq F(\theta) + k|F(\eta)|$$

and

$$F(f - f(x)v) = F((f - f(x)v)_B) \geq -F(\theta_B) - kF(\eta_B) \geq -F(\theta) - k|F(\eta)|.$$

Hence $|F(f - f(x)v)| \leq F(\theta) + k|F(\eta)|$. Then (2.8) follows.

The case $B = \emptyset$ is similar. The theorem is proved. $\qquad\square$

Remark 2.1.1. Condition (2.5) is necessary in order to obtain (2.8). Indeed, if we use only condition (2.7), then by applying functional F we get only

$$|F(f) - f(x)| \leq |f(x)| \cdot |F(v) - 1| + kF(|\eta|) + F(\theta).$$

We have $|F(\eta)| \leq F(|\eta|)$, and in general we do not have equality.

In the sequel we apply Theorem 2.1.1 in the case where D is an interval I. First we give some auxiliary definitions and results.

Definition 2.1.1. *A linear positive functional $F : V \to \mathbb{R}$, where V is a linear subspace of $\mathcal{F}(I)$ is called* **admissible related to the point** *$x \in I$, if $\Pi_2 \subset V$ and one of the following conditions is satisfied:*

i) For any $g \in V$ we have $g \cdot \chi_{I \cap (x, \infty)} \in V$ and $g \cdot \chi_{I \cap (-\infty, x)} \in V$.

ii) The functional F is induced by a positive Borel measure (see (1.10), where $V = \mathcal{L}_\mu(I)$).

iii) $V = C(I)$.

Remark 2.1.2. Therefore the functionals defined on the space $C(I)$ are admissible related to any point $x \in I$. Also the functionals of the form (1.10) when $\Pi_2 \subset V$ are admissible related to any point $x \in I$. In the case where the interval I is not finite, the class of integral functionals of the form (1.10) is larger than the class of the functionals given on $C(I)$.

In view of a greater generality we use admissible functionals related to a point, although the most important cases are given by the functionals defined on the space $C(I)$, or more generally by the functionals induced by a positive Borel measure. See Remark 1.1.2.

We introduce the following notation.

Definition 2.1.2. *For $f \in \mathcal{F}(I)$ and $a, b, x \in I$, $a \neq b$ set*

$$\Delta(f; a, x, b) := \frac{b-x}{b-a} f(a) + \frac{x-a}{b-a} f(b) - f(x). \tag{2.11}$$

Definition 2.1.3. *For $\Psi \in \mathcal{F}[0, \infty)$, $a, b, x \in I$, $a < x < b$ and $h > 0$ set*

$$\delta_h(\Psi; a, x, b) := \frac{b-x}{b-a} \Psi\left(\frac{x-a}{h}\right) + \frac{x-a}{b-a} \Psi\left(\frac{b-x}{h}\right). \tag{2.12}$$

Lemma 2.1.1. *Let $f \in \mathcal{F}(I)$, $a, b \in I$, $a \neq b$. Define the function*

$$g(t) := -\Delta(f; a, t, b), \quad t \in I. \tag{2.13}$$

We have:

i) $g(a) = 0 = g(b)$.

ii) $\Delta(f; \alpha, u, \beta) = \Delta(g; \alpha, u, \beta)$, for all $\alpha, \beta, u \in I$, $\alpha \neq \beta$.

iii) if Ω_2 is a second order modulus on the subspace V and $f \in V$, then $g \in V$ and $\Omega_2(f, h) = \Omega_2(g, h)$, for all $h > 0$.

Proof. The point i) is immediate. The points ii) and iii) hold, because we can write $g = f + l$, with $l \in \Pi_1$. □.

Remark 2.1.3. If F is a functional induced by a positive Borel measure such that $F(e_0) = 1$, $F(e_1) = x$, $F(|e_1 - xe_0|) \neq 0$, where $x \in I$, then relation (2.10) can be rewritten in the following equivalent mode:

$$F(f) - f(x) = \iint\limits_{(I^-) \times (I^+)} \frac{t - s}{M} \cdot \Delta(f; s, x, t) \, d(\mu_1 \times \mu_2)(s, t), \quad f \in \mathcal{L}_\mu(I),$$

$$(2.14)$$

where $I^- := I \cap (-\infty, x)$, $I^+ := I \cap (x, \infty)$, $M := \frac{1}{2} F(|e_1 - xe_0|)$, μ_1 and μ_2 are the restrictions of the measure μ on the intervals I^- and I^+, respectively and $\mu_1 \times \mu_2$ is their product on the space $(I^-) \times (I^+)$.

The main result of this section is the following.

Theorem 2.1.2. *Let $F : V \to \mathbb{R}$, $V \subset \mathcal{F}(I)$, be a linear positive functional that is admissible related to a point $x \in I$. Let also Ω_1 be a first order modulus and Ω_2 be a second order modulus, both of them defined on a subspace $W \subset \mathcal{F}(I)$. Let $f \in V \cap W$ and $h > 0$. Suppose that there exists $\Psi \in \mathcal{F}[0, \infty)$, $\Psi(t) \geq 0$, $t \geq 0$ such that $\Psi\left(\left|\frac{e_1 - xe_0}{h}\right|\right) \in V$ and*

$$|\Delta(f; t_1, x, t_2)| \leq \delta_h(\Psi; t_1, x, t_2)\Omega_2(f, h), \text{ for all } t_1, t_2 \in I, \ t_1 < x < t_2. \quad (2.15)$$

i) In the particular case where $F(e_0) = 1$, $F(e_1) = x$ and x is an interior point of I, we have

$$|F(f) - f(x)| \leq F\left(\Psi\left(\left|\frac{e_1 - xe_0}{h}\right|\right)\right)\Omega_2(f, h). \quad (2.16)$$

ii) In the general case, if we suppose in addition to (2.15) that

$$|f(t) - f(x)| \leq h^{-1}\Omega_1(f, h)|x - t| + \Psi\left(\left|\frac{t - x}{h}\right|\right)\Omega_2(f, h), \text{ for all } t \in I,$$

$$(2.17)$$

then one has

$$|F(f) - f(x)| \leq |F(e_0) - 1| \cdot |f(x)| + |F(e_1 - xe_0)| h^{-1}\Omega_1(f, h)$$
$$+ F\left(\Psi\left(\left|\frac{e_1 - xe_0}{h}\right|\right)\right)\Omega_2(f, h). \quad (2.18)$$

Proof. We have to consider only the case where $\Omega_1(f, h) < \infty$ and $\Omega_2(f, h) < \infty$. We make the proof according to condition i), ii), or iii) from Definition 2.1.1, which is satisfied by the functional F.

If condition i) holds, then we apply Theorem 2.1.1 taking $D := I$, $A := I \cap (-\infty, x)$, $B := I \cap (x, \infty)$, $v := e_0$, $\eta := e_1 - xe_0$, $k := h^{-1}\Omega_1(f, h)$ and $\theta(t) := \Psi\left(\left|\frac{t - x}{h}\right|\right)\Omega_2(f, h)$, $t \in I$.

If condition ii) is satisfied, then obviously $f \cdot \chi_{I \cap (x, \infty)} \in V$ and $f \cdot \chi_{I \cap (-\infty, x)} \in V$, for any $f \in V$. Hence condition i) holds.

If condition iii) is satisfied, then from Remark 1.1.2, the extension of the functional F can be expressed in the form (1.10). Therefore this case is reduced to the previous case. $\qquad \square$

2.2 Estimates with moduli ω_2^λ and $\omega_2^\star$

2.2.1 Introduction. Auxiliary results

The classical second order modulus is given by

$$\omega_2(f, h)$$
$$:= \sup\left\{\left|f(u) - 2f\left(\frac{u+v}{2}\right) + f(v)\right|, \ u, v \in I, \ |u - v| \le 2h\right\}, \qquad (2.19)$$

where $f \in \mathcal{F}(I)$ and $h > 0$. Here, as well as in the case of the next moduli, we accept the supremum to be equal to ∞.

The first estimate for general linear positive operators, using modulus ω_2 and with precise constants, was given by H. Gonska, see [36]. We mention also the paper by Zhuk [144]. In the paper of Gonska and Kovacheva [44], one obtains an estimate which, in terms of functionals, can be expressed as the following: If $F : C[a, b] \to \mathbb{R}$ is a linear positive functional, then

$$|F(f) - f(x)| \le |F(e_0) - 1| \cdot \|f\| + \frac{2}{h} \cdot |F(e_1 - xe_0)| \cdot \omega_1(f, h) \qquad (2.20)$$

$$+ \left[\frac{3}{4} \cdot (1 + F(e_0)) + \frac{3}{4} \cdot |f(e_0) - 1|\right.$$

$$\left. + \frac{3}{2h} \cdot |F(e_1 - xe_0)| + \frac{3}{4h^2} \cdot F((e_1 - xe_0)^2)\right] \omega_2(f, h),$$

for $f \in C[a, b]$, $x \in [a, b]$ and $0 < h \le \frac{1}{2}(b - a)$.

We consider a generalized second order modulus with a parameter as follows.

Definition 2.2.1. [91] *For* $0 \le \lambda \le \frac{1}{2}$, $f \in \mathcal{F}(I)$, $h > 0$, *put*

$$\omega_2^\lambda(f, h) := \sup\left\{|\Delta(f; u, tu + (1 - t)v, v)|, \ u, v \in I, \ u \ne v,\right.$$

$$\left. |u - v| \le 2h, \frac{1}{2} - \lambda \le t \le \frac{1}{2} + \lambda\right\}. \qquad (2.21)$$

In this definition, the most important cases are $\lambda = 0$ and $\lambda = 1/2$.

Remark 2.2.1. We have for $f \in \mathcal{F}(I)$, $h > 0$:

$$\omega_2^0(f, h) = \frac{1}{2} \cdot \omega_2(f, h), \quad f \in \mathcal{F}(I), \ h > 0, \qquad (2.22)$$

$$\omega_2^{\lambda_1}(f, h) \le \omega_2^{\lambda_2}(f, h), \ \text{if } \lambda_1 \le \lambda_2, \qquad (2.23)$$

$$\omega_2^{1/2}(f, h) = \sup\left\{|\Delta(f; u, y, v)|, \ u, v \in I, \ u \ne v,\right.$$

$$\left. u \le y \le v, \ v - u \le 2h\right\}. \qquad (2.24)$$

The modulus $\omega_2^{1/2}$ was used in a slightly modified form in our papers [82], [83] and by J. Adell and J. de la Cal in [2]. A variant of the modulus $\omega_2^{1/2}$ is given below.

Definition 2.2.2. *For* $f \in \mathcal{F}(I)$, $h > 0$, *set*

$$\omega_2^\star(f, h) := \sup \Big\{ |\Delta(f; u, y, v)|, \ u, v \in I, \ u \neq v, \ u \leq y \leq v,$$

$$y - u \leq h, \ v - y \leq h \Big\}. \tag{2.25}$$

Remark 2.2.2. Obviously we have

$$\omega_2^\star(f, h) \leq \omega_2^{1/2}(f, h), \quad f \in \mathcal{F}(I), \ h > 0. \tag{2.26}$$

★

We give some auxiliary results.

Lemma 2.2.1. *We have*

$$|\Delta(f; y, u, z)| \leq \frac{2}{1 + 2\lambda} \cdot \omega_2^\lambda(f, h), \tag{2.27}$$

for all functions $f \in \mathcal{F}_b(I)$ *and all* $h > 0$, $0 \leq \lambda \leq \frac{1}{2}$, $y, z \in I$, $y < z \leq y + 2h$, $y \leq u \leq z$.

Proof. Fix f, h, y, z as in the statement of the lemma. Assume that $\omega_2^\lambda(f, h) < \infty$. First consider the case $0 \leq \lambda < \frac{1}{2}$. Let the function $g(t) := -\Delta(f; y, t, z)$, $t \in I$. We have $g(y) = 0 = g(z)$ and $\omega_2^\lambda(f, h) = \omega_2^\lambda(g, h)$.

Let $\varepsilon > 0$ be arbitrary. There exists $u_\varepsilon \in [y, z]$ such that $|g(u_\varepsilon)| > M - \varepsilon$, where $M := \sup\{|g(t)|, \ t \in [y, z]\}$. In the case where $y + \left(\frac{1}{2} - \lambda\right)(z - y) \leq u_\varepsilon \leq y + \left(\frac{1}{2} + \lambda\right)(z - y)$, we have $M \leq |g(u_\varepsilon)| + \varepsilon \leq \omega_2^\lambda(f, h) + \varepsilon$. In the case $y \leq u_\varepsilon \leq y + \left(\frac{1}{2} - \lambda\right)(z - y)$ define

$$v := y + \frac{2(u_\varepsilon - y)}{1 - 2\lambda}.$$

We have $M < |g(u_\varepsilon)| + \varepsilon = |-\Delta(g; y, u_\varepsilon, v) + \frac{1}{2}(1 - 2\lambda)g(v)| + \varepsilon \leq \omega_2^\lambda(f, h) + \frac{1}{2}(1 - 2\lambda)M + \varepsilon$. From this one obtains

$$M < \frac{2}{1 + 2\lambda}(\omega_2^\lambda(f, h) + \varepsilon).$$

Using the symmetry, this inequality is true also in the case $y + \left(\frac{1}{2} + \lambda\right)(z - y) \leq u_\varepsilon \leq z$. Since $\varepsilon > 0$ is arbitrary, relation (2.27) is proved for each $0 \leq \lambda < \frac{1}{2}$. In the case $\lambda = \frac{1}{2}$, (2.27) follows directly from relation (2.21). $\qquad\square$

Lemma 2.2.2. *Assume that $g \in \mathcal{F}_b(I)$, and let $\alpha, \beta \in I$ be such that $\beta = \alpha + 2h$, $h > 0$ and $g(\alpha) = 0 = g(\beta)$. Then the inequality*

$$|g(t)| \leq \left(v^2 + 2v + \frac{2}{1 + 2\lambda} \right) \omega_2^\lambda(g, h) \tag{2.28}$$

holds true for all $t \in I$, $t = \beta + vh$, $v \in (0, 1]$ and $\lambda \in \left[0, \frac{1}{2} \right]$.

Proof. Suppose $\omega_2^\lambda(f, h) < \infty$. First, consider $0 \leq \lambda < \frac{1}{2}$. Let $v \in (0, 1]$, and $t := \beta + vh$. Put $q := \frac{1}{2} - \lambda$ and let $k \geq 0$ be the unique integer such that

$$2q^{k+1} < v \leq 2q^k. \tag{2.29}$$

Consider three cases.

Case (1): $2q^{k+1} < v \leq \dfrac{2q^k(1 - q)}{1 - q^k}$, and $k \geq 1$.

$$\tag{2.30}$$

Set:

$$y_j := \alpha + \frac{(1 - q)q^{k-j+1}}{1 - q^{k+1}} \cdot (2 + v)h, \quad \text{for } 1 \leq j \leq k + 1.$$

We have the identity

$$g(t) = \frac{vq^k}{2(1 - q^k)} \left[\Delta(g; y_1, y_{k+1}, t) + \sum_{j=1}^{k} q^{-j} \Delta(g; \alpha, y_{k+1-j}, y_{k+2-j}) \right]$$

$$+ \frac{1}{2}(2 + v)\Delta(g; y_1, \beta, t).$$

From the right inequality in (2.30) it follows that

$$0 < t - y_1 = \frac{1 - q^k}{1 - q^{k+1}} \cdot (2 + v)h \leq 2h,$$

and

$$y_{k+1} - \alpha = \frac{1 - q}{1 - q^{k+1}} \cdot (2 + v)h \leq 2h.$$

Also we have

$$\frac{y_{k+1} - y_1}{t - y_1} = \frac{1}{2} + \lambda \quad \text{and} \quad \frac{y_j - \alpha}{y_{j+1} - \alpha} = \frac{1}{2} - \lambda, \ (1 \leq j \leq k).$$

By using Lemma 2.2.1 one obtains

$$|g(t)| \leq \left[\frac{vq^k}{2(1 - q^k)} \sum_{j=0}^{k} q^{-j} + \frac{2 + v}{1 + 2\lambda} \right] \omega_2^\lambda(g, h)$$

$$= \left[\frac{2}{1 + 2\lambda} + v \left(2 + 2q^{k+1} - \frac{T}{2(1 - q)(1 - q^k)} \right) \right] \omega_2^\lambda(g, h),$$

where

$$T := \left((2 - 4q^k) + (q^k - q^{2k+1}) + 3(q^{k+1} - q^{2k+1}) \right)(1 - 2q) + 2q^{k+2}(1 - 2q^k) > 0.$$

Since $2q^{k+1} < v$, we get (2.28).

Case (2): $\dfrac{2q^k(1-q)}{1-q^k} < v \le 2q^k$ and $k \ge 1.$

$$(2.31)$$

Define $z_j := t - vq^{-j}h,\ 0 \le j \le k$. Note that $z_0 = \beta$. We have

$$g(t) = \frac{1}{1-q^k}\left[\sum_{j=0}^{k-1} q^j \Delta(g; z_{j+1}, z_j, t) + q^k \Delta(g; \alpha, z_k, \beta) \right].$$

Since $v \le 2q^k$ we have: $z_k = (2 + v(1 - q^{-k}))h + \alpha \ge (2 + 2q^k(1 - q^{-k}))h + \alpha > \alpha,$
$z_k < t - vh = \beta$ and $0 < t - z_k = q^{-k}vh \le 2h$. Also, we have

$$\frac{z_j - z_{j+1}}{t - z_{j+1}} = \frac{1}{2} - \lambda, \ (0 \le j \le k - 1).$$

From Lemma 2.2.1 and using the left inequality in (2.31) one obtains

$$|g(t)| \le \left(\frac{1}{1-q} + \frac{q^k}{1-q^k} \cdot \frac{2}{1+2\lambda} \right) \omega_2^\lambda(g, h)$$

$$= \left(\frac{2}{1+2\lambda} + \frac{q^k}{(1-q)(1-q^k)} \right) \omega_2^\lambda(g, h)$$

$$\le \left(\frac{2}{1+2\lambda} + \frac{4q^k(1-q)}{1-q^k} \right) \omega_2^\lambda(g, h) < \left(\frac{2}{1+2\lambda} + 2v + v^2 \right) \omega_2^\lambda(g, h).$$

Case (3): $2q < v \le 1.$

$$(2.32)$$

Put $z := t - 2h$. We have

$$g(t) = \frac{2}{2-v}\left(\frac{v}{2} \cdot g(z) - g(\beta) + \frac{2-v}{2} \cdot g(t) \right) - \frac{v}{2-v} \cdot g(z)$$

$$= \frac{2}{2-v} \cdot \Delta(g; z, \beta, t) + \frac{v}{2-v} \cdot \Delta(g; \alpha, z, \beta).$$

Since $1 \ge v > 2q$ we get $|\Delta(g; z, \beta, t)| \le \omega_2^\lambda(g, h)$. By applying Lemma 2.2.1 one obtains

$$|g(t)| \le \left(\frac{2}{2-v} + \frac{v}{2-v} \cdot \frac{2}{1+2\lambda} \right) \omega_2^\lambda(g, h)$$

$$= \left(\frac{2}{1+2\lambda} + 2v - \frac{S}{(2-v)(1+2\lambda)} \right) \omega_2^\lambda(g, h)$$

where $S := -2(1 + 2\lambda)v^2 + 8\lambda v + 2 - 4\lambda \geq 0$, since $\frac{4\lambda - 2}{2 + 4\lambda} \leq v \leq 1$.

Therefore (2.28) is proved for each $0 \leq \lambda < \frac{1}{2}$. Using the inequality $\omega_2^\lambda(g, h) \leq \omega_2^*(g, h)$, for all $\lambda \in [0, \frac{1}{2})$, and passing to the limit $\lambda \to \frac{1}{2}$, (2.28) follows also for $\lambda = \frac{1}{2}$. $\qquad\square$

Lemma 2.2.3. *If $f \in \mathcal{F}_b(I)$, $a \in I$, $b \in I$, $b = a + 2h$ and $f(a) = 0 = f(b)$, then the inequality*

$$|f(t)| \leq \left(\frac{1 - 2\lambda}{1 + 2\lambda} + h^{-2}\left(t - \frac{a + b}{2}\right)^2\right) \omega_2^\lambda(f, h) \qquad (2.33)$$

holds for all $t \in I$, $t \geq b$ and $\lambda \in \left[0, \frac{1}{2}\right]$.

Proof. Put $c_j := b + jh$, for any integer j. Let $t > b$ be fixed, and denote by k the unique integer such that $k \geq 0$, and $c_k < t \leq c_{k+1}$. Let us also denote $\alpha := c_{k-2}$, $\beta := c_k$ and $v := h^{-1}(t - \beta)$. Define the function $g \in F_b(I)$ given by $g(y) := \Delta(f; \alpha, y, \beta)$, $(y \in I)$. We have $\omega_2^\lambda(f, h) = \omega_2^\lambda(g, h)$, $g(\alpha) = 0 = g(\beta)$ and

$$|f(t)| \leq |g(t)| + \frac{1}{2h} \cdot |(t - \beta)f(\alpha) + (\alpha - t)f(\beta)| \leq |g(t)| + |f(\beta)|$$
$$+ \frac{v}{2} \cdot |f(\beta) - f(\alpha)|.$$

From Lemma 2.2.2 we obtain

$$|g(t)| \leq \left(\frac{2}{1 + 2\lambda} + 2v + v^2\right) \omega_2^\lambda(g, h).$$

Also, we have

$$f(c_k) = \sum_{j=1}^{k} 2j\Delta(f; c_{k-1-j}, c_{k-j}, c_{k+1-j}) + k\Delta(f; c_{-2}, c_{-1}, c_0)$$
$$+ \left(\frac{k}{2} + 1\right) f(b) - \frac{k}{2} \cdot f(a).$$

Hence $|f(\beta)| \leq (k^2 + 2k)\omega_2^\lambda(f, h)$. Using the identity

$$f(c_k) - f(c_{k-2}) = 2\Delta(f; c_{-2}, c_{-1}, c_0) + 4 \sum_{j=1}^{k-1} \Delta(f; c_{j-2}, c_{j-1}, c_j)$$
$$+ 2\Delta(f; c_{k-2}, c_{k-1}, c_k),$$

one obtains $|f(\beta) - f(\alpha)| \leq 4k\omega_2^\lambda(f, h)$. From these relations follows

$$|f(t)| \leq \left(\frac{1 - 2\lambda}{1 + 2\lambda} + (1 + k + v)^2\right) \omega_2^\lambda(f, h)$$
$$= \left(\frac{1 - 2\lambda}{1 + 2\lambda} + h^{-2}\left(t - \frac{a + b}{2}\right)^2\right) \omega_2^\lambda(f, h).$$

The case $t = b$ is immediate. The lemma is proved. $\qquad\square$

Lemma 2.2.4. *Let $f \in \mathcal{F}(I)$, $x, y \in I$ and the number $h > 0$, such that $y \geq x + h$. Suppose that $|f(x + h) - f(x)| \leq \omega_2^*(f, h)$. Then*

$$|f(y) - f(x + h)| \leq h^{-2}(y - x)^2 \omega_2^*(f, h). \tag{2.34}$$

Proof. Define $c_i = x + ih$, $i = 0, 1, \ldots$. Denote $k = [(y - x)/h]$, where $[\cdot]$ is the integer part. Hence $k \geq 1$. From the identity

$$f(c_k) - f(c_{k-1}) = \sum_{i=2}^{k}[f(c_i) - 2f(c_{i-1}) + f(c_{i-2})] + f(c_1) - f(c_0), \tag{2.35}$$

(where the sum is 0 for $k = 1$), we obtain

$$|f(c_k) - f(c_{k-1})| \leq (2k - 1)\omega_2^*(f, h). \tag{2.36}$$

Also, from the identity

$$f(c_k) - f(c_1) = \sum_{i=2}^{k}(k - i + 1)[f(c_i) - 2f(c_{i-1})$$
$$+ f(c_{i-2})] + (k - 1)(f(c_1) - f(c_0)), \tag{2.37}$$

we obtain

$$|f(c_k) - f(c_1)| \leq \left(2\sum_{i=2}^{k}(k - i + 1) + k - 1\right)\omega_2^*(f, h)$$
$$= (k^2 - 1) \cdot \omega_2^*(f, h). \tag{2.38}$$

Let $t \in [k, k + 1)$ be such that $y = x + th$. Let us prove $|f(y) - f(x + h)| \leq t^2 \omega_2^*(f, h)$. If $t = k$ this follows directly from (2.38). Consider now the case $t > k$. We have $|\Delta(f; c_{k-1}, c_k, y)| \leq \omega_2^*(f, h)$, that is

$$\left|\frac{t - k}{t - k + 1} \cdot f(c_{k-1}) + \frac{1}{t - k + 1} \cdot f(y) - f(c_k)\right| \leq \omega_2^*(f, h).$$

Hence

$$|f(y) - f(c_k)| \leq (t - k)|f(c_k) - f(c_{k-1})| + (t - k + 1)\omega_2^*(f, h). \tag{2.39}$$

Finally, from (2.36), (2.38) and (2.39) we have

$$|f(y) - f(x + h)| \leq |f(y) - f(c_k)| + |f(c_k) - f(c_1)|$$
$$\leq \left[(t - k)(2k - 1) + (t - k + 1) + (k^2 - 1)\right]\omega_2^*(f, h)$$
$$\leq t^2 \omega_2^*(f, h). \quad \square \tag{2.40}$$

2.2.2 Main results

The following theorem was proved, for the most part, in [91], [97], [111], see also [87].

Theorem 2.2.1. *Let $F : V \to \mathbb{R}$, $V \subset \mathcal{F}(I)$, be a linear positive functional that is admissible related to a point $x \in I$. Let $\lambda \in \left[0, \frac{1}{2}\right]$, $b \in [0, 1)$ and $p \in [1, \infty)$. We have*

$$|F(f) - f(x)| \leq |F(e_0) - 1| \cdot |f(x)| + |F(e_1 - xe_0)| \cdot h^{-1}\omega_1(f, h)$$
$$+ \left[\frac{2}{1 + 2\lambda} \cdot F(e_0) + \frac{1}{(1 - b)^2} \cdot F\left(\left(\left|\frac{e_1 - xe_0}{h}\right|^p - b\right)^2\right)\right] \omega_2^\lambda(f, h) \quad (2.41)$$

for any $h > 0$ such that $length(I) \geq 2h$ and any $f \in V \cap \mathcal{F}_b(I)$.

Conversely, let $\lambda \in \left[0, \frac{1}{2}\right)$, $b \in [0, 1)$ and $p \in [1, \infty)$ and suppose that there are the constants $A, B, C, D \geq 0$, such that the inequality

$$|F(f) - f(x)| \leq A \cdot |f(x)| \cdot |F(e_0) - 1| + B \cdot |F(e_1 - xe_0)| \cdot h^{-1}\omega_1(f, h)$$
$$+ \left[C \cdot F(e_0) + D \cdot F\left(\left(\left|\frac{e_1 - xe_0}{h}\right|^p - b\right)^2\right)\right] \omega_2^\lambda(f, h) \quad (2.42)$$

is satisfied (only) for all linear positive functionals F of the form (1.11), for all $x \in I$, all $f \in C(I)$ and all $h > 0$ such that $length(I) \geq 2h$. Then we must have $A \geq 1$, $B \geq 1$; for $b = 0$ we have also $C \geq \frac{2}{1 + 2\lambda}$; and for $p = 1$ and $b = 0$ we have also $D \geq 1$.

Proof. We have to consider only the case where $\omega_1(f, h) < \infty$, which implies that $\omega_2^\lambda(f, h) < \infty$.

The proof of the direct part of the theorem is based on Theorem 2.1.2 (ii) in which we take $\Omega_1 := \omega_1$, $\Omega_2 := \omega_2^\lambda$ and the function

$$\Psi(t) = \Psi_{\lambda, p, b}(t) := \frac{2}{1 + 2\lambda} + \left(\frac{t^p - b}{1 - b}\right)^2, \quad t \geq 0. \quad (2.43)$$

Let us prove the corresponding relation (2.15), namely

$$|\Delta(f; t_1, x, t_2)| \leq \delta_h(\Psi_{\lambda, p, b}; t_1, x, t_2)\omega_2^\lambda(f, h),$$
$$\text{for all } t_1 < x < t_2, \ t_1, t_2 \in I. \quad (2.44)$$

Case (1) : $t_2 - t_1 \leq 2h$. From Lemma 2.2.1 we have

$$|\Delta(f; t_1, x, t_2)| \leq \frac{2}{1 + 2\lambda} \cdot \omega_2^\lambda(f, h) \leq \delta_h(\Psi_{\lambda, p, b}; t_1, x, t_2)\omega_2^\lambda(f, h).$$

Case (2) : $x - t_1 < h$ and $t_2 - t_1 > 2h$. Put $a := t_1$, $b := a + 2h$, $r := (x - t_1)h^{-1}$ and $s := (t_2 - x)h^{-1}$. We can assume that $f(a) = 0 = f(b)$, since otherwise we can replace f by the function $g(t) := -\Delta(f; a, x, b)$, see Lemma 2.1.1. With this assumption, one obtains from Lemma 2.2.1:

$$|f(x)| = |\Delta(f; a, x, b)| \le \frac{2}{1 + 2\lambda} \cdot \omega_2^\lambda(f, h).$$

From Lemma 2.2.3 we get $|f(t_2)| \le \left[\frac{1-2\lambda}{1+2\lambda} + (r + s - 1)^2\right] \cdot \omega_2^\lambda(f, h)$. Then we have

$$\begin{aligned}
|\Delta(f; t_1, x, t_2)| &\le \frac{r}{r + s} |f(t_2) - f(x)| + \frac{s}{r + s} |f(t_1) - f(x)| \\
&\le \frac{r}{r + s} \left(\frac{2}{1 + 2\lambda} + \frac{1 - 2\lambda}{1 + 2\lambda} + (r + s - 1)^2\right) \omega_2^\lambda(f, h) \\
&\quad + \frac{s}{r + s} \cdot \frac{2}{1 + 2\lambda} \cdot \omega_2^\lambda(f, h) \\
&\le \left[\frac{2}{1 + 2\lambda} + \frac{r}{r + s} (1 + (r + s - 1)^2)\right] \omega_2^\lambda(f, h).
\end{aligned}$$

On the other hand we have

$$\delta_h(\Psi_{\lambda, p, b}; t_1, x, t_2) = \frac{2}{1 + 2\lambda} + \frac{r}{r + s} \cdot \left(\frac{s^p - b}{1 - b}\right)^2 + \frac{s}{r + s} \cdot \left(\frac{r^p - b}{1 - b}\right)^2.$$

It remains to prove that $F_{r,s,b}(p) \ge 0$, for any $0 < r \le 1$, $0 \le b < 1$, $s > 2 - r$ and $p \ge 2$, where

$$F_{r,s,b}(p) := r \left(\frac{s^p - b}{1 - b}\right)^2 + s \left(\frac{r^p - b}{1 - b}\right)^2 - r(1 + (r + s - 1)^2).$$

We then have $\frac{d}{dp} F_{r,s,b}(p) > 0$, $p \ge 1$, which is equivalent to $(s^p - b)s^{p-1} \ln s + (r^p - b)r^{p-1} \ln r > 0$. Indeed, since this inequality is obvious for $b = 1$ it suffices to show it for $b = 0$, that is $s^{2p-1} \ln s + r^{2p-1} \ln r > 0$. But this inequality is implied by the inequality $s \ln s + r \ln r > 0$, which is implied itself by the immediate inequality $(2 - r) \ln(2 - r) + r \ln r > 0$.

Therefore $F_{r,s,b}(p) \ge F_{r,s,b}(1)$. For r, s, b as above, define $G_{r,b}(s) := F_{r,s,b}(1)$. For fixed r, b we have

$$\begin{aligned}
\frac{d}{ds} G_{r,b}(s) &= 2r(s - b)(1 - b)^{-2} + (r - b)^2(1 - b)^{-2} - 2r(r + s - 1) \\
&\ge 2rs + (r - b)^2(1 - b)^{-2} - 2rs - 2r^2 + 2r > 0.
\end{aligned}$$

Consequently it suffices to prove $G_{r,b}(2 - r) \ge 0$, which is equivalent to

$$r \left(\frac{2 - r - b}{1 - b}\right)^2 + (2 - r) \left(\frac{r - b}{1 - b}\right)^2 - 2r \ge 0.$$

Put $t := \frac{1-r}{1-b}$. The inequality above becomes the inequality

$$r(1 + t)^2 + (2 - r)(1 - t)^2 - 2r \ge 0,$$

which is immediate, if we verify it for $r = 0$ and $r = 1$. Then (2.44) holds in Case 2.

Case $(2')$: $t_2 - x < h$ and $t_2 - t_1 > 2h$. It is symmetrical to Case (2).

Case (3) : $t_2 - x \geq h$ and $x - t_1 \geq h$. Put $a := x - h$ and $b := x + h$. We can assume that $f(a) = 0 = f(b)$. From Definition 2.2.1 we get $|f(x)| = \left| \frac{1}{2} f(a) - f(x) + \frac{1}{2} f(b) \right| \leq \omega_2^\lambda(f, h)$. Then using Lemma 2.2.3 we obtain

$$|f(t_2) - f(x)| \leq \omega_2^\lambda(f, h) + \left(\frac{1 - 2\lambda}{1 + 2\lambda} + h^{-2}(t_2 - x)^2 \right) \omega_2^\lambda(f, h)$$

$$= \left(\frac{2}{1 + 2\lambda} + h^{-2}(t_2 - x)^2 \right) \omega_2^\lambda(f, h).$$

By taking into account the symmetry, we have also

$$|f(t_1) - f(x)| \leq \left(\frac{2}{1 + 2\lambda} + h^{-2}(t_1 - x)^2 \right) \omega_2^\lambda(f, h).$$

It follows that

$$|\Delta(f; t_1, x, t_2)| \leq \left(\frac{2}{1 + 2\lambda} + h^{-2}(t_2 - x)(x - t_1) \right) \omega_2^\lambda(f, h). \tag{2.45}$$

On the other hand, since $\Psi_{\lambda, p, b}(t) \geq \frac{2}{1+2\lambda} + t^2$, for $t \geq 1$, we obtain (2.44).

It remains to prove (2.17). By the symmetry it is enough to take $t > x$. Consider three cases.

Case (a) : $t \geq x + 2h$. Let l be a linear function such that the function $g := f + l$ satisfies the condition $g(x) = 0 = g(x + 2h)$. From Lemma 2.1.1 we obtain the relations $\Delta(f; x, t, x + 2h) = \Delta(g; x, t, x + 2h)$ and $\omega_2^\lambda(f, h) = \omega_2^\lambda(g, h)$. By applying Lemma 2.2.3 to the function $t \rightarrow \Delta(g; x, t, x + 2h)$, $(t \in I)$ for $a := x$ and $b := x + 2h$ we obtain

$$|f(t) - f(x)| = \left| \frac{t - x}{2h} \cdot (f(x + 2h) - f(x)) - \Delta(f; x, t, x + 2h) \right|$$

$$\leq \frac{|t - x|}{2h} \cdot \omega_1(f, 2h) + \left(\frac{1 - 2\lambda}{1 + 2\lambda} + \left(\frac{t - x - h}{h} \right)^2 \right) \omega_2^\lambda(f, h).$$

Finally, the inequality

$$|f(t) - f(x)| \leq h^{-1} |t - x| \omega_1(f, h) + \Psi_{\lambda, p, b} \left(\left| \frac{t - x}{h} \right| \right) \omega_2^\lambda(f, h)$$

follows from the immediate inequality

$$(u - 1)^2 \leq 1 + \left(\frac{u^p - b}{1 - b} \right)^2, \quad u \geq 1, \ b \in [0, 1), \ p \geq 1.$$

Case (b) : $x + h \leq t < x + 2h$. We have

$$f(t) - f(x) = \frac{t - x}{h} (f(x + h) - f(x)) + \frac{t - x}{h} \cdot \Delta(f; x, x + h, t).$$

From Lemma 2.2.1 and the inequality

$$\frac{2u}{1+2\lambda} \leq \frac{2}{1+2\lambda} + \left(\frac{u^p - b}{1-b}\right)^2, \quad u \geq 1,$$

it follows that

$$|f(t) - f(x)| \leq \frac{|t-x|}{h} \cdot \omega_1(f,h) + \frac{|t-x|}{h} \cdot \frac{2}{1+2\lambda} \cdot \omega_2^\lambda(f,h)$$

$$\leq h^{-1} \cdot |t-x| \omega_1(f,h) + \Psi_{\lambda,p,b}\left(\left|\frac{t-x}{h}\right|\right) \omega_2^\lambda(f,h).$$

Case (c) : $x < t < x+h$. Since length $(I) \geq 2h$ we have either $x+h \in I$ or $t - h \in I$. Using the symmetry we consider only the case $x + h \in I$. It follows that

$$f(t) - f(x) = \frac{t-x}{h} \cdot (f(x+h) - f(x)) - \Delta(f; x, t, x+h).$$

From this relation and from Lemma 2.2.1 one obtains immediately (2.17).

Therefore (2.17) is proved in all cases. By applying Theorem 2.1.2, relation (2.41) holds.

Now we proceed to prove the converse part.

If we take $f := e_0$ and F of the form (1.11), such that $F(e_0) \neq 1$, one obtains from (2.42) that $A \geq 1$.

Choosing in (2.42) $I := [0,1]$, $x \in (0,1)$, $F : \mathcal{F}[0,1] \to \mathbb{R}$ defined by $F(f) := f(1)$, $f \in C[0,1]$ and $f := e_1$, $h := \frac{1}{2}$ one obtains $1 - x \leq B(1-x)$, that is $B \geq 1$.

Let $0 \leq \lambda < \frac{1}{2}$, $b = 0$ and $p \in [1,\infty)$. If we take in (2.42) $I := [0,1]$, x an arbitrary point in $(0,1)$ and the functional $F : \mathcal{F}[0,1] \to \mathbb{R}$, defined by

$$F(f) := (1-x)f(0) + xf(1), \quad f \in \mathcal{F}[0,1],$$

we obtain for all $0 < h \leq \frac{1}{2}$ and $f \in \mathcal{F}[0,1]$,

$$|\Delta(f; 0, x, 1)| \leq (C + Dh^{-2p}((1-x)x^{2p} + x(1-x)^{2p}))\omega_2^\lambda(f,h). \quad (2.46)$$

We choose in (2.46) $h := \frac{1}{2}$, $0 < x \leq \frac{1}{2} - \lambda$ and the function f defined by $f(t) := tx^{-1}$, for $0 \leq t \leq x$, and $f(t) := (1-t)(1-x)^{-1}$, for $x \leq t \leq 1$. We have

$$\omega_2^\lambda\left(f, \frac{1}{2}\right) = \sup\left\{\left|\Delta\left(f; 0, u\left(\frac{1}{2} - \lambda\right), u\right)\right|, \quad \frac{2x}{1-2\lambda} \leq u \leq 1\right\}$$

$$= \frac{1+2\lambda}{2(1-x)},$$

and (2.46) becomes

$$1 \leq \left(C + 4^p D(x(1-x)^{2p} + (1-x)x^{2p})\right) \cdot \frac{1+2\lambda}{2(1-x)}.$$

By passing to the limit $x \to 0$, it follows that $C \geq \frac{2}{1+2\lambda}$.

Also, by taking $p = 1$ and $f := e_2$ in (2.46) and using the relation $\omega_2^\lambda(e_2, h) = h^2$, $h > 0$, one obtains $x(1 - x) \le C \cdot h^2 + D \cdot x(1 - x)$. By passing to the limit $h \to 0$ it follows that $D \ge 1$.

Note that, for any $f \in \mathcal{F}_b(I)$, any $h > 0$ such that $\omega_2^{1/2}(f, h) < \infty$ and any $\varepsilon > 0$ there is $0 \le \lambda < \frac{1}{2}$ such that $\omega_2^\lambda(f, h) > \omega_2^{1/2}(f, h) - \varepsilon$. Then we obtain for $\lambda = \frac{1}{2}$, the optimal values $A = 1$, $B = 1$, $C = 1$ and $D = 1$. The theorem is completely proved.   □

Remark 2.2.3. The optimality, for the general case, of the constants appearing in estimate (2.41) does not assure their optimality for particular functionals and particular values of h. A way to improve the estimates is to use additional parameters. The parameters p and b were be introduced in estimate (2.41), in order to obtain good estimates for the Bernstein operators, see Section 4.1.

Remark 2.2.4. For the particular functionals F, for which $F(e_0) = 1$, $F(e_1) = x$, the optimal constants C and D remain $C = \frac{2}{1+2\lambda}$, and $D = 1$, since the functional used in the proof of Theorem 2.2.1 for showing the optimality of values satisfies the above conditions. For the other second order moduli, discussed in the next sections, for these particular functionals the corresponding optimal values are smaller than for the general case.

Theorem 2.2.2. *Let $F : V \to \mathbb{R}$, $V \subset \mathcal{F}(I)$, be a linear positive functional. Suppose that F is admissible related to an interior point x of I and $F(e_0) = 1$, $F(e_1) = x$. Let $\lambda \in \left[0, \frac{1}{2}\right]$. Let $h > 0$ such that length $(I) \le 2h$. Then we have*

$$|F(f) - f(x)| \le \frac{2}{1 + 2\lambda} \cdot \omega_2^\lambda(f, h), \quad f \in V \cup \mathcal{F}_b(I). \tag{2.47}$$

Proof. Let $h_0 := \frac{1}{2}\text{length}(I)$. We have $\omega_2^\lambda(f, h) = \omega_2^\lambda(f, h_0)$ for all $h \ge h_0$. Therefore we obtain, for all $h \ge h_0$,

$$|F(f) - f(x)| \le \left(\frac{2}{1 + 2\lambda} + h^{-2p}F(|e_1 - xe_0|^{2p})\right)\omega_2^\lambda(f, h_0).$$

If we allow $h \to \infty$ in this inequality, we obtain (2.47).   □

The next corollary contains simple types of estimates using the modulus ω_2.

Corollary 2.2.1. *Let $F : V \to \mathbb{R}$, $V \subset \mathcal{F}(I)$ be a positive linear functional, that is admissible related to a point x of I. Let $f \in V \cap \mathcal{F}_b(I)$ and $h > 0$. We have:*
i) If length $(I) \ge 2h$, and $s \ge 2$, then

$$|F(f) - f(x)| \le |F(e_0) - 1| \cdot |f(x)| + |F(e_1 - xe_0)| \cdot h^{-1}\omega_1(f, h)$$
$$+ \left(F(e_0) + \frac{1}{2}h^{-s}F(|e_1 - xe_0|^s)\right)\omega_2(f, h). \tag{2.48}$$

ii) If $F(e_0) = 1$, $F(e_1) = x$, x is an interior point of I and length $(I) \le 2h$, then

$$|F(f) - f(x)| \le \omega_2(f, h). \tag{2.49}$$

Corollary 2.2.2. *We have*

$$\omega_2^\lambda(f,h) \le \left(\frac{2}{1+2\lambda} + \left(\frac{h}{\rho}\right)^2\right)\omega_2^\lambda(f,\rho) \tag{2.50}$$

for all $f \in \mathcal{F}_b(I)$, $\lambda \in \left[0,\frac{1}{2}\right]$, $0 < \rho \le h$, such that length $(I) \ge 2\rho$.

Proof. Let f, λ, ρ, h as in the hypothesis. Take $u, v \in I$, $u \ne v$, $|u-v| \le 2h$, and $t \in \left[\frac{1}{2}-\lambda, \frac{1}{2}+\lambda\right]$. Choose $x := tu + (1-t)v$. Then we can apply Theorem 2.2.1 for $p = 1$, $b = 0$ and the functional $F : \mathcal{F}_b(I) \to \mathbb{R}$ defined by $F(f) := tf(u) + (1-t)f(v)$. $\square$

In the sequel we deduce estimates for ω_2^*.

Theorem 2.2.3. *Let $F : V \to \mathbb{R}$, $V \subset \mathcal{F}(I)$, be a linear positive functional that is admissible related to a point $x \in I$. Let $f \in V \cap \mathcal{F}(I)$, $h > 0$ and $s \ge 2$. We have*
i) If length $(I) \ge 2h$, then

$$|F(f) - f(x)| \le |F(e_0) - 1| \cdot |f(x)| + |F(e_1 - xe_0)| \cdot h^{-1}\omega_1(f,h)$$
$$+ \left(F(e_0) + h^{-s}F(|e_1 - xe_0|^s)\right)\omega_2^*(f,h). \tag{2.51}$$

ii) If $F(e_0) = 1$, $F(e_1) = x$, x is an interior point of I and length $(I) \le 2h$, we have

$$|F(f) - f(x)| \le \omega_2^*(f,h). \tag{2.52}$$

Proof. We can restrict ourselves to the case $\omega_1(f,h) < \infty$. The theorem follows by applying Theorem 2.1.2 (ii) in which we take $\Omega_1 := \omega_1$, $\Omega_2 := \omega_2^*$ and the function

$$\Psi(t) := 1 + t^s, \ t \ge 0. \tag{2.53}$$

In order to prove the corresponding relation (2.15), namely

$$|\Delta(f; t_1, x, t_2)| \le (1 + h^{-2}(t_2 - x)(x - t_1))\omega_2^*(f,h),$$
$$t_1 < x < t_2, \ t_1, t_2 \in I, \tag{2.54}$$

let $t_1 < x < t_2$ be points of I. Define the points $a, b \in I$, in the following mode. Put $a := t_1$, if $x - t_1 \le h$ and $a := x - h$, conversely. Also, put $b := t_2$, if $t_2 - x \le h$ and $b := x + h$, conversely. We can assume that $f(a) = 0 = f(b)$, since otherwise we can replace f by the function $g(t) := -\Delta(f; a, x, b)$, (Lemma 2.1.1). Hence we have $|f(x)| = |\Delta(f; a, x, b)| \le \omega_2^*(f,h)$. We consider the following cases.

Case (1) : $a = t_1$ and $b = t_2$. Then $|\Delta(f; t_1, x, t_2)| \le \omega_2^*(f,h)$. Hence (2.54) is true.

Case (2) : $a = t_1$ and $b = x + h$. Then $|f(x+h) - f(x)| \le \omega_2^*(f,h)$ and from Lemma 2.2.4 we obtain

$$|f(t_2) - f(x)| \le |f(t_2) - f(x+h)| + |f(x+h) - f(x)|$$
$$\le (1 + h^{-2}(t_2 - x)^2)\omega_2^*(f,h)$$
$$\le (1 + h^{-s}(t_2 - x)^s)\omega_2^*(f,h) = \Psi\left(\left|\frac{t_2 - x}{h}\right|\right)\omega_2^*(f,h).$$

Also
$$|f(t_1) - f(x)| = |f(x)| \le \omega_2^*(f, h) \le \Psi\left(\left|\frac{t_1 - x}{h}\right|\right)\omega_2^*(f, h).$$

From these, one obtains immediately (2.54).

Case $(2')$: $a = x - h$ and $b = t_2$. This case is symmetrical to Case 2.

$\overline{\text{Case (3)}}$: $a = x - h$ and $b = x + h$. Analogous to Case 2, we obtain $|f(t_2) - f(x)| \le \Psi\left(\left|\frac{t_2-x}{h}\right|\right)\omega_2^*(f, h)$. Using the symmetry we then obtain, $|f(t_1) - f(x)| \le \Psi\left(\left|\frac{t_1-x}{h}\right|\right)\omega_2^*(f, h)$. Consequently one obtains (2.54).

It remains to prove (2.17). By the symmetry it is enough to take $t > x$. Consider three cases.

Case (a) : $t \ge x + 2h$. Let l be a linear function such that the function $g := f + l$ satisfies the condition $g(x) = 0 = g(x+2h)$. Lemma 2.1.1 gives $\Delta(f; x, t, x+2h) = \Delta(g; x, t, x+2h)$ and $\omega_2^\lambda(f, h) = \omega_2^\lambda(g, h)$. Also $|g(x+h)| \le \omega_2^*(f, h)$. By applying Lemma 2.2.4 we obtain $|\Delta(g; x, t, x + 2h)| = |g(t)| = |g(t) - g(x + 2h)| \le \left(\frac{t-x-h}{h}\right)^2 \omega_2^*(f, h)$. Consequently

$$|f(t) - f(x)| = \left|\frac{t - x}{2h} \cdot (f(x + 2h) - f(x)) - \Delta(f; x, t, x + 2h)\right|$$

$$\le \frac{|t - x|}{2h} \cdot \omega_1(f, 2h) + \left(\frac{t - x - h}{h}\right)^2 \omega_2^*(f, h)$$

$$\le h^{-1}|t - x|\,\omega_1(f, h) + \Psi\left(\left|\frac{t - x}{h}\right|\right)\omega_2^*(f, h).$$

Case (b) : $x + h \le t < x + 2h$. We have

$$f(t) - f(x) = \frac{t - x}{h}(f(x + h) - f(x)) + \frac{t - x}{h} \cdot \Delta(f; x, x + h, t).$$

It follows that

$$|f(t) - f(x)| \le \frac{|t - x|}{h} \cdot \omega_1(f, h) + \frac{|t - x|}{h} \cdot \omega_2^*(f, h)$$

$$\le h^{-1} \cdot |t - x|\,\omega_1(f, h) + \Psi\left(\left|\frac{t - x}{h}\right|\right)\omega_2^\lambda(f, h).$$

Case (c) : $x < t < x + h$. Since length $(I) \ge 2h$ we have either $x + h \in I$ or $t - h \in I$. Using the symmetry we consider only the case $x + h \in I$. It follows that

$$f(t) - f(x) = \frac{t - x}{h} \cdot (f(x + h) - f(x)) - \Delta(f; x, t, x + h).$$

From this relation one obtains immediately (2.17).

Therefore (2.17) is proved in all cases. By applying Theorem 2.1.2, the proof of point i) is finished. Relation (2.52) follows from (2.51) similarly as follows relation (2.47) from relation (2.41), see the proof of Theorem 2.2.2. $\qquad\square$

Remark 2.2.5. By taking into account relation (2.26) it follows that the estimate (2.51) is stronger than the estimate (2.41), for $\lambda = \frac{1}{2}$, $p = 1$ and $b = 0$, since the constants that appear in these two estimates are equal. This fact assures the optimality of the constants in estimate (2.51).

Example 2.2.1. Let the functional $F : V \to \mathbb{R}$, $F(f) = \lim_{t \to x} f(t)$, where V is the subspace of $\mathcal{F}(I)$ of functions admitting a finite limit in a given point $x \in I$. From Corollary 2.2.1 follows the inequality

$$| \lim_{t \to x} f(t) - f(x)| \leq \omega_2(f, h), \tag{2.55}$$

for any $f \in V$ and any $h > 0$. If we consider the case where x is an end point of I we can see that the constant 1 in front of the term $\omega_2(f, h)$ is optimal.

2.3 Estimates with modulus ω_2^d

2.3.1 Introduction. Auxiliary results

In the category of the second order moduli of continuity we can include the modulus $(f, h) \mapsto h\omega_1(f', h)$, $f \in \mathcal{D}(I)$. The coefficient h in front of the first order modulus of the derivative is required by the condition of normalization, see Definition 1.1.3

Estimates with the first modulus of the derivative were obtained by many authors. We mention for instance, E. Censor [21], R. DeVore [27], B. Mond and R. Vasudevan [70], Varshney and P. Singh [136], H. Gonska [37], C. Badea, I. Badea, H. Gonska [9].

The deficiency of the first modulus of the derivative, due to the fact that it is not applicable to nondifferentiable functions, will be remedied by constructing two extensions of it for the class of arbitrary functions. The main example of these extended second order moduli was introduced in [88], in the following mode.

Definition 2.3.1. *For $f \in \mathcal{F}(I)$ and the number $h > 0$ put*

$$\omega_2^d(f, h)$$
$$:= h \sup \left\{ \left| \frac{f(x + t_1) - f(x)}{t_1} - \frac{f(y + t_2) - f(y)}{t_2} \right|, \right. \tag{2.56}$$
$$t_1 > 0, \ t_2 > 0,$$
$$\left. x, x + t_1, y, y + t_2 \in I, \ \max\{x + t_1, y + t_2\} - \min\{x, y\} \leq h \right\}.$$

In (2.56) we admit that the supremum may be ∞.
The following theorem is immediate and we omit this proof.

Theorem 2.3.1. *For any function $f \in \mathcal{D}(I)$ and any number $h > 0$ we have*

$$\omega_2^d(f, h) = h \, \omega_1(f', h). \tag{2.57}$$

We state some auxiliary results for the modulus ω_2^d.

Lemma 2.3.1. *If* $g \in \mathcal{F}(I)$, $h > 0$, $a, b \in I$, $b = a + h$, $g(a) = 0 = g(b)$, $s \in (0, 1)$, *then*

$$| g(a + sh) | \leq s(1 - s)\, \omega_2^d(g, h). \tag{2.58}$$

Proof. The inequality (2.58) follows from the identity

$$g(a + sh) = s(1 - s)h \left(\frac{g(a + sh) - g(a)}{sh} - \frac{g(b) - g(a + sh)}{(1 - s)h} \right). \quad \square$$

Lemma 2.3.2. *If* $g \in \mathcal{F}(I)$, $h > 0$, $a, b \in I$, $b = a + h$, $g(a) = 0 = g(b)$, $q \in (0, 1]$, *then*

$$| g(b + qh) - g(a + qh) | \leq q\, \omega_2^d(g, h). \tag{2.59}$$

Proof. In order to prove the relation in (2.59), let the integer value be $m \geq 1$. From the identity

$$
\begin{aligned}
&g(b + qh) - g(a + qh) \\
&= \sum_{k=1}^{m-1} \left(g\left(b + \frac{k}{m} \cdot qh \right) - g\left(b + \frac{k-1}{m} \cdot qh \right) \right. \\
&\qquad \left. - g\left(a + \frac{k+1}{m} \cdot qh \right) + g\left(a + \frac{k}{m} \cdot qh \right) \right) \\
&\quad + \left(g(b + qh) - g\left(b + \frac{m-1}{m} \cdot qh \right) - g\left(b + \frac{1}{m} \cdot qh \right) + g(b) \right) \\
&\quad - g\left(a + \frac{1}{m} \cdot qh \right) \\
&\quad + \left(g\left(b + \frac{1}{m} \cdot qh \right) - 2g(b) + g\left(b - \frac{1}{m} \cdot qh \right) \right) \\
&\quad - g\left(a + \frac{m-q}{m} \cdot h \right),
\end{aligned}
$$

using Lemma 2.3.1 one obtains

$$
\begin{aligned}
&| g(b + qh) - g(a + qh) | \\
&\leq \left(q \cdot \frac{m-1}{m} + \frac{q}{m} + \frac{q}{m}\left(1 - \frac{q}{m} \right) + \frac{q}{m} + \frac{q}{m}\left(1 - \frac{q}{m} \right) \right) \omega_2^d(g, h).
\end{aligned}
$$

Since m can be arbitrarily chosen, (2.59) follows. $\quad \square$

Theorem 2.3.2. *We have*

$$\omega_2(f, h) \leq \omega_2^d(f, h), \qquad f \in \mathcal{F}(I), \ h > 0 \tag{2.60}$$

and there is no constant $C > 0$ *such that*

$$\omega_2^d(f, h) \leq C\, \omega_2(f, h), \qquad f \in C^1(I), \ h > 0. \tag{2.61}$$

Proof. Let $\alpha, \beta \in I$, $\beta = \alpha + 2\rho$, $0 < \rho \le h$. Consider the function $g := f + l$ where $l \in \Pi_1$ is chosen such that $g(\alpha) = 0 = g(\alpha + \rho)$. We have $\omega_2^d(f, \rho) = \omega_2^d(g, \rho)$ and $f(\beta) - 2f(\alpha + \rho) + f(\alpha) = g(\beta) - 2g(\alpha + \rho) + g(\alpha)$. Using (2.59), for the choices $a := \alpha$, $b := \alpha + \rho$, $h := \rho$ and $q := 1$, one obtains $| g(\beta) - 2g(\alpha + \rho) + g(\alpha) | = |g(\beta)| \le \omega_2^d(g, \rho)$. From this follows (2.60).

The second part of the theorem can be obtained by using, for any arbitrary $\varepsilon > 0$, continuously differentiable functions f for which there are $a \in I$, such that $f(a) = 1$ and $f(x) = 0$, $x \in I$, $| x - a | > \varepsilon$. $\qquad\square$

Lemma 2.3.3. *Let $f \in \mathcal{F}(I)$, $h > 0$, $a, b \in I$, $b = a + h$, $f(a) = 0 = f(b)$, $s \in (0, 1)$, $q \in (0, 1]$ and the integer $n \ge 0$. Then*

$$| f(b + (n + q)h) - f(a + sh) |$$
$$\le \left(\frac{1}{2} \cdot n^2 + \left(\frac{1}{2} + q \right) n + qs - s^2 + \max\{q, s\} \right) \omega_2^d(f, h). \qquad (2.62)$$

Proof. First, we consider $n = 0$. In the case $s > q$, from the identity

$$f(b + qh) - f(a + sh)$$
$$= \frac{q(s - q)}{s} h \left(\frac{f(b + qh) - f(b)}{qh} - \frac{f(a + sh) - f(a + qh)}{(s - q)h} \right)$$
$$+ \frac{q}{s}(f(b + qh) - f(a + qh)) - \frac{s - q}{s} \cdot f(a + sh),$$

and using Lemmas 2.3.1 and 2.3.2, we arrive at

$$|f(b + qh) - f(a + sh)| \le \left[\frac{q(s - q)}{s} + \frac{q^2}{s} + (s - q)(1 - s) \right] \omega_2^d(f, h)$$
$$= (s - s^2 + qs)\omega_2^d(f, h).$$

In the case $q > s$, from the identity

$$f(b + qh) - f(a + sh)$$
$$= \frac{(q - s)(1 - q)}{1 - s} h \left(\frac{f(b + qh) - f(b + sh)}{(q - s)h} - \frac{f(b) - f(a + qh)}{(1 - q)h} \right)$$
$$+ \frac{q - s}{1 - s}(f(b + qh) - f(a + qh))$$
$$+ \frac{1 - q}{1 - s}(f(b + sh) - f(a + sh)) - \frac{q - s}{1 - s} \cdot f(a + sh),$$

and using Lemmas 2.3.1 and 2.3.2, we arrive at

$$|f(b + qh) - f(a + sh)|$$
$$\le \left[\frac{(q - s)(1 - q)}{1 - s} + \frac{(q - s)q}{1 - s} + \frac{(1 - q)s}{1 - s} + (q - s)s \right] \omega_2^d(f, h)$$
$$= (q + qs - s^2)\omega_2^d(f, h).$$

In the case $s = q$, the relation in (2.62) results from (2.59).

Now, let $n \geq 1$. We have

$$f(b + (n + q)h) - f(a + sh)$$

$$= \sum_{k=1}^{n} (n - k + 1) \left(f(b + (k + q)h) \right.$$

$$- 2 f(b + (k - 1 + q)h) + f(b + (k - 2 + q)h))$$

$$+ n(f(b + qh) - f(a + qh)) + (f(b + qh) - f(a + sh)).$$

Using (2.60),(2.59) and (2.62, for $n = 0$), the lemma is proved. □

Finally, we present another possibility of extension of the first modulus of the derivative,

Definition 2.3.2. [88] *For $f \in \mathcal{F}(I)$ and the number $h > 0$, put*

$$\omega_2^e(f, h) := h \sup \left\{ \frac{|f(x + t_1 + t_2) - f(x + t_1) - f(x + t_2) + f(x)|}{t_1}, \right.$$

$$\left. t_1 > 0, \ t_2 > 0, \ t_1 + t_2 \leq h, \ x, \ x + t_1 + t_2 \in I \right\}. \qquad (2.63)$$

In (2.63) we admit that the supremum may be ∞. One obtains the same definition if the denominator t_1 is replaced by t_2 or by $\min\{t_1, t_2\}$. The following theorems are immediate.

Theorem 2.3.3. *For any function $f \in \mathcal{D}(I)$ and any number $h > 0$ we have*

$$\omega_2^e(f, h) = h \, \omega_1(f', h). \qquad (2.64)$$

Theorem 2.3.4. *We have*

$$\omega_2^e(f, h) \leq \omega_2^d(f, h), \qquad f \in \mathcal{F}(I), \ h > 0. \qquad (2.65)$$

A partial reciprocal result is given in the next theorem.

Theorem 2.3.5. *We have*

$$\omega_2^d(f, h) = \omega_2^e(f, h), \qquad f \in \mathcal{F}_b(I), \ h > 0. \qquad (2.66)$$

Proof. Let us prove the inverse inequality to (2.65). Fix f and h. Consider two cases.

Case 1: : f is continuous. If $\omega_2^e(f, h) = \infty$ then (2.64) is obvious. Consider now that $\omega_2^e(f, h) < \infty$. Choose $\varepsilon > 0$ arbitrarily. Let $x, y \in I$, $t_1 > 0$, $t_2 > 0$ such that $x + t_1, \ x + t_2 \in I$, $\max\{x + t_1, y + t_2\} - \min\{x, y\} \leq h$.

Let $n \in \mathbb{N}$, such that $t_1 \leq n t_2$ and denote $\rho := t_1/n$. For such a number n there are the unique numbers $m \in N$, $r \in [0, 1)$, depending on n such that $t_2 = (m + r)\rho$.

Since f is continuous we can take n sufficiently great such that

$$| (f(x + t_1) - f(x))/t_1 - (f(y + t_2) - f(y))/t_2 |$$

$$\leq | (f(x + n\rho) - f(x))/(n\rho) - (f(y + m\rho) - f(y))/(m\rho) | + \varepsilon.$$

But we have

$$| (f(x+n\rho) - f(x))/(n\rho) - (f(y+m\rho) - f(y))/(m\rho) |$$

$$\leq \frac{1}{mn} \sum_{i=1}^{n} \sum_{j=1}^{m} | (f(x+i\rho) - f(x+(i-1)\rho)$$

$$- f(y+j\rho) + f(y+(j-1)\rho) | /\rho$$

$$\leq \omega_2^e(f,h).$$

Hence $\omega_2^d(f,h) \leq \omega_2^e(f,h)$.

Case 2 : f has at least a point of discontinuity. Consider, for a choice, that f has a discontinuity from the right at $a \in I$. Denote

$$A := \{l \in \bar{\mathbb{R}} \mid (\exists)(x_n)_n, \ x_n \in I \cap (a, a+h), \ n \in \mathbb{N},$$

$$\lim_{n\to\infty} x_n = a, \ \lim_{n\to\infty} f(x_n) = l\},$$

and let $\alpha, \beta \in \bar{\mathbb{R}}$, $\alpha := \inf A$, $\beta := \sup A$. We have $\beta > f(a)$ or $f(a) > \alpha$. Consider only the case $\beta > f(a)$. Consider a sequence $(x_n)_n$ as in A for $l = \beta$. There is $n_0 \in \mathbb{N}$, such that $2x_n - a \in I$, $n \geq n_0$. We have

$$\limsup_{n\to\infty}(f(2x_n - a) - 2f(x_n) + f(a)) \leq (f(a) - \beta).$$

Then

$$\lim_{n\to\infty} (f(2x_n - a) - 2f(x_n) + f(a))/(x_n - a) = -\infty.$$

Therefore $\omega_2^e(f,h) = \infty$ and hence (2.64) is true. $\square$

In the next subsection we shall use only the modulus ω_2^d.

2.3.2 Main results

Most of the results given above in this section are obtained by combining the results given in [87], [88], [91] and [104].

In estimates of form (2.1) with the modulus ω_2^d a different situation appears by comparing them with the estimates made with the modulus ω_2^λ, as regards the optimality of the constants. So, in the case of the modulus ω_2^d, the optimal values for C and D in the corresponding estimate (2.1), are not optimal for functionals F with the properties $F(e_0) = 1$ and $F(e_1) = x, x \in I$. First we consider this particular type of functionals. In the statements of the corollaries the moments will be denoted as in (1.17).

Theorem 2.3.6. *Let $F : V \to \mathbb{R}$, $V \subset \mathcal{F}(I)$, be a linear positive functional. Suppose that F is admissible related to an interior point x of I. Suppose also that $F(e_0) = 1$, $F(e_1) = x$. Let $r \geq 1$. We have*

$$|F(f) - f(x)| \leq \left(\frac{1}{8r} + \frac{r}{2} h^{-2} F((e_1 - xe_0)^2) \right) \omega_2^d(f,h), \qquad (2.67)$$

for all $f \in V$ and $h > 0$.

Conversely, if there are constants C, $D \geq 0$, such that the inequality

$$|F(f) - f(x)| \leq \left(C + D \cdot h^{-2} F((e_1 - xe_0)^2) \right) \omega_2^d(f, h) \qquad (2.68)$$

is satisfied (only) for all linear positive functionals F of the form (1.11), for all $x \in I$, all $f \in C^1(I)$ and all $h > 0$, then it follows that $D \geq \frac{1}{2}$ and if $D = \frac{r}{2}, r \geq 1$, then it also follows that $C \geq \frac{1}{8r}$.

Proof. Consider the direct part. Let $h > 0$ and $f \in V$, with $\omega_2^d(f, h) < \infty$. Let x be an interior point of I. We apply Theorem 2.1.2 (i), by choosing $\Omega_2 := \omega_2$ and the function $\Psi(y) = \frac{1}{8r} + \frac{r}{2} y^2$, $y \in [0, \infty)$. We have to prove the corresponding relation (2.15), namely

$$|\Delta(f; t_1, x, t_2)| \qquad (2.69)$$
$$\leq \left(\frac{1}{8r} + \frac{r}{2} h^{-2}(t_2 - x)(x - t_1) \right) \omega_2^d(f, h), \quad t_1 < x < t_2, \ t_1, t_2 \in I.$$

We distinguish among three cases.

<u>Case 1:</u> $t_2 - t_1 \leq h$. We put $a := t_1, b := t_2, \ \rho := t_2 - t_1$ and $s := (x - a)\rho^{-1}$. We can suppose that $f(a) = 0 = f(b)$, since otherwise we can replace function f by a function $g := f + l$, where $l \in \Pi_1$ is taken such that $g(a) = 0 = g(b)$. Then $\omega_2^d(f, h) = \omega_2^d(g, h)$ and $\Delta(f; t_1, x, t_2) = \Delta(g; t_1, x, t_2)$.

From the condition $f(a) = 0 = f(b)$, using Lemma 2.3.1 one gets

$$|\Delta(f; t_1, x, t_2)| = |f(x)| \leq s(1 - s)\omega_2^d(f, \rho) \leq s(s - 1)\rho h^{-1}\omega_2^d(f, h).$$

It remains to prove

$$\frac{1}{8r} + \frac{r}{2} h^{-2}(t_2 - x)(x - t_1) \geq s(1 - s)\rho. \qquad (2.70)$$

If we put $u := \rho h^{-1}$, since $0 < u \leq 1$ and $s \in (0, 1)$ it follows that

$$\frac{1}{8r} + \frac{r}{2} h^{-2}(t_2 - x)(x - t_1) - s(1 - s)\rho h^{-1}$$
$$= \frac{1}{8r} + \frac{r}{2} s(1 - s)u^2 - s(1 - s)u$$
$$= \frac{1}{8r} (1 - 2ru\sqrt{s(1 - s)})^2 + \frac{u}{2}\sqrt{s(1 - s)}(1 - 2\sqrt{s(1 - s)}) \geq 0.$$

Therefore relation (2.70) is true and consequently relation (2.69) is valid.

<u>Case 2:</u> $x - t_1 < \frac{h}{2}$ and $t_2 - t_1 > h$. Put $a := t_1, b := a + h$ and $s := h^{-1}(x - a)$. We can write t_2 in the form $t_2 = b + (n + q)h$, where $n \in \mathbb{N} \cup \{0\}$ and $q \in (0, 1]$. We can suppose as in Case 1, that $f(a) = 0 = f(b)$. From Lemma 2.3.1, we have

$$|f(t_1) - f(x)| \leq s(1 - s)\omega_2^d(f, h)$$

and from Lemma 2.3.3,

$$|f(t_2) - f(x)| \le \left(\frac{1}{2}n^2 + \left(\frac{1}{2} + q\right)n + qs - s^2 + \max\{q, s\}\right)\omega_2^d(f, h).$$

Consequently it follows that

$$|\Delta(f; t_1, x, t_2)| \le \frac{x - t_1}{t_2 - t_1}|f(t_2) - f(x)| + \frac{t_2 - x}{t_2 - t_1}|f(t_1) - f(x)|$$

$$\le \left[\frac{s}{1 + n + q}\left(\frac{n^2}{2} + \frac{2q + 1}{2}n + qs - s^2 + \max\{q, s\}\right)\right.$$

$$\left. + \frac{1 + n + q - s}{1 + n + q}s(1 - s)\right]\omega_2^d(f, h)$$

$$= \frac{1}{1 + n + q}\left(\frac{s}{2}n^2 + \left(\frac{3}{2}s + sq - s^2\right)n + qs + s - 2s^2 + s\max\{q, s\}\right)\omega_2^d(f, h).$$

On the other hand we have

$$\frac{1}{8r} + \frac{r}{2}h^{-2}(t_2 - x)(x - t_1) = \frac{1}{8r} + \frac{r}{2}(1 + n + q - s)s$$

$$= \frac{1}{1 + n + q}\left[\frac{r}{2}sn^2 + \left(\frac{1}{8r} + rs + rsq - \frac{r}{2}s^2\right)n\right.$$

$$\left. + \frac{1}{8r} + \frac{1}{8r}q + \frac{r}{2}s + rsq + \frac{r}{2}sq^2 - \frac{r}{2}s^2 - \frac{r}{2}qs^2\right].$$

Therefore, in order to prove inequality (2.69) it is enough to show the validity of the inequality

$$T(n, r, q, s) \ge 0, \text{ for } n \ge 0, \ r \ge 1, \ q \in (0, 1] \ s \in \left(0, \frac{1}{2}\right),$$

where

$$T(n, r, q, s) := T_1(r, s)n^2 + T_2(r, q, s)n + T_3(r, q, s), \text{ and}$$

$$T_1(r, s) := \frac{r}{2}s - \frac{s}{2},$$

$$T_2(r, q, s) := \frac{1}{8r} + sr + rsq - \frac{r}{2}s^2 - \frac{3}{2}s - sq + s^2,$$

$$T_3(r, q, s) := \frac{1}{8r} + \frac{1}{8r}q + \frac{r}{2}s + rsq + \frac{r}{2}sq^2 - \frac{r}{2}s^2$$

$$- \frac{r}{2}qs^2 - qs - s + 2s^2 - s\max\{q, s\}.$$

We have obviously, $T_1(r, s) \ge 0$. Afterwards

$$T_2(r, q, s) = \frac{1}{8r}(1 - 2rs)^2 + sq(r - 1) + s(1 - s)(r - 1) \ge 0.$$

It remains to prove $T_3(r, q, s) \ge 0$. We consider some cases.

Case a: $0 < q \le s < \frac{1}{2}$, $r \ge 1$. It is sufficient to show that:

i) $T_3(r, 0, s) \geq 0$, for $0 < s < \dfrac{1}{2}$, $r \geq 1$,

ii) $\dfrac{\partial}{\partial q} T_3(r, q, s) \geq 0$, for $0 \leq q < s < \dfrac{1}{2}$, $r \geq 1$.

If $r \geq 2$, obviously, $T_3(r, 0, s) \geq 0$ and if $1 \leq r < 2$ we have

$$T_3(r, 0, s) \geq \frac{1}{8r} + \frac{1}{4}\left(\frac{r}{2} - 1\right) = \frac{1}{8r}(r - 1)^2 \geq 0.$$

Thus i) is true. Also, putting $r := 1 + v$, $v \geq 0$, we have

$$\frac{\partial}{\partial q} T_3(r, q, s) = \frac{1}{8r} + rs + rsq - \frac{r}{2}s^2 - s$$

$$\geq \frac{1}{8r}(1 + 8sr^2 - 4r^2s^2 - 8rs)$$

$$= \frac{1}{8r}\left((1 - 4s^2) + 8vs(1 - s) + 4v^2s(1 - s) + 4v^2s\right) \geq 0. \qquad (2.71)$$

Hence ii) is true.

Case b: $0 < s < \frac{1}{2}$, $s < q \leq 1 \leq r$. If we define $U(r, q, s) := rT_3(r, q, s)$, it is enough to show the validity of the relations

i) $U(1, q, s) \geq 0$, for $0 \leq s < \dfrac{1}{2}$, $s < q \leq 1$,

ii) $\dfrac{\partial}{\partial r} U(r, q, s) \geq 0$, for $0 \leq s < \dfrac{1}{2}$, $s < q \leq 1 \leq r$.

We have

$$U(1, q, s) = \left(\frac{3}{2} - \frac{q}{2}\right)s^2 + \left(\frac{1}{2}q^2 - q - \frac{1}{2}\right)s + \frac{1}{8} + \frac{q}{8}.$$

For a fixed $q \in (0, 1]$ the discriminant Δ of the polynomial of degree 2: $s \mapsto U(1, q, s)$ is

$$\Delta = \left(\frac{1}{2} - q - \frac{1}{2}\right)^2 - 4\left(\frac{3}{2} - \frac{q}{2}\right)\left(\frac{1}{8} + \frac{q}{8}\right) = \frac{1}{4}(q - 1)^2((q - 1)^2 - 3) \leq 0.$$

Then i) is true. Also, for $0 \leq s < \frac{1}{2}$, $s < q \leq 1 \leq r$ we have

$$\frac{\partial}{\partial r} U(r, q, s) = r(s(1 - s) + 2sq + sq(q - s)) - 2qs - s + 2s^2$$

$$\geq s(1 - s) + 2sq + sq(q - s) - 2qs - s + 2s^2$$

$$= s^2 + sq(q - s) \geq 0.$$

Hence ii) is true too. Then Case 2 is completely proved.

Case 2': $t_2 - x < \frac{h}{2}$ and $t_2 - t_1 > h$ is symmetrical to Case 2.

Case 3: $x - t_1 \geq \frac{h}{2}$ and $t_2 - x \geq \frac{h}{2}$. We put $a := x - \frac{h}{2}$, $b := x + \frac{h}{2}$ and we can assume as in Case 1 that $f(a) = 0 = f(b)$. From Lemma 2.3.3, for $s = \frac{1}{2}$ we get for any $t \in I$, $t = b + (n + q)h$, $n \in \mathbb{N} \cup \{0\}$, $q \in (0, 1]$:

$$|f(t) - f(x)| \le \left(\frac{1}{2}n^2 + \left(\frac{1}{2}+q\right)n + \frac{q}{2} - \frac{1}{4} + \max\left\{q, \frac{1}{2}\right\}\right)\omega_2^d(f, h)$$

$$\le \left(\frac{1}{2}n^2 + \left(\frac{1}{2}+q\right)n + \frac{q}{2} + \frac{1}{4} + \frac{1}{2}q^2\right)\omega_2^d(f, h)$$

$$= \left(\frac{1}{8} + \frac{1}{2}\left(n + \frac{1}{2} + q\right)^2\right)\omega_2^d(f, h).$$

Therefore there holds

$$|f(t) - f(x)| \le \left(\frac{1}{8} + \frac{1}{2}h^{-2}(t - x)^2\right)\omega_2^d(f, h), \tag{2.72}$$

for any $t \in I$, $t > b$.

Because of the symmetry, the estimate in (2.72) is also true for any $t \in I$, $t < a$. For any $u \ge \frac{1}{4}$ and $r \ge 1$ we have

$$\left(\frac{1}{8r} + \frac{r}{2}u\right) - \left(\frac{1}{8} + \frac{u}{2}\right) = \frac{1}{8r}(r - 1)(4ru - 1) \ge 0.$$

Then, by taking $u = h^{-2}(t - x)^2$ we obtain from relation (2.72), and from the symmetrical relation of it, that

$$|f(t) - f(x)| \le \left(\frac{1}{8r} + \frac{r}{2}h^{-2}(t - x)^2\right)\omega_2^d(f, h), \text{ for } t \in I, \ |t - x| \ge \frac{h}{2}. \tag{2.73}$$

Now the inequality (2.73) and the inequality

$$|\Delta(f; t_1, x, t_2)| \le \frac{t_2 - x}{t_2 - t_1}|f(t_1) - f(x)| + \frac{x - t_1}{t_2 - t_1}|f(t_2) - f(x)|$$

lead to (2.69).

The direct part of the theorem is proved. For the converse part we consider the following example. Let $I := [0, 1]$ $x \in (0, 1)$ and $F : \mathcal{F}[0, 1] \to \mathbb{R}$, defined by

$$F(f) := (1 - x)f(0) + xf(1), \quad f \in \mathcal{F}[0, 1].$$

We have $F(e_0) = 1$, $F(e_1) = x$ and $F((e_1 - xe_0)^2) = x(1 - x)$. Also $\omega_2^d(f, h) = h\omega_1(f', h)$, for $f \in C^1[0, 1]$ and $h > 0$. Suppose that there are real numbers $C \ge 0$ and $D \ge 0$ such that the inequality

$$|F(f) - f(x)| \le (C + Dx(1 - x)h^{-2})\omega_2^d(f, h) \tag{2.74}$$

holds for any $f \in C^1[0, 1]$, and any $h > 0$.

By choosing $f = e_2$ in (2.74) it follows that

$$x(1 - x) \le 2Ch^2 + 2Dx(1 - x), \text{ for all } h > 0.$$

If we allow $h \to 0+$ in this inequality we get $D \ge \frac{1}{2}$.

Take now $D \ge \frac{1}{2}$ and put $D = \frac{r}{2}, r \ge 1$. We show that we must have $C \ge \frac{1}{8r}$. For $0 < \varepsilon < \frac{1}{2}$ define the function $f_\varepsilon \in C^1[0, 1]$, given by:

$$f_\varepsilon(t) := \begin{cases} \frac{1}{2}t, & 0 \le t \le \frac{1}{2} - \varepsilon, \\[2mm] \frac{1}{2}t - \frac{1}{4\varepsilon}\left(t - \frac{1}{2} + \varepsilon\right)^2, & \frac{1}{2} - \varepsilon \le t \le \frac{1}{2} + \varepsilon, \\[2mm] \frac{1}{2}(1-t), & \frac{1}{2} + \varepsilon \le t \le 1. \end{cases} \qquad (2.75)$$

We have $f_\varepsilon\left(\frac{1}{2}\right) = \frac{1}{4} - \frac{\varepsilon}{4}$ and $\omega_2^d(f_\varepsilon, h) = h$, for $h \ge 2\varepsilon$. By taking $f := f_\varepsilon$, $x := \frac{1}{2}$ and $h := r$ in (2.74) we obtain

$$\frac{1}{4} - \frac{\varepsilon}{4} \le \left(C + \frac{1}{8r}\right)r, \text{ for } 0 < \varepsilon < \frac{1}{2}.$$

Allowing $\varepsilon \to 0+$ we get $C \ge \frac{1}{8r}$. The theorem is completely proved. □

Since in estimate (2.67) the number $r \ge 1$ is arbitrary, we can choose for r the best value for fixed F, h and x. So we get the following corollaries.

Corollary 2.3.1. *Let F and x be as in Theorem 2.3.6. We have*

$$|F(f) - f(x)| \le \begin{cases} \left(\frac{1}{8} + \frac{1}{2}h^{-2}m_2\right)\omega_2^d(f, h), & 0 < h \le 2\sqrt{m_2}, \\[2mm] \frac{1}{2}\sqrt{m_2}\,h^{-1}\omega_2^d(f, h), & h > 2\sqrt{m_2} \end{cases} \qquad (2.76)$$

for all $f \in V$, $h > 0$.

Proof. If $m_2 = 0$, then $F(f) = f(x)$ and hence (2.76) is true. If $m_2 > 0$, we take into account that the minimum of the function $r \mapsto \frac{1}{8r} + \frac{r}{2}m_2$, $r \ge 1$ is $r := \frac{h}{2\sqrt{m_2}}$ in the case $h \ge 2\sqrt{m_2}$ and $r := 1$ in the case $0 < h \le 2\sqrt{m_2}$. □

In the case of differentiable functions we have

Corollary 2.3.2. *If F and x are as in Theorem 2.3.6, then we have*

$$|F(f) - f(x)| \le \begin{cases} \left(\frac{h}{8} + \frac{1}{2}h^{-1}m_2\right)\omega_1(f', h), & 0 < h \le 2\sqrt{m_2}, \\[2mm] \frac{1}{2}\sqrt{m_2}\,\omega_1(f', h), & h > 2\sqrt{m_2} \end{cases} \qquad (2.77)$$

for $f \in V \cap \mathcal{D}(I)$, $h > 0$.

Remark 2.3.1. The estimates (2.76) and (2.77) in the case $h > 2\sqrt{m_2}$ can be obtained from the corresponding estimates given for $h = 2\sqrt{m_2}$.

Now we consider the general case.

Theorem 2.3.7. *Let $F : V \to \mathbb{R}$, $V \subset \mathcal{F}(I)$, be a linear positive functional that is admissible related to a point $x \in I$. Let $r \ge 1$. The inequality*

$$|F(f) - f(x)| \le |f(x)| \cdot |F(e_0) - 1| + |F(e_1 - xe_0)| \cdot h^{-1}\omega_1(f, h) \qquad (2.78)$$
$$+ \left(\frac{1}{2r+4} \cdot F(e_0) + \frac{r}{2} \cdot h^{-2}F((e_1 - xe_0)^2)\right)\omega_2^d(f, h)$$

holds for $f \in V$ and $h > 0$, such that length $(I) \geq 2h$.

Conversely, if the inequality

$$|F(f) - f(x)| \leq A \cdot |f(x)| \cdot |F(e_0) - 1| + B \cdot |F(e_1 - xe_0)| \cdot h^{-1}\omega_1(f, h)$$
$$+ (C \cdot F(e_0) + D \cdot h^{-2}F((e_1 - xe_0)^2))\omega_2^d(f, h) \quad (2.79)$$

is satisfied (only) for all linear positive functionals F of the form (1.11), for all $x \in I$, all $f \in C^1(I)$ and all $h > 0$ such that length $(I) \geq 2h$, then we must have $A \geq 1$, $B \geq 1$, $D \geq \frac{r}{2}$ and moreover, if $D = \frac{r}{2}$, $r \geq 1$ and $B = 1$, then we must have also $C \geq \frac{1}{2r+4}$.

Proof. For the direct part of the theorem, let us consider that $\omega_1(f, h) < \infty$, $\omega_2^d(f, h) < \infty$. We prove it by choosing $\Omega_1 := \omega_1$, $\Omega_2 := \omega_2^d$ and the function $\Psi(y) := \frac{1}{2r+4} + \frac{r}{2}y^2$, $y \in [0, \infty)$, in Theorem 2.1.2. If x is an interior point of I, then the corresponding relation (2.15) becomes

$$|\Delta(f; t_1, x, t_2)| \leq \left(\frac{1}{2r+4} + \frac{r}{2}h^{-2}(t_2 - x)(x - t_1)\right)\omega_2^d(f, h), \quad (2.80)$$
$$t_1 < x < t_2, \; t_1, t_2 \in I.$$

Since we have $\frac{1}{2r+4} \geq \frac{1}{8r}$, relation (2.80) follows from relation (2.69).

It remains to prove the corresponding inequality (2.17). Suppose for a choice that $t > x$. We distinguish between two cases.

Case (1): $t > x + h$. Applying relation (2.80) we get successively

$$|f(t) - f(x)| = \frac{t - x}{h}|f(x + h) - f(x) + \Delta(f; x, x + h, t)|$$
$$\leq \frac{t - x}{h} \cdot (\omega_1(f, h) + |\Delta(f; x, x + h, t)|)$$
$$\leq \frac{t - x}{h}\left[\omega_1(f, h) + \left(\frac{1}{2 + 4r} + \frac{r}{2} \cdot \frac{t - x - h}{h}\right)\omega_2^d(f, h)\right]$$
$$\leq h^{-1}|t - x|\omega_1(f, h) + \Psi\left(\left|\frac{t - x}{h}\right|\right)\omega_2^d(f, h).$$

Case (2): $x < t \leq x + h$. Since length $(I) \geq 2h$, we have either $x + h \in I$ or $t - h \in I$. Consider only the case $x + h \in I$. There holds

$$f(t) - f(x) = \frac{t - x}{h} \cdot (f(x + h) - f(x)) - \Delta(f; x, t, x + h).$$

We have:

$$|\Delta(f; x, t, x + h)| = \left|\frac{(t - x)(x + h - t)}{h^2} \cdot h\left[\frac{f(x + h) - f(t)}{x + h - t} - \frac{f(t) - f(x)}{t - x}\right]\right|$$
$$\leq \frac{(t - x)(x + h - t)}{h^2} \cdot \omega_2^d(f, h).$$

Hence

$$|f(t) - f(x)| \le h^{-1}|t - x|\omega_1(f, h) + \left[\frac{t-x}{h}\left(1 - \frac{t-x}{h}\right)\right]\omega_2^d(f, h).$$

Put $y := \frac{t-x}{h}$. Using the inequality $y(1 - y) \le \frac{1}{2r+4} + \frac{r}{2} \cdot y^2$, $y \in \mathbb{R}$, $r \ge 1$ the corresponding relation (2.17) follows. Therefore the direct part of the theorem is proved.

For the inverse part we make appropriate choices.

If we take $f := e_0$ and F of type (1.11) such that $F(e_0) \ne 1$, one obtains from (2.79) that $A \ge 1$.

By choosing $I := [0, 1]$, $x \in (0, 1)$, the functional F, given by $F(g) := g(1)$, $g \in \mathcal{F}[0, 1]$, the function $f := e_1$, and $h := \frac{1}{2}$, one obtains from (2.79) that $1 - x \le B(1 - x)$, that is $B \ge 1$.

If we take $I := [0, 1]$, $x \in (0, 1)$ and the functional F, defined by

$$F(f) := (1 - x)f(0) + xf(1), \quad f : [0, 1] \to \mathbb{R},$$

we obtain from (2.79) for all $0 < h \le \frac{1}{2}$ and $f : [0, 1] \to \mathbb{R}$:

$$|\Delta(f; 0, x, 1)| \le (C + Dh^{-2}x(1 - x))\omega_2^d(f, h). \tag{2.81}$$

We have $\omega_2^d(e_2, h) = 2h^2$, for all $h > 0$. By taking $f := e_2$ in (2.81) one obtains $x(1 - x) \le 2C \cdot h^2 + 2D \cdot x(1 - x)$, and by passing to the limit $h \to 0$, $D \ge \frac{1}{2}$ follows.

Finally suppose that (2.79) holds with $B = 1$, $D = \frac{r}{2}$, $r \ge 1$ and for certain values $A \in \mathbb{R}$, $C \in \mathbb{R}$, for all I, x, F, f and h that verify the conditions in the theorem. Denote $y = \frac{1}{r+2}$. Choose $I := [-1, 1]$, $x := 0$, $h := 1$, and the functional F defined by $F(f) := f(y)$, for $f \in \mathcal{F}[-1, 1]$. Let $0 < \varepsilon < \frac{1}{4}y$ and the function $f_0 \in C^1[-1, 1]$ be given by

$$f_0(t) := \begin{cases} 0, & -1 \le t \le -\varepsilon, \\ \frac{(t+\varepsilon)^2}{2\varepsilon}, & -\varepsilon < t \le 0, \\ t + \frac{\varepsilon}{2}, & 0 < t \le y - \varepsilon, \\ y - \frac{1}{2\varepsilon}(t - y)^2, & y - \varepsilon < t \le y, \\ y, & y < t \le 1. \end{cases}$$

We have $\omega_1(f_0, 1) = y$ and $\omega_2^d(f_0, 1) = \omega_1(f_0', 1) = 1$. With these data, relation (2.79) becomes

$$y \le y^2 + C + \frac{r}{2}y^2.$$

From this it follows that $C \ge \frac{1}{2r+4}$. The theorem is completely proved. $\qquad\square$

Since in estimate (2.78), the number $r \ge 1$ is arbitrary, we can choose for r the best value, for fixed F, h and x. So we get the following corollaries.

Corollary 2.3.3. *Let F and x be as in Theorem 2.3.7. Suppose $m_0 > 0$ and $m_2 > 0$. We have*

$$|F(f) - f(x)| \le |m_0 - 1| \cdot |f(x)| + m_1 h^{-1} \omega_1(f, h)$$

$$+ \begin{cases} \left(\frac{1}{6} m_0 + \frac{1}{2} h^{-2} m_2\right) \omega_2^d(f, h), & 0 < h \le 3\sqrt{\frac{m_2}{m_0}}, \\ \left(h^{-1}\sqrt{m_0 m_2} - h^{-2} m_2\right) \omega_2^d(f, h), & h > 3\sqrt{\frac{m_2}{m_0}}, \end{cases} \tag{2.82}$$

if length $(I) \ge 2h$ *and* $f \in V$.

Proof. We take into account that the minimum of the function $r \mapsto \frac{m_0}{2r+4} + \frac{r}{2} \cdot h^{-2} m_2$, $r \ge 1$ is $r := h\sqrt{\frac{m_0}{m_2}} - 2$ in the case $h \ge 3\sqrt{\frac{m_2}{m_0}}$ and $r := 1$ in the case $0 < h \le 3\sqrt{\frac{m_2}{m_0}}$. $\qquad\square$

In the case of differentiable functions we have

Corollary 2.3.4. *Let F and x be as in Theorem 2.3.7. Suppose $m_0 > 0$ and $m_2 > 0$. We have*

$$|F(f) - f(x)| \le |m_0 - 1| \cdot |f(x)| + m_1 h^{-1} \omega_1(f, h)$$

$$+ \begin{cases} \left(\frac{1}{6} h m_0 + \frac{1}{2} h^{-1} m_2\right) \omega_1(f', h), & 0 < h \le 3\sqrt{\frac{m_2}{m_0}}, \\ \left(\sqrt{m_0 m_2} - h^{-1} m_2\right) \omega_1(f', h), & h > 3\sqrt{\frac{m_2}{m_0}}, \end{cases} \tag{2.83}$$

if length $(I) \ge 2h$ *and* $f \in V \cap \mathcal{D}(I)$.

Remark 2.3.2. The estimates (2.82) and (2.83) in the case $h > 3\sqrt{\frac{m_2}{m_0}}$ can be obtained from the corresponding estimates given for $h = 3\sqrt{\frac{m_2}{m_0}}$.

For differentiable functions another type of estimate is obtained by replacing the term $h^{-1} \omega_1(f', h)$ with the smaller term $|f'(x)|$. In the following theorem, for any real number a we denote by $[a]$ and $]a[$ the integers defined in Section 1.1.

Theorem 2.3.8. *Let $F : V \to \mathbb{R}$, $V \subset \mathcal{F}(I)$, be a linear positive functional that is admissible related to a point $x \in I$. Let $r \in (1, \infty)$. We have*

$$|F(f) - f(x)| \le |F(e_0) - 1| \cdot |f(x)| + |F(e_1 - xe_0)| \cdot |f'(x)|$$

$$+ (C_r h F(e_0) + \frac{r}{2} h^{-1} F((e_1 - xe_0)^2)) \omega_1(f', h), \tag{2.84}$$

for $f \in V \cap \mathcal{D}(I)$ and $h > 0$, where

$$C_r := \begin{cases} \max\left\{ P_r\left(\left[\frac{2-r}{2(r-1)}\right]\right), \; P_r\left(\left]\frac{2-r}{2(r-1)}\right[+ 1\right)\right\}, & 1 < r < 2, \\ \frac{1}{2r}, & r \ge 2, \end{cases} \tag{2.85}$$

and P_r is the polynomial $P_r(u) := \frac{1}{2r}(u + 1)(u + 1 - ru)$, $u \in \mathbb{R}$.

Conversely, if there are the numbers $A, B, C, D \ge 0$ such that the inequality

$$|F(f) - f(x)| \leq A \cdot |F(e_0) - 1| \cdot |f(x)| + B \cdot |F(e_1 - xe_0)| \cdot |f'(x)|$$
$$+ (C\, h F(e_0) + D\, h^{-1} F((e_1 - xe_0)^2)) \omega_1(f', h) \qquad (2.86)$$

holds (only) for all linear positive functionals F of the form (1.11), for all $x \in I$, all $f \in C^1(I)$ and all $h > 0$, then we must have $A \geq 1$, $B \geq 1$, $D > \frac{1}{2}$ and if $D = \frac{r}{2}$, $r > 1$, then we must have also $C \geq C_r$.

Proof. For the direct part we consider $\omega_1(f', h) < \infty$ and apply Theorem 2.1.2 (ii), choosing $\Omega_1(f, h) := h|f'(x)|$, $f \in \mathcal{D}(I)$, $h > 0$, $\Omega_2 := \omega_2^d$ and the function $\Psi(y) = C_r + \frac{r}{2} y^2$, $y \in [0, \infty)$, $r > 1$. Let x be an interior point of I. The corresponding relation (2.15), namely

$$|\Delta(f; t_1, x, t_2)| \qquad (2.87)$$
$$\leq \left(C_r + \frac{r}{2} h^{-2}(t_2 - x)(x - t_1) \right) \omega_2^d(f, h), \quad t_1 < x < t_2, \; t_1, t_2 \in I,$$

follows from relation (2.69). Indeed, since $\max\limits_{u \in \mathbb{R}} P_r(u) = P_r\left(\frac{2-r}{2(r-1)} \right)$, we obtain that

$$C_r = \max\limits_{n \in \mathbb{N} \cup \{0\}} P_r(n).$$

Hence $C_r \geq P_r(0) = \frac{1}{2r} \geq \frac{1}{8r}$.

It remains to prove the corresponding relation (2.17). Let $x, t \in I$. Suppose for a choice that $t \geq x$. There are $n \in \mathbb{N} \cup \{0\}$ and $q \in [0, 1)$, such that $t - x = h(n + q)$. We have successively,

$$|f(t) - f(x)| \leq |f'(x)| \, |t - x| + \left| \int_x^t (f'(u) - f'(x))\, du \right|$$

$$\leq |f'(x)| \, |t - x| + \sum_{j=0}^{n-1} \int_{x+jh}^{x+(j+1)h} |f'(u) - f'(x)|\, du$$

$$+ \int_{x+nh}^t |f'(u) - f'(x)|\, du$$

$$\leq |f'(x)| \, |t - x| + \left[\sum_{j=0}^{n-1} (j+1)h + (n+1)qh \right] \omega_1(f', h)$$

$$= |f'(x)| \, |t - x| + \left(\frac{n(n+1)}{2} + (n+1)q \right) h \omega_1(f', h).$$

For $r > 1$ consider the function

$$\varphi_r(n, q) := \frac{n(n+1)}{2} + (n+1)q - \frac{r}{2}(n+q)^2, \quad n \in \mathbb{N} \cup \{0\}, \; q \in [0, 1).$$

Take

$$T_r := \sup\limits_{\substack{n \in \mathbb{N} \cup \{0\} \\ q \in [0,1)}} \varphi_r(n, q).$$

We have $T_r \geq \varphi_r(0,0) = 0$. Using $\frac{\partial \varphi_r}{\partial q}(n,q) = n + 1 - r(n+q)$, we conclude that for $nr \geq n+1$, we have $\varphi_r(n,q) \leq \varphi_r(n,0) = \frac{n(n+1)}{2} - \frac{r}{2}n^2 \leq 0 \leq T_r$. Then

$$
\begin{aligned}
T_r = \sup_{\substack{n \in \mathbb{N} \cup \{0\}, \, q \in [0,1) \\ rn < n+1}} \varphi_r(n,q) &= \sup_{\substack{n \in \mathbb{N} \cup \{0\} \\ rn < n+1}} \sup_{q \in [0,1)} \varphi_r(n,q) \\
= \sup_{\substack{n \in \mathbb{N} \cup \{0\} \\ rn < n+1}} \varphi_r\left(n, \frac{n+1-rn}{r}\right) &= \sup_{\substack{n \in \mathbb{N} \cup \{0\} \\ rn < n+1}} P_r(n).
\end{aligned}
$$

If $r \geq 2$ we have $\max_{n \in \mathbb{N} \cup \{0\}} P_r(n) = P_r(0)$. If $1 < r < 2$ we have $\frac{1}{r-1} - \frac{2-r}{2(r-1)} > 1$. Hence $\left[\frac{2-r}{2(r-1)}\right], \left]\frac{2-r}{2(r-1)}\right[+ 1 \in \{n \in \mathbb{N} \cup \{0\}, \, rn < n+1\}$. Consequently $T_r = \sup_{\substack{n \in \mathbb{N} \cup \{0\} \\ rn < n+1}} P_n(n) = \sup_{n \in \mathbb{N} \cup \{0\}} P_r(n) = C_r$. Therefore we get

$$
|f(t) - f(x)| \leq |f'(x)| \cdot |t - x| + \left(C_r + \frac{r}{2}\left(\frac{t-x}{h}\right)^2\right) h\omega_1(f',h),
$$

i.e., the corresponding form of relation (2.17). The direct part is finished.

Consider now that inequality (2.86) holds. The optimality of the constant $A = 1$ follows if we take $F = 0$ and $f := e_0$. Also we obtain $B \geq 1$ if we take $I := [0,1]$, $x := 0$, $F(g) := g(1)$, $g \in \mathcal{F}[0,1]$, $f := e_1$ and $h = 1$.

Finally, let $0 < \varepsilon < 1$, $n \in \mathbb{N} \cup \{0\}$, $q \in [0,1)$. Consider the interval $I := [0, n+q]$, $x := 0$ and the functional $F(g) := g(n+q)$, $g \in \mathcal{F}(I)$. Let $\Psi \in C^1[0,\infty)$, defined in the following mode:

$$
\Psi(t) := \begin{cases} k - 1 + \frac{t-k+1}{\varepsilon}, & t \in [k-1, k-1+\varepsilon], \; k \in \mathbb{N}, \\ k, & t \in [k-1+\varepsilon, k], \; k \in \mathbb{N}. \end{cases}
$$

Then consider $f \in C^1(I)$, $f(t) := \int_0^t \Psi(u)\,du$, $t \in I$. Take $h = 1$. We have $\omega_1(f',1) = \omega_1(\Psi,1) = 1$. From relation (2.86) we obtain

$$
f(n+q) \leq C + D(n+q)^2.
$$

But

$$
f(n+q) = \int_0^{n+q} f'(t)\,dt = \int_0^{n+q} \Psi(t)\,dt \geq \frac{n(n+1)}{2} + (n+1)q - \varepsilon(n+1).
$$

Since $\varepsilon > 0$ can be taken arbitrarily we obtain

$$
\frac{n(n+1)}{2} + (n+1)q \leq C + D(n+q)^2.
$$

Since $n \in \mathbb{N} \cup \{0\}$ and $q \in [0,1)$ can be taken arbitrarily, we obtain $D > \frac{1}{2}$. Let us take $D = \frac{r}{2}$ with $r > 1$. We must have $C \geq \varphi_r(n,q)$, for $n \in \mathbb{N} \cup \{0\}$, $q \in [0,1)$. Since $T_r = C_r$, it follows that $C \geq C_r$. $\qquad\square$

Corollary 2.3.5. *Let F and x be as in Theorem 2.3.8. Suppose $m_0 > 0$ and $m_2 > 0$.
We have*

$$|F(f) - f(x)| \leq |m_0 - 1| \cdot |f(x)| + m_1 |f'(x)| \tag{2.88}$$

$$+ \begin{cases} \left(\frac{m_0}{8} \left(1 + 2\sqrt{\frac{m_2}{m_0}}\, h^{-1} \right)^2 \right) h\omega_1(f', h), & 0 < h \leq \frac{\sqrt{6}}{1+\sqrt{6}} \sqrt{\frac{m_2}{m_0}}, \\[2mm] \left(\frac{h}{3} m_0 + \frac{3}{4} h^{-1} m_2 \right) \omega_1(f', h), & \frac{\sqrt{6}}{1+\sqrt{6}} \sqrt{\frac{m_2}{m_0}} < h \leq \frac{3}{2} \sqrt{\frac{m_2}{m_0}}, \\[2mm] \sqrt{m_0 m_2}\, \omega_1(f', h), & h > \frac{3}{2} \sqrt{\frac{m_2}{m_0}}, \end{cases}$$

for $f \in V \cap \mathcal{D}(I)$ and $h > 0$.

Proof. We apply Theorem 2.3.8. We take into account that $C_r := \max\limits_{n \in \mathbb{N} \cup \{0\}} P_r(n)$. If
$r \geq \frac{3}{2}$, we obtain $C_r = P_r(0) = \frac{1}{2r}$. For $1 < r < \frac{3}{2}$ we have $C_r \leq \max\limits_{u \in \mathbb{R}} P_r(u) =$
$P_r\left(\frac{2-r}{2(r-1)} \right) = \frac{r}{8(r-1)}$. Let us define

$$T_1(r) := \frac{1}{2r} m_0 + \frac{r}{2} m_2 h^{-2}, \quad T_2(r) := \frac{r}{8(r-1)} m_0 + \frac{r}{2} m_2 h^{-2}$$

and

$$T(r) := \begin{cases} T_1(r), & r > \frac{3}{2}, \\ T_2(r), & 1 < r \leq \frac{3}{2}. \end{cases}$$

We have $\min\limits_{r>1} T_1(r) = T_1\left(h\sqrt{\frac{m_0}{m_2}} \right)$ and $\min\limits_{r>1} T_2(r) = T_2\left(1 + \frac{h}{2}\sqrt{\frac{m_0}{m_2}} \right)$. We distinguish among three cases.

<u>Case 1.</u> $h \leq \sqrt{\frac{m_2}{m_0}}$. We have:

$$\min\limits_{r>1} T(r) = \min\left\{ T_1\left(\frac{3}{2} \right), T_2\left(1 + \frac{h}{2}\sqrt{\frac{m_0}{m_2}} \right) \right\}$$

$$= \begin{cases} T_2\left(1 + \frac{h}{2}\sqrt{\frac{m_0}{m_2}} \right), & 0 < h \leq \frac{\sqrt{6}}{1+\sqrt{6}} \sqrt{\frac{m_2}{m_0}}, \\[2mm] T_1\left(\frac{3}{2} \right), & \frac{\sqrt{6}}{1+\sqrt{6}} \sqrt{\frac{m_2}{m_0}} < h \leq \sqrt{\frac{m_2}{m_0}}. \end{cases}$$

<u>Case 2.</u> $\sqrt{\frac{m_2}{m_0}} < h \leq \frac{3}{2}\sqrt{\frac{m_2}{m_0}}$. We obtain:

$$\min\limits_{r>1} T(r) = \min\left\{ T_1\left(\frac{3}{2} \right), T_2\left(\frac{3}{2} \right) \right\} = T_1\left(\frac{3}{2} \right).$$

<u>Case 3.</u> $h \geq \frac{3}{2}\sqrt{\frac{m_2}{m_0}}$. We obtain

$$\min\limits_{r>1} T(r) = \min\left\{ T_1\left(h\sqrt{\frac{m_0}{m_2}} \right), T_2\left(\frac{3}{2} \right) \right\} = T_1\left(h\sqrt{\frac{m_0}{m_2}} \right).$$

From the results given in these three cases one obtains (2.88). $\qquad \square$

Remark 2.3.3. The estimate (2.88) in the case $h > \frac{3}{2} \sqrt{\frac{m_2}{m_0}}$ can be obtained from the estimate given for $h = \frac{3}{2} \sqrt{\frac{m_2}{m_0}}$.

2.4 Estimates with modulus ω_2^{dd}

The results given in this section were obtained in [104]. We consider the following normalized second order modulus.

Definition 2.4.1. *For $f \in \mathcal{F}(I)$ and $h > 0$ set*

$$\omega_2^{dd}(f, h) := h^2 \sup\{|\,[f; t_1, x, t_2]\,|;\ t_1 < x < t_2,\ t_1, t_2 \in I,\ t_2 - t_1 \le h\} \quad (2.89)$$

where $[f; t_1, x, t_2]$ is the divided difference of the function f on the points t_1, x, t_2, (see (1.5)).

We admit that the supremum in (2.89) may be infinity.

In order to obtain an estimate for the modulus ω_2^{dd} we first prove some lemmas.

Lemma 2.4.1. *For any function $f \in \mathcal{F}(I)$, and any points $a < x < b$ of I, such that $b - a \le h$, we have*

$$|\,\Delta(f; a, x, b)\,| \le (b - x)(x - a)h^{-2}\omega_2^{dd}(f, h). \quad (2.90)$$

Proof. We take into account that $\Delta(f; a, x, b) = (b - x)(x - a)[a, x, b; f]$. $\square$

Lemma 2.4.2. *Let there be the function $f \in \mathcal{F}(I)$, the points $s < u < x < t$ of I, and the number $h > 0$. If the inequalities*

$$|\Delta(f; s, u, x)| \le (x - u)(u - s)h^{-2}\omega_2^{dd}(f, h)$$

and

$$|\Delta(f; u, x, t)| \le (t - x)(x - u)h^{-2}\omega_2^{dd}(f, h)$$

hold, then the inequality:

$$|\Delta(f; s, x, t)| \le (t - x)(x - s)h^{-2}\omega_2^{dd}(f, h)$$

is true.

Proof. We have:

$$
\begin{aligned}
&|\Delta(f; s, x, t)| \\
&= \left| \frac{(t - x)(x - s)}{(t - s)(x - u)}\Delta(f; s, u, x) + \frac{(x - s)(t - u)}{(t - s)(x - u)}\Delta(f; u, x, t) \right| \\
&\le \left(\frac{(t - x)(x - s)}{(t - s)(x - u)}(x - u)(u - s) \right. \\
&\quad \left. + \frac{(x - s)(t - u)}{(t - s)(x - u)}(t - x)(x - u) \right) h^{-2}\omega_2^{dd}(f, h) \\
&= (t - x)(x - s)h^{-2}\omega_2^{dd}(f, h). \quad \square
\end{aligned}
$$

Lemma 2.4.3. *For any function* $f \in \mathcal{F}(I)$, *any points* $s < x < t$ *of* I *and any number* $h > 0$, *we have*

$$|\Delta(f; s, x, t)| \leq (t - x)(x - s)h^{-2}\omega_2^{dd}(f, h). \tag{2.91}$$

Proof. By taking into account Lemma 2.4.1, it is enough to consider only the case $t - s \geq h$. We prove relation (2.91) by induction with regard to $m \in \mathbb{N}$, $m \geq 6$, if $t - s \leq m\frac{h}{6}$. For $m = 6$ this follows from Lemma 2.4.1. Suppose now that the statement is true for $m \geq 6$ and prove it for $m + 1$. Let $s < x < t$ such that $m\frac{h}{6} < t - s \leq (m + 1)\frac{h}{6}$. Suppose, for a choice, that $x - s \geq t - x$. Consider the points $u := s + \frac{1}{3}(x - s)$ and $v := s + \frac{2}{3}(x - s)$. Hence $v - s \leq m\frac{h}{6}$, $x - u \leq m\frac{h}{6}$ and $t - v \leq m\frac{h}{6}$. By using the induction hypothesis we obtain the following three inequalities: $|\Delta(f; s, u, v)| \leq (v - u)(u - s)h^{-2}\omega_2^{dd}(f, h)$, $|\Delta(f; u, v, x)| \leq (x - v)(v - u)h^{-2}\omega_2^{dd}(f, h)$ and $|\Delta(f; v, x, t)| \leq (t - x)(x - v)h^{-2}\omega_2^{dd}(f, h)$. By using Lemma 2.4.2, from the first two it follows that $|\Delta(f; s, v, x)| \leq (x - v)(v - s)h^{-2}\omega_2^{dd}(f, h)$, and next, by using also the third one we get $|\Delta(f; s, x, t)| \leq (t - x)(x - s)h^{-2}\omega_2^{dd}(f, h)$.   □

Lemma 2.4.4. *Let* $f \in \mathcal{F}(I)$, $x \in I$, $t \in I$, $h > 0$, *such that* $length(I) \geq 2h$, *and* $r \geq 1$. *Then we have*

$$|f(t) - f(x)| \leq h^{-1}\omega_1(f, h)|t - x| + \left(\frac{1}{4(1 + r)} + r\left(\frac{t - x}{h}\right)^2\right)\omega_2^{dd}(f, h). \tag{2.92}$$

Proof. Suppose, for a choice, that $t > x$. We distinguish between two cases.

   Case (1): $t > x + h$. Applying relation (2.91) we get

$$
\begin{aligned}
|f(t) - f(x)| &= \frac{t - x}{h}|f(x + h) - f(x) + \Delta(f; x, x + h, t)| \\
&\leq \frac{t - x}{h}\left[\omega_1(f, h) + \frac{t - x - h}{h}\omega_2^{dd}(f, h)\right] \\
&\leq h^{-1}\omega_1(f, h)|t - x| + \left(\frac{1}{4(1 + r)} + r\left(\frac{t - x}{h}\right)^2\right)\omega_2^{dd}(f, h).
\end{aligned}
$$

   Case (2): $x < t \leq x + h$. Since $length(I) \geq 2h$, we have either $x + h \in I$ or $t - h \in I$. Consider only the case $x + h \in I$. We have

$$
\begin{aligned}
|f(t) - f(x)| &= \left|\frac{t - x}{h} \cdot (f(x + h) - f(x)) - \Delta(f; x, t, x + h)\right| \\
&\leq h^{-1}\omega_1(f, h)|t - x| + \frac{(x + h - t)(t - x)}{h^2}\omega_2^{dd}(f, h).
\end{aligned}
$$

Put $u = \frac{t - x}{h}$. From the inequality $u(1 - u) \leq \frac{1}{4(1 + r)} + ru^2$, $u \in \mathbb{R}$, $r \geq 1$, we get relation (2.92).   □

Theorem 2.4.1. *Let* $F : V \to \mathbb{R}$, $V \subset \mathcal{F}(I)$, *be a linear positive functional that is admissible related to a point* $x \in I$. *Let* $r \geq 1$. *The inequality*

$$|F(f) - f(x)| \leq |f(x)| \cdot |F(e_0) - 1| + |F(e_1 - xe_0)| \cdot h^{-1}\omega_1(f, h) \qquad (2.93)$$

$$+ \left(\frac{1}{4(r+1)} \cdot F(e_0) + r \cdot h^{-2} F((e_1 - xe_0)^2) \right) \omega_2^{dd}(f, h)$$

holds for $f \in V$ and $h > 0$, such that length $(I) \geq 2h$.

Conversely, if the inequality

$$|F(f) - f(x)| \leq A \cdot |F(e_0) - 1| \cdot |f(x)| + B \cdot |F(e_1 - xe_0)|$$
$$\cdot h^{-1}\omega_1(f, h) + (C \cdot F(e_0) + D \cdot h^{-2} F((e_1 - xe_0)^2))\omega_2^{dd}(f, h) \qquad (2.94)$$

is satisfied only for all linear positive functionals F of the form (1.11), for all $x \in I$, all $f \in C^2(I)$ and all $h > 0$ such that length $(I) \geq 2h$, then we must have $A \geq 1$, $B \geq 1$, $D \geq 1$ and if $D = r$, $r \geq 1$ and $B = 1$, then we must have also $C \geq \frac{1}{4(r+1)}$.

Proof. We can consider without any loss of generality that $\omega_1(f, h) < \infty$, $\omega_2^{dd}(f, h) < \infty$. The direct part of the theorem can be obtained from Theorem 2.1.2, if we take $\Omega_1 := \omega_1$, $\Omega_2 := \omega_2^{dd}$ and the function $\Psi(y) := \frac{1}{4(r+1)} + ry^2$, $y \in [0, \infty)$. Indeed, the corresponding relation (2.15) follows from Lemma 2.4.3 and the corresponding relation (2.17) follows from Lemma 2.4.4.

For the inverse part of the theorem, the necessity of the inequalities $A \geq 1$ and $B \geq 1$ can be obtained similarly as in the proof of Theorem 2.3.7. Also if we take $I := [0, 1]$, $x \in (0, 1)$ and the functional F, defined by

$$F(f) := (1 - x)f(0) + xf(1), \quad f \in \mathcal{F}[0, 1],$$

we get from (2.94), for all $0 < h \leq \frac{1}{2}$ and $f : [0, 1] \to \mathbb{R}$,

$$|\Delta(f; 0, x, 1)| \leq (C + Dh^{-2}x(1 - x))\omega_2^{dd}(f, h). \qquad (2.95)$$

We have $\omega_2^{dd}(e_2, h) = h^2$, for all $h > 0$. By taking $f := e_2$ in (2.95) one obtains $x(1 - x) \leq C \cdot h^2 + D \cdot x(1 - x)$, and by passing to the limit $h \to 0$ it follows that $D \geq 1$.

Finally suppose that (2.94) holds with $B = 1$, $D = r$, $r \geq 1$ and for certain values $A \in \mathbb{R}$, $C \in \mathbb{R}$ for all I, x, F, f and h that verify the conditions in the theorem. Put $y := \frac{1}{2(r+1)}$. Choose : $I := [0, 2]$, $x := 0$, $h := 1$, the functional F defined by $F(f) := f(y)$, $f \in \mathcal{F}[0, 2]$. Let the function $f(t) = 2t - t^2$, $t \in [0, 2]$. We have $\omega_1(f, 1) = 1$. Since $f''(t) = -2$, $t \in [0, 2]$, we have $\omega_2^{dd}(f, 1) = 1$. With these data relation (2.94) becomes

$$2y - y^2 \leq y + C + ry^2.$$

From this it follows that $C \geq \frac{1}{4(r+1)}$. The theorem is completely proved. $\square$

If in the previous theorem we choose the best value for the parameter r we obtain:

Corollary 2.4.1. *Let F and x be as in Theorem 2.4.1. Suppose $m_0 > 0$ and $m_2 > 0$. Then*

$$|F(f) - f(x)| \le |m_0 - 1| \cdot |f(x)| + m_1 h^{-1} \omega_1(f, h)$$

$$+ \begin{cases} \left(\frac{1}{8} m_0 + h^{-2} m_2\right) \omega_2^{dd}(f, h), & \text{if } 0 < h \le 4\sqrt{\frac{m_2}{m_0}}, \\ \left(h^{-1}\sqrt{m_0 m_2} - h^{-2} m_2\right) \omega_2^{dd}(f, h), & \text{if } h > 4\sqrt{\frac{m_2}{m_0}}, \end{cases} \qquad (2.96)$$

if length $(I) \ge 2h$ and $f \in V$.

Proof. Relation (2.96) is obtained from (2.93), by taking $r = 1$, if $h \le 4\sqrt{\frac{m_2}{m_0}}$ and $r = \frac{1}{2} h \sqrt{\frac{m_0}{m_2}} - 1$, if $h > 4\sqrt{\frac{m_2}{m_0}}$. $\qquad \square$

Remark 2.4.1. The estimate (2.96) in the case $h > 4\sqrt{\frac{m_2}{m_0}}$ can be obtained from the estimate given for $h = 4\sqrt{\frac{m_2}{m_0}}$.

In a particular case we obtain:

Theorem 2.4.2. *Let $F : V \to \mathbb{R}$, $V \subset \mathcal{F}(I)$, be a linear positive functional. Suppose that F is admissible related to an interior point x of I. Suppose that $F(e_0) = 1$, $F(e_1) = x$. The inequality*

$$|F(f) - f(x)| \le h^{-2} F((e_1 - x e_0)^2) \omega_2^{dd}(f, h) \qquad (2.97)$$

holds true for any $f \in V$ and $h > 0$.
  Conversely, if the inequality

$$|F(f) - f(x)| \le A \cdot h^{-2} F((e_1 - x e_0)^2)) \omega_2^{dd}(f, h) \qquad (2.98)$$

is satisfied only for all linear positive functionals F of the form (1.11), for all $x \in I$, $f = e_2$ and all $h > 0$, then we must have $A \ge 1$.

Proof. Let $f \in V$ such that $\omega_2^{dd}(f, h) < \infty$. The direct part of the theorem follows from Theorem 2.1.2 (i), if we take $\Omega_2 := \omega_2^{dd}$ and the function $\Psi(y) := y^2$, $y \in [0, \infty)$. Indeed, the corresponding relation (2.15) follows from Lemma 2.4.3.
  For the inverse part of the theorem, we take $I := [0, 1]$, $x \in (0, 1)$ and the functional F, defined by

$$F(f) := (1 - x) f(0) + x f(1), \quad f : [0, 1] \to \mathbb{R},$$

then we obtain from (2.98) for all $0 < h \le \frac{1}{2}$ and $f : [0, 1] \to \mathbb{R}$,

$$|\Delta(f; 0, x, 1)| \le A h^{-2} x (1 - x) \omega_2^{dd}(f, h). \qquad (2.99)$$

We have $\omega_2^{dd}(e_2, h) = h^2$, for all $h > 0$. By taking $f := e_2$ in (2.99) one obtains $x(1 - x) \le A \cdot x(1 - x)$, that is $A \ge 1$. $\qquad \square$

The estimates with modulus ω_2^{dd} can be expressed in another form, if we use the following notation:

$$M_2(f) := \inf_{h > 0} \sup\{|\,[t_1, x, t_2; f]\,|; \ t_1 < x < t_2, \ t_1, t_2 \in I, \ t_2 - t_1 \le h\}, \qquad (2.100)$$

where, $f \in \mathcal{F}(I)$. So, we get

Corollary 2.4.2. *Let F and x be as in Theorem 2.4.1. Suppose $m_0 > 0$ and $m_2 > 0$. We have*

$$|F(f) - f(x)|$$
$$\leq |m_0 - 1| \cdot |f(x)| + m_1 h^{-1} \omega_1(f, h) + \left(\frac{m_0}{8} h^2 + m_2 \right) M_2(f), \quad (2.101)$$

for $f \in V$ and $0 < h \leq 4\sqrt{\frac{m_2}{m_0}}$ such that length $(I) \geq 2h$.

Corollary 2.4.3. *Let F and x be as in Theorem 2.4.2. We have*

$$|F(f) - f(x)| \leq m_2 M_2(f), \quad (2.102)$$

for $f \in V$.

In the case of functions having a bounded second derivative, since $h^{-2} \omega_2^{dd}(f, h) \leq \frac{1}{2} \| f'' \|$, we get:

Corollary 2.4.4. *Let F and x be as in Theorem 2.4.1. Suppose $m_0 > 0$ and $m_2 > 0$. We have*

$$|F(f) - f(x)| \leq |m_0 - 1| \cdot |f(x)| + \left(m_1 + \frac{1}{2} m_2 \right) \| f'' \|, \quad (2.103)$$

if $f \in V$ has a bounded second derivative.

Corollary 2.4.5. *Let F and x like in Theorem 2.4.2. We have*

$$|F(f) - f(x)| \leq \frac{1}{2} m_2 \| f'' \|, \quad (2.104)$$

if $f \in V$ has a bounded second derivative.

2.5 Estimates with Ditzian–Totik modulus

On the interval $[0, 1]$ the Ditzian–Totik moduli of the first and second orders are defined with the aid of the weight function $\varphi(x) := \sqrt{x(1 - x)}$, $x \in [0, 1]$, in the mode

$$\omega_1^\varphi(f, h) := \sup \left\{ |f(u) - f(v)|, \ u, v \in [0, 1], \right. \quad (2.105)$$

$$\left. |v - u| \leq h\varphi\left(\frac{u + v}{2} \right) \right\},$$

$$\omega_2^\varphi(f, h) := \sup \left\{ \left| f(u) - 2f\left(\frac{u + v}{2} \right) + f(v) \right|, \right. \quad (2.106)$$

$$\left. u, v \in I, \ |u - v| \leq 2h\varphi\left(\frac{u + v}{2} \right) \right\},$$

$f \in \mathcal{F}[0, 1]$, $h > 0$.

These moduli, as well as their higher order variants, are studied by Ditzian and Totik in the monograph [30]. A first estimate with the Ditzian–Totik modulus for the general linear positive operators that preserve linear functions was obtained by H. Gonska and G. Tachev, see [46].

An estimate with the second order Ditzian–Totik modulus for linear positive operators reproducing linear functions, and based on the results in the next subsection, is given in [35]. Our main result, given in Subsection 2.5.2, is an improvement of this one.

2.5.1 Auxiliary results

Lemma 2.5.1. *Let $x \in (0, 1)$ and $h > 0$. The following inequalities are equivalent:*

$$x + h\varphi(x) \le 1 \tag{2.107}$$

and

$$x \le \frac{1}{1 + h^2}. \tag{2.108}$$

Also the corresponding strict inequalities are equivalent.

Proof. The lemma is immediate. $\qquad\square$

Lemma 2.5.2. *For any $0 < z < 1$ and any $h > 0$ there is a unique $w \in (z, 1)$ such that*

$$z + h\varphi(w) = w. \tag{2.109}$$

Denote this point w by $\Theta_h(z)$. The function $\Theta_h : (0, 1) \to (0, 1)$ is strictly increasing.

Proof. After a simple calculus, the relation in (2.109) is equivalent with

$$P(w) = 0,$$

where

$$P(w) = (1 + h^2)w^2 - (2z + h^2)w + z^2.$$

We have:

$$P(z) = -h^2 z(1 - z) < 0, \quad P(1) = (1 - z)^2 > 0.$$

Since P is a polynomial of degree 2 there exists a unique $w \in (z, 1)$ such that $P(w) = 0$. From Lemma 2.5.1 it follows by symmetry that we have $w - h\varphi(w) > 0$, if and only if $w > \frac{h^2}{1+h^2}$. Therefore, we have $\Theta_h(z) > \frac{h^2}{1+h^2}$, for all $z \in (0, 1)$. The fact that Θ_h is increasing on $(0, 1)$ is equivalent with the fact that the function $p(w) := w - h\varphi(w)$, $w \in (0, 1)$, is increasing on the interval $\left(\frac{h^2}{1+h^2}, 1\right)$. We have $p'(w) = 1 - h\frac{1-2w}{2\varphi(w)}$. The unique root of the equation $p'(w) = 0$ is $w_0 = \frac{\sqrt{1+h^2}-1}{2\sqrt{1+h^2}} < \frac{h^2}{1+h^2}$. Then $p'(w) > 0$, $w \in \left(\frac{h^2}{1+h^2}, 1\right)$. $\qquad\square$

Lemma 2.5.3. *Let $h > 0$, $y \in \left[\frac{1}{2}, \frac{1}{1+h^2}\right)$ and $z = y + h\varphi(y)$. If we represent*

$$y = \sin^2 \alpha, \quad z = \sin^2 \beta, \quad \frac{\pi}{4} \leq \alpha < \beta < \frac{\pi}{2}, \tag{2.110}$$

then

$$\beta - \alpha \geq \frac{h}{2}. \tag{2.111}$$

Proof. We have

$$\varphi(y) = \sin \alpha \cos \alpha, \quad \text{and} \quad \beta = \arcsin \sqrt{\sin^2 \alpha + h \sin \alpha \cos \alpha}. \tag{2.112}$$

Consider the function

$$F(\rho) = \arcsin \sqrt{\sin^2 \alpha + \rho \sin \alpha \cos \alpha}, \quad \rho \in [0, h]. \tag{2.113}$$

We have

$$F'(\rho) = \frac{\sin \alpha \cos \alpha}{2\sqrt{\cos^2 \alpha - \rho \sin \alpha \cos \alpha}\sqrt{\sin^2 \alpha + \rho \sin \alpha \cos \alpha}}$$

$$= \frac{\sqrt{\sin \alpha \cos \alpha}}{2\sqrt{(1 - \rho^2) \sin \alpha \cos \alpha + \rho(\cos^2 \alpha - \sin^2 \alpha)}}, \quad \rho \in [0, h].$$

Note that, $F(\rho)$ and $F'(\rho)$ are correctly defined, since $\beta < \frac{\pi}{2}$.

Since $\alpha \geq \frac{\pi}{4}$ it follows that F' is increasing on $[0, h]$, i.e., F is convex on $[0, h]$.
Consequently we have $\beta = F(h) \geq F(0) + hF'(0) = \alpha + \frac{h}{2}$. $\qquad\square$

In the following definition we use Lemma 2.5.1 and Lemma 2.5.2.

Definition 2.5.1. *Let $h > 0$ and $x \in \left(0, \frac{1}{1+h^2}\right)$. We define the* canonical sequence *of the pair (x, h) by recurrence in the following way. Firstly put*

$$u_0 = x, \quad y_1 = u_0 + h\varphi(u_0), \quad u_1 = \Theta_h(y_1). \tag{2.114}$$

Consider now that there are constructed the terms

$$x = u_0 < y_1 < u_1 < \cdots < u_{i-1} < y_i < u_i, \quad i \geq 1.$$

We have the following cases:

i) If $u_i < \frac{1}{1+h^2}$, then define

$$y_{i+1} := u_i + h\varphi(u_i) \quad \text{and} \quad u_{i+1} = \Theta_h(y_{i+1}). \tag{2.115}$$

ii) If $u_i = \frac{1}{1+h^2}$, then define

$$y_{i+1} := u_i + h\varphi(u_i) = 1, \tag{2.116}$$

and y_{i+1} is the last term of the canonical sequence.

iii) If $u_i > \frac{1}{1+h^2}$, then u_i is the last term of the canonical sequence.

Remark 2.5.1. *If $y_i < u_i < y_{i+1}$ are terms of the canonical sequence of a pair (x, h), then*

$$y_{i+1} - u_i = u_i - y_i = h\varphi(u_i). \tag{2.117}$$

Remark 2.5.2. *From Definition 2.5.1 we deduce that for a pair (x, h) as in Definition 2.5.1, one of the following cases is possible.*
Case (a) The canonical sequence is of the form

$$x = u_0 < y_1 < \cdots < u_{p-1} < y_p < u_p,$$

and $u_p \in \left(\frac{1}{1+h^2}, 1\right)$, $p \geq 1$.
Case (b) The canonical sequence is of the form

$$x = u_0 < y_1 < \cdots < u_{p-1} < y_p = 1, \quad p \geq 2.$$

Case (c) The canonical sequence is infinite:

$$x = u_0 < y_1 < u_1 < \cdots < 1.$$

The following lemma shows that in fact Case (c) is impossible.

Lemma 2.5.4. *For any $h > 0$ and $x \in \left(0, \frac{1}{1+h^2}\right)$, the canonical sequence of the pair (x, h) is finite, that is one of the cases (a) or (b) of Remark 2.5.2, is true.*

Proof. With the notation in Definition 2.5.1, we show firstly that there is an index $j_0 \geq 1$ such that

$$u_{j_0} > \frac{1}{2}. \tag{2.118}$$

Indeed, if $u_j < \frac{1}{2}$, then

$$u_j = u_j - y_j + \sum_{i=2}^{j}(y_i - y_{i-1}) + y_1 - x + x$$

$$= h\varphi(u_j) + \sum_{i=2}^{j} 2h\varphi(u_{i-1}) + h\varphi(u_0) + x$$

$$> 2jh\varphi(x) + x.$$

From this it follows that there is j_0 satisfying (2.118).
If $u_{j_0} \geq \frac{1}{1+h^2}$, then Lemma 2.5.4 is proved. In the converse case, let us represent

$$u_i = \sin^2 \alpha_i, \ (i \geq j_0), \ y_i = \sin^2 \beta_i, \ (i \geq j_0 + 1). \tag{2.119}$$

We have

$$\alpha_{j_0} < \beta_{j_0+1} < \alpha_{j_0+1} < \beta_{j_0+2} < \cdots < \frac{\pi}{2}. \tag{2.120}$$

By taking into account Lemma 2.5.3, it follows that

$$\alpha_i > \alpha_{j_0} + \frac{h}{2}(i - j_0), \quad i \geq j_0 + 1. \tag{2.121}$$

From (2.120) and (2.121) we obtain the statement of the lemma.   □

Lemma 2.5.5. *Let $h > 0$. Let x and t be such that*

$$0 < x < x + h\varphi(x) < t \leq 1.$$

Then, by considering the canonical sequence of the pair (x, h):

$$x = u_0 < y_1 < u_1 < \cdots \quad (finite), \tag{2.122}$$

one of the following conditions is true:

(I) There is $m \in \mathbb{N}$ such that

$$u_0 < y_1 < \cdots < y_m \leq t$$

and $t = y_m$ or $t \in (y_m, u_m]$. (Remark that if $t > y_m$, then u_m exists.)
(II)There is $m \in \mathbb{N}$ such that

$$u_0 < y_1 < \cdots < y_m < u_m < t$$

and $u_m > \frac{1}{1+h^2}$ or $t < y_{m+1}$. (Remark that if $u_m \leq \frac{1}{1+h^2}$, then y_{m+1} exists.)

Proof. The proof is a consequence of Lemma 2.5.4 and Remark 2.5.2.   □

Lemma 2.5.6. *(i) Let $0 < t \leq 1$. The function $u \mapsto \frac{t-u}{u(1-u)}$, $u \in (0, t)$, is decreasing.*

(ii) Let $0 \leq s < 1$. The function $u \mapsto \frac{(u-s)^2}{u(1-u)}$, is decreasing on the interval $(0, s]$ and is increasing on the interval $[s, 1)$.

Proof. (i) Let $\Psi_1(u) = \frac{t-u}{u(1-u)}$, $u \in (0, t)$. We have $\Psi_1(u) = \frac{t}{u} - \frac{1-t}{1-u}$ and $\Psi_1'(u) = -\frac{t}{u^2} - \frac{1-t}{(1-u)^2} < 0$.

(ii) Let $\Psi_2(u) = \frac{(u-s)^2}{u(1-u)}$, $u \in (0, 1)$. We take into account that $\Psi_2'(u) = \frac{(u-s)(u+s-2us)}{[u(1-u)]^2} \geq 0$.   □

Lemma 2.5.7. *Let $f \in B[a, b]$, where $[a, b] \subset [0, 1]$, $a < b$. For any $x \in [a, b]$ we have*

$$|\Delta(f; a, x, b)| \leq \omega_2^\varphi \left(f, \frac{b - a}{2\varphi\left(\frac{a+b}{2}\right)} \right), \tag{2.123}$$

and consequently we have

$$|\Delta(f; a, x, b)| \leq \omega_2^\varphi(f, h), \quad \text{if } b - a \leq 2h\varphi\left(\frac{a+b}{2}\right), \quad \text{and } h > 0. \tag{2.124}$$

Proof. Denote

$$M = \sup_{x \in [a,b]} |\Delta(f; a, x, b)|.$$

For $\varepsilon > 0$ arbitrarily chosen there is $x_\varepsilon \in (a, b)$ such that

$$|\Delta(f; a, x_\varepsilon, b)| > M - \varepsilon.$$

We can choose a linear function l such that the function $g := f + l$ has the properties $g(a) = 0$ and $g(b) = 0$. Note that we have

$$\omega_2^\varphi(f, h) = \omega_2^\varphi(g, h), \quad \text{for any } h > 0,$$

and

$$g(x) = \Delta(g; a, x, b) = \Delta(f; a, x, b), \quad \text{for any } x \in [a, b].$$

Therefore $|g(x_\varepsilon)| > M - \varepsilon$. By using the symmetry we can consider only the case $x_\varepsilon \leq \frac{a+b}{2}$. Denote $y = 2x_\varepsilon - a$ and $h_* = \frac{b-a}{2\varphi\left(\frac{a+b}{2}\right)}$. From Lemma 2.5.6 (ii), with s:=a,

we get $x_\varepsilon - a \leq h_* \varphi(x_\varepsilon)$. Hence

$$|g(a) - 2g(x_\varepsilon) + g(y)| \leq \omega_2^\varphi(f; h_*).$$

Then we have

$$\begin{aligned}
M &\geq |g(y)| = |g(y) - 2g(x_\varepsilon) + g(a) + 2g(x_\varepsilon)| \\
&\geq 2|g(x_\varepsilon)| - |g(y) - 2g(x_\varepsilon) + g(a)| \\
&\geq 2(M - \varepsilon) - \omega_2^\varphi(f; h_*).
\end{aligned}$$

Thus

$$M \leq \omega_2^\varphi(f; h_*) + 2\varepsilon.$$

Since $\varepsilon > 0$ was arbitrarily chosen we get (2.123) and consequently (2.124).

$\square$

Lemma 2.5.8. *Let $h > 0$ and $[a, b] \subset [0, 1]$ such that $b - a \leq 2h\varphi\left(\frac{a+b}{2}\right)$. Then for any points $a \leq u < v \leq b$, we have $v - u \leq 2h\varphi\left(\frac{u+v}{2}\right)$.*

Proof. From Lemma 2.5.6 (ii), it follows that the function $y \mapsto \frac{2(y-a)}{\varphi(y)}$, $y \in \left(a, \frac{a+b}{2}\right]$, is increasing. If we take $y = \frac{a+t}{2}$, it follows that the function $t \mapsto \frac{t-a}{\varphi\left(\frac{a+t}{2}\right)}$, $t \in (a, b]$, is also increasing. Hence

$$\frac{v - a}{\varphi\left(\frac{a+v}{2}\right)} \leq \frac{b - a}{\varphi\left(\frac{a+b}{2}\right)} \leq 2h.$$

In a symmetrical mode, the function $y \mapsto \frac{2(v-y)}{\varphi(y)}$, $y \in \left[\frac{a+v}{2}, v\right)$ is decreasing. If we take $y = \frac{v+t}{2}$, we obtain that the function $t \mapsto \frac{v-t}{\varphi\left(\frac{v+t}{2}\right)}$, $t \in (a, v]$, is also decreasing. Hence

$$\frac{v - u}{\varphi\left(\frac{u+v}{2}\right)} \leq \frac{v - a}{\varphi\left(\frac{a+v}{2}\right)} \leq 2h. \qquad \square$$

Lemma 2.5.9. *For $h > 0$ and $x \in \left(0, \frac{1}{1+h^2}\right)$, consider the canonical sequence of pair (x, h) given in (2.122). Let $f \in B[0, 1]$. We have*

i) If $y_i < u_i < y_{i+1}$, $(i \geq 1)$, are consecutive terms of the sequence (2.122), then

$$|\Delta(f; y_i, u_i, y_{i+1})| \leq \frac{1}{2}\omega_2^\varphi(f, h). \tag{2.125}$$

ii) If $u_i < y_{i+1} < u_{i+1}$, $(i \geq 0)$, are consecutive terms of the sequence (2.122), then

$$|\Delta(f; u_i, y_{i+1}, u_{i+1})| \leq \omega_2^\varphi(f, h). \tag{2.126}$$

Proof. i) We have

$$|\Delta(f; y_1, u_1, y_2)| = \left|\frac{1}{2}f(y_1) - 2f\left(\frac{y_1 + y_2}{2}\right) + f(y_2)\right| \leq \frac{1}{2}\omega_2^\varphi(f, h).$$

ii) It follows by Lemma 2.5.7, if we show that

$$u_{i+1} - u_i \leq 2h\varphi\left(\frac{u_i + u_{i+1}}{2}\right). \tag{2.127}$$

But this inequality is equivalent with

$$\varphi(u_i) + \varphi(u_{i+1}) \leq 2\varphi\left(\frac{u_i + u_{i+1}}{2}\right),$$

which is true, since φ is concave.   □

Lemma 2.5.10. *Let $h \in (0, 1]$ and let x and t be such that*

$$0 < x - h\varphi(x) < x < x + h\varphi(x) \leq t \leq 1.$$

If $f \in B[0, 1]$ satisfies the condition $f(x - h\varphi(x)) = 0 = f(x + h\varphi(x))$, then we have

$$f(t) \leq \left[\frac{1}{2} + \frac{3}{2}\left(\frac{t - x}{h\varphi(x)}\right)^2\right]\omega_2^\varphi(f, h). \tag{2.128}$$

Proof. From the hypothesis, we get

$$f(x) = -\frac{1}{2}(f(x - h\varphi(x)) - 2f(x) + f(x + h\varphi(x))) \geq -\frac{1}{2}\omega_2^\varphi(f, h). \tag{2.129}$$

Let x, t, h be as in the hypothesis of the lemma. Consider the canonical sequence (2.122)) of the pair (x, h) and denote by $M(x, t, h)$ the unique indices $m \geq 1$ for which one of the conditions (I) or (II) in Lemma 2.5.5 holds. We prove by induction with regard to m the following proposition:

$P(m)$: "For any function $f \in B[0, 1]$ and any numbers x, t, h that satisfy the conditions in the hypothesis of the lemma, if $M(x, t, h) = m$, then relation (2.128)) holds."

For $m = 1$ we distinguish between two cases.

Case A. Condition (I) holds, i.e., $x < y_1 \leq t$ and $t = y_1$, or $t \in (y_1, u_1]$. If $t = y_1$, relation (2.128) is obvious. Let $t > y_1$. Since $[x, t] \subseteq [u_0, u_1]$, using Lemmas 2.5.7–2.5.9, we obtain

$$|\Delta(f; x, y_1, t)| \leq \omega_2^\varphi(f, h).$$

Consequently we have

$$f(t) = \frac{t - x}{y_1 - x} \Delta(f; x, y_1, t) - \frac{t - y_1}{y_1 - x} f(x)$$

$$\leq \left(\frac{t - x}{y_1 - x} + \frac{1}{2} \frac{t - y_1}{y_1 - x} \right) \omega_2^\varphi(f, h)$$

$$= \left(-\frac{1}{2} + \frac{3}{2} \frac{t - x}{h\varphi(x)} \right) \omega_2^\varphi(f, h)$$

$$\leq \left(\frac{1}{2} + \frac{3}{2} \left(\frac{t - x}{h\varphi(x)} \right)^2 \right) \omega_2^\varphi(f, h).$$

Case B. Condition (II) holds, i.e., $x < y_1 < u_1 < t$ and $u_1 > \frac{1}{1+h^2}$ or $t < y_2$. First, we prove that

$$t - y_1 \leq 2h\varphi\left(\frac{t + y_1}{2} \right). \tag{2.130}$$

Indeed we have to consider two cases.

Case 1: $u_1 \in \left(\frac{1}{1+h^2}, 1 \right)$. By using the concavity of the function $t \mapsto 2h\varphi\left(\frac{t+y_1}{2} \right) - t + y_1, t \in [y_1, 1]$, in order to prove (2.130) it suffices to see it for $t = y_1$ and $t = 1$ The inequality (2.130) is obvious for $t = y_1$ and for $t = 1$ it can be rewritten after a simple computation on the equivalent form

$$y_1 \geq \frac{1 - h^2}{1 + h^2},$$

that is

$$u_1 - h\varphi(u_1) \geq \frac{1 - h^2}{1 + h^2}. \tag{2.131}$$

The function $u \mapsto u - h\varphi(u)$ is increasing on the interval $\left(\frac{h^2}{1+h^2}, 1 \right)$, see the proof of Lemma 2.5.2. It follows that in order to prove (2.131) it is sufficient to show that

$$\frac{1}{1 + h^2} - h\varphi\left(\frac{1}{1 + h^2} \right) \geq \frac{1 - h^2}{1 + h^2}. \tag{2.132}$$

But (2.132) is an equality for any $h \in (0, 1]$.

Case 2. There is y_2 and $t \leq y_2$. We have $y_2 - y_1 = 2h\varphi(u_1)$. Since $[y_1, t] \subseteq [y_1, y_2]$, by using Lemma 2.5.8 we obtain (2.130).

Now from (2.130) and Lemma 2.5.7 we arrive at

$$|\Delta(f; y_1, u_1, t)| \leq \omega_2^\varphi(f, h). \tag{2.133}$$

We use the identity

$$f(s) = \frac{(s - y_1)(u_1 - x)}{(u_1 - y_1)(y_1 - x)} \Delta(f; x, y_1, u_1) + \frac{s - y_1}{u_1 - y_1} \Delta(f; y_1, u_1, s)$$
$$- \frac{s - y_1}{y_1 - x} f(x) + \frac{s - x}{y_1 - x} f(y_1), \quad s \in (u_1, 1]. \tag{2.134}$$

From relations (2.134) for $s := t$, (2.126), (2.129) and (2.133), we obtain

$$f(t) \leq \left[\frac{(t - y_1)(u_1 - x)}{(u_1 - y_1)(y_1 - x)} + \frac{t - y_1}{u_1 - y_1} + \frac{1}{2} \frac{t - y_1}{y_1 - x} \right] \omega_2^\varphi(f, h). \tag{2.135}$$

It suffices to show that

$$\frac{(t - y_1)(u_1 - x)}{(u_1 - y_1)(y_1 - x)} + \frac{t - y_1}{u_1 - y_1} + \frac{1}{2} \frac{t - y_1}{y_1 - x} \leq \frac{1}{2} + \frac{3}{2} \left(\frac{t - x}{h\varphi(x)} \right)^2.$$

Put $\alpha := y_1 - x$, $\beta := u_1 - y_1$, $\gamma := t - u_1$. The inequality above can be rewritten in the equivalent form

$$4\alpha^2 \gamma \leq 3\beta(\beta + \gamma)(\alpha + \beta + \gamma). \tag{2.136}$$

From Lemma 2.5.6 (i), we have

$$\frac{t - u_1}{\varphi^2(u_1)} \leq \frac{t - y_1}{\varphi^2(y_1)}, \quad \text{i.e.,} \quad (h\varphi(y_1))^2 \gamma \leq \beta^2(\beta + \gamma).$$

Since the function φ^2 is concave, we have

$$\varphi^2(y_1) \geq \frac{\alpha}{\alpha + \beta} \varphi^2(u_1) + \frac{\beta}{\alpha + \beta} \varphi^2(x) = h^{-2}\alpha\beta.$$

Hence $4\alpha^2 \gamma \leq 4\frac{\alpha}{\beta}(h\varphi(y_1))^2 \gamma \leq 4\alpha\beta(\beta+\gamma)$. Consequently, inequality (2.136) holds if $\alpha \leq 3(\beta+\gamma)$. Suppose now that $\alpha > 3(\beta+\gamma)$. For $\alpha \leq h\varphi(y_1)$, inequality (2.136) is true. Indeed, we have

$$4\alpha^2 \gamma \leq 4(h\varphi(y_1))^2 \gamma \leq 4\beta^2(\beta + \gamma) < 3\beta(\beta + \gamma)(\alpha + \beta + \gamma).$$

It remains to prove relation (2.136) in the case $3(\beta + \gamma) < \alpha$ and $h\varphi(y_1) < \alpha$. From the hypothesis of Case B, it follows that $\gamma \leq \beta$. We obtain

$$t - y_1 < h\varphi(y_1).$$

Indeed we have

$$(t - y_1)^2 = (\beta + \gamma)^2 < \frac{1}{3}\alpha(\beta + \gamma) \leq \frac{2}{3}\alpha\beta \leq \frac{2}{3}(h\varphi(y_1))^2.$$

Put $v := 2y_1 - t$. Since $h\varphi(y_1) < \alpha$, it follows $v \in (x, y_1)$. From Lemma 2.5.7 we have

$$f(v) = -\Delta(f; x - h\varphi(x), v, x + h\varphi(x)) \geq -\omega_2^{\varphi}(f, h).$$

Consequently, $f(t) = (f(t) - 2f(y_1) + f(v)) - f(v) \leq 2\omega_2^{\varphi}(f, h)$. On the other hand we have

$$\frac{1}{2} + \frac{3}{2}\left(\frac{t - x}{h\varphi(x)}\right)^2 \geq 2.$$

Therefore proposition $P(1)$ is completely proved. Suppose now that $P(m - 1)$, $m \geq 2$ is true and prove $P(m)$.

Let f, x, t, h be as in the hypothesis. We can choose a linear function l, such that the function $g := f + l$ satisfies the following condition: $g(y_1) = g(y_2) = 0$. We get $\Delta(f; y_1, y_2, t) = \Delta(g; y_1, y_2, t)$ and $\omega_2^{\varphi}(f, h) = \omega_2^{\varphi}(g, h)$. Note that $M(u_1, t, h) = m - 1$, because the canonical sequence of the pair (u_1, h), until t is $u_1 < y_2 < \cdots < y_m \leq t$ or $u_1 < y_2 < \cdots < y_m < u_m < t$. Then, by applying proposition $P(m - 1)$ with g instead of f and u_1 instead of x, we get $\Delta(f; y_1, y_2, t) = \Delta(g; y_1, y_2, t)$ and $\omega_2^{\varphi}(f, h) = \omega_2^{\varphi}(g, h)$. If we apply the hypothesis of induction with the point u_1 instead of x, we get

$$g(t) \leq \left[\frac{1}{2} + \frac{3}{2}\left(\frac{t - u_1}{u_1 - y_1}\right)^2\right]\omega_2^{\varphi}(g, h).$$

Using identity (2.134) for $s := y_2$ and relations (2.125), (2.126) and (2.129), we obtain

$$f(y_2) \leq \left[\frac{(y_2 - y_1)(u_1 - x)}{(u_1 - y_1)(y_1 - x)} + \frac{1}{2}\frac{y_2 - y_1}{u_1 - y_1} + \frac{1}{2}\frac{y_2 - y_1}{y_1 - x}\right]\omega_2^{\varphi}(f, h).$$

Put $\alpha := y_1 - x, \beta := u_1 - y_1, \rho := t - u_1$. Note that $\rho \geq \beta$. From the relation above we obtain

$$f(t) = \frac{t - y_1}{y_2 - y_1}\Delta(f; y_1, y_2, t) + \frac{t - y_1}{y_2 - y_1}f(y_2)$$

$$= g(t) + \frac{t - y_1}{y_2 - y_1}f(y_2)$$

$$\leq \left[\frac{1}{2} + \frac{3}{2}\left(\frac{\rho}{\beta}\right)^2 + \frac{\rho + \beta}{2\beta}\left(2\frac{\beta + \alpha}{\alpha} + 1 + \frac{\beta}{\alpha}\right)\right]\omega_2^{\varphi}(f, h)$$

$$= \left[\frac{1}{2} + \frac{3}{2}\left[\left(\frac{\rho}{\beta}\right)^2 + 1 + \frac{\beta}{\alpha} + \frac{\rho}{\alpha} + \frac{\rho}{\beta}\right]\right]\omega_2^{\varphi}(f, h).$$

It suffices to have

$$\left(\frac{\rho}{\beta}\right)^2 + 1 + \frac{\beta}{\alpha} + \frac{\rho}{\alpha} + \frac{\rho}{\beta} \leq \left(\frac{\alpha + \beta + \rho}{\alpha}\right)^2.$$

This inequality is immediate for any $0 < \alpha \leq \beta, 0 < \rho$. It remains the case where $0 < \beta < \alpha$ and $\beta \leq \rho$. From Lemma 2.5.6 (i) we have

$$\frac{t - u_1}{\varphi^2(u_1)} \leq \frac{t - x}{\varphi^2(x)}, \quad \text{i.e.,} \quad \alpha^2\rho \leq \beta^2(\alpha + \beta + \rho).$$

This implies that

$$\left(\frac{\rho}{\beta}\right)^2 + 1 + \frac{\beta}{\alpha} + \frac{\rho}{\alpha} + \frac{\rho}{\beta} \le \frac{\rho(\alpha + \beta + \rho)}{\alpha^2} + 1 + \frac{\beta}{\alpha} + \frac{\beta(\alpha + \beta + \rho)}{\alpha^2} + \frac{\rho}{\alpha}$$

$$= \left(\frac{\alpha + \beta + \rho}{\alpha}\right)^2. \tag{2.137}$$

The proof is finished. □

2.5.2 Main result

Our main result is

Theorem 2.5.1. *If* $F : V \to \mathbb{R}$, $V \subset \mathcal{F}[0, 1]$, *is a linear positive functional that is admissible related to a point* $x \in (0, 1)$, *then we have*

$$|F(f) - f(x)| \le |F(e_0) - 1| \cdot |f(x)| + \frac{|F(e_1 - xe_0)|}{2h\varphi(x)} \cdot \omega_1^\varphi(f, 2h) \tag{2.138}$$

$$+ \left[F(e_0) + \frac{3}{2}\frac{F((e_1 - xe_0)^2)}{(h\varphi(x))^2}\right] \omega_2^\varphi(f, h),$$

for all $f \in V \cap B[0, 1]$ *and* $h > 0$, *such that,* $h \in \left(0, \frac{1}{2}\right]$.

Proof. Let f, x, h be fixed as in the premises. Suppose that $\omega_1^\varphi(f, 2h) < \infty$, $\omega_2^\varphi(f, h) < \infty$. We apply Theorem 2.1.2 by choosing Ω_1, defined by $\Omega_1(g, h) := \frac{1}{2\varphi(x)} \cdot \omega_1^\varphi(g, 2h)$, $g \in B[0, 1]$, $h > 0$, $\Omega_2 := \omega_2^\varphi$ and the function

$$\Psi(y) = 1 + \frac{3}{2}\left(\frac{y}{\varphi(x)}\right)^2, \ y \in [0, \infty). \tag{2.139}$$

First, let us prove the corresponding inequality (2.15), namely

$$|\Delta(f; s, x, t)| \le \left(1 + \frac{3}{2} \cdot \frac{(t - x)(x - s)}{(h\varphi(x))^2}\right) \cdot \omega_2^\varphi(f, h), \ s < x < t, \ s, t \in [0, 1]. \tag{2.140}$$

We have some cases.

Case 1: $t - s \le 2h\varphi\left(\frac{t+s}{2}\right)$. Then by using Lemma 2.5.7 we obtain

$$|\Delta(f; s, x, t)| \le \omega_2^\varphi(f, h) \le \left(1 + \frac{3}{2} \cdot \frac{(t - x)(x - s)}{(h\varphi(x))^2}\right) \cdot \omega_2^\varphi(f, h).$$

Case 2: $t - x > h\varphi(x)$ and $x - s > h\varphi(x)$. We can choose a polynomial l of degree 1, such that the function $g = f + l$ satisfies the conditions

$$g(x - h\varphi(x)) = 0 = g(x + h\varphi(x)). \tag{2.141}$$

We have

$$\Delta(f; s, x, t) = \Delta(g; s, x, t) \quad \text{and} \quad \omega_2^\varphi(f, h) = \omega_2^\varphi(g, h). \tag{2.142}$$

Suppose for a choice that

$$\Delta(f; s, x, t) \geq 0. \tag{2.143}$$

We have $g(x) \geq -\frac{1}{2}\omega_2^\varphi(f, h)$. By applying Lemma 2.5.10 we get

$$g(t) - g(x) \leq \left[1 + \frac{3}{2}\left(\frac{t - x}{h\varphi(x)}\right)^2\right] \omega_2^\varphi(f, h), \tag{2.144}$$

and by using the symmetry we have also

$$g(s) - g(x) \leq \left[1 + \frac{3}{2}\left(\frac{x - s}{h\varphi(x)}\right)^2\right] \omega_2^\varphi(f, h). \tag{2.145}$$

Therefore we obtain

$$|\Delta(f; s, x, t)| = \Delta(g; s, x, t) = \frac{t - x}{t - s}(g(s) - g(x)) + \frac{x - s}{t - s}(g(t) - g(x))$$
$$\leq \left(1 + \frac{3}{2} \cdot \frac{(t - x)(x - s)}{(h\varphi(x))^2}\right) \cdot \omega_2^\varphi(f, h).$$

<u>Case 3</u>: $t - s > 2h\varphi\left(\frac{t+s}{2}\right)$ and $x - s \leq h\varphi(x)$. We can construct two points y and w that satisfy the following two conditions:

$$s < x \leq y < w < t, \tag{2.146}$$
$$y - s = w - y = h\varphi(y). \tag{2.147}$$

Indeed, first consider the case $s > 0$. Define $y := \Theta_h(s)$ and $w = y + h\varphi(y)$. Therefore (2.147) is true. There exists a unique h_1, $0 < h_1 \leq h$ such that $x - s = h_1\varphi(x)$, that is $x = \Theta_{h_1}(s)$. Also since $y - s = h\varphi(y) \geq h_1\varphi(y)$, there is a unique s_1, $s \leq s_1 < y$ such that $y - s_1 = h_1\varphi(y)$, that is $y = \Theta_{h_1}(s_1)$. Since $s \leq s_1$, from Lemma 2.5.2 it follows that $x \leq y$.

Put now $v = \frac{t+s}{2}$. Since $v - s > h\varphi(v)$, there is a unique s_2, $s < s_2 < v$ such that $v - s_2 = h\varphi(v)$, that is $v = \Theta_h(s_2)$. Because $y = \Theta_h(s)$, we conclude from Lemma 2.5.2 that $y < v$. From this we get $w < t$. Hence relation (2.146) is true.

Second, consider the case $s = 0$. Since $x \leq h\varphi(x)$, from the symmetrical result in Lemma 2.5.1 it follows that $x \in \left(0, \frac{h^2}{1+h^2}\right]$. Define $y = \frac{h^2}{1+h^2}$ and $w = y + h\varphi(y)$. We have $y - h\varphi(y) = 0$ and thus (2.147) is true. In order to prove (2.146) it remains to show $w < t$. Let $v = \frac{t}{2}$. The hypothesis of Case 3, i.e., $v > h\varphi(v)$, is equivalent with $v > \frac{h^2}{1+h^2}$, i.e., $v > y$. Consequently $t > w$.

Let us choose a linear function l such that the function $g = f + l$ satisfies the conditions

$$g(s) = 0 = g(w).$$

The function g satisfies (2.142). Suppose for a choice that (2.143) is true. We have $g(y) \geq -\frac{1}{2}\omega_2^{\varphi}(f, h)$, and by applying Lemma 2.5.10 for the point y, we arrive at

$$g(t) \leq \left[\frac{1}{2} + \frac{3}{2}\left(\frac{t-y}{h\varphi(y)}\right)^2\right]\omega_2^{\varphi}(f, h). \tag{2.148}$$

From Lemma 2.5.7 we obtain

$$g(x) \geq -\omega_2^{\varphi}(f, h). \tag{2.149}$$

Hence

$$\begin{aligned}
|\Delta(f; s, x, t)| &= \Delta(g; s, x, t) \\
&\leq \frac{t-x}{t-s}(g(s) - g(x)) + \frac{x-s}{t-s}(g(t) - g(x)) \\
&\leq \left[\frac{t-x}{t-s} + \frac{x-s}{t-s}\left(\frac{3}{2} + \frac{3}{2}\left(\frac{t-y}{h\varphi(y)}\right)^2\right)\right]\omega_2^{\varphi}(f, h) \\
&= \left[1 + \frac{x-s}{t-s}\left(\frac{1}{2} + \frac{3}{2}\left(\frac{t-y}{h\varphi(y)}\right)^2\right)\right]\omega_2^{\varphi}(f, h). \tag{2.150}
\end{aligned}$$

From Lemma 2.5.6 (i) we deduce $\frac{t-x}{(h\varphi(x))^2} \geq \frac{t-y}{(h\varphi(y))^2}$. It follows that

$$\left[1 + \frac{3}{2}\frac{(t-x)(x-s)}{(h\varphi(x))^2}\right]\omega_2^{\varphi}(f, h) \geq \left[1 + \frac{3}{2}\frac{(t-y)(x-s)}{(h\varphi(y))^2}\right]\omega_2^{\varphi}(f, h).$$

In order to prove (2.140) it is enough to show that

$$\frac{3}{2}\frac{(t-y)(x-s)}{(h\varphi(y))^2} \geq \frac{x-s}{t-s}\left[\frac{1}{2} + \frac{3}{2}\left(\frac{t-y}{h\varphi(y)}\right)^2\right].$$

But this inequality is immediate, by taking into account that $y - s = h\varphi(y)$ and $t - y > h\varphi(y)$.

Case 3′: $t - s > 2h\varphi\left(\frac{t+s}{2}\right)$ and $t - x \leq h\varphi(x)$. This case is similar to the symmetrical Case 3.

The proof of relation (2.140) is complete. We prove now the corresponding relation (2.17), namely, for any $t \in [0, 1]$:

$$|f(t) - f(x)| \leq \frac{|t-x|}{2h\varphi(x)} \cdot \omega_1^{\varphi}(f, 2h) + \left[1 + \frac{3}{2}\left(\frac{t-x}{h\varphi(x)}\right)^2\right]\omega_2^{\varphi}(f, h). \tag{2.151}$$

Using the symmetry we consider that $x \leq \frac{1}{2}$. We distinguish between some cases.

Case a: $x \pm h\varphi(x) \in [0, 1]$. The function $g(t) := \Delta(f; x - h\varphi(x), t, x + h\varphi(x))$, $t \in [0, 1]$ satisfies the conditions $g(x - h\varphi(x)) = 0 = g(x + h\varphi(x))$ and $\omega_2^{\varphi}(g, h) = \omega_2^{\varphi}(f, h)$. Consequently, from Lemma 2.5.10 we deduce

$$|\Delta(f; x - h\varphi(x), t, x + h\varphi(x))| \le \left(\frac{1}{2} + \frac{3}{2}\left(\frac{t-x}{h\varphi(x)}\right)^2\right)\omega_2^\varphi(f, h).$$

We have

$$|f(t) - f(x)| = \left|\Delta(f; x - h\varphi(x), x, x + h\varphi(x))\right.$$
$$-\Delta(f; x - h\varphi(x), t, x + h\varphi(x))$$
$$\left.+\frac{t-x}{2h\varphi(x)}\Big(f(x + h\varphi(x)) - f(x - h\varphi(x))\Big)\right|$$
$$\le \frac{1}{2}\omega_2^\varphi(f, h) + \left(\frac{1}{2} + \frac{3}{2}\left(\frac{t-x}{h\varphi(x)}\right)^2\right)\omega_2^\varphi(f, h) + \frac{|t-x|}{2h\varphi(x)} \cdot \omega_1^\varphi(f, 2h).$$

From this we get (2.151). For the following two cases we put

$$y := \Theta_h(x) \quad \text{and} \quad z := y + h\varphi(y).$$

From the symmetrical result given in Lemma 2.5.1 we have $x < \frac{h^2}{1+h^2}$. The relation $x + h\varphi(y) = y$ is equivalent, see the proof of Lemma 2.5.2, to the equality $P(y) = 0$, where $P(y) = (1 + h^2)y^2 - (2x + h^2)y + x^2$. We have $P(x) < 0$. Since $h \le \frac{1}{2}$ and $x < \frac{h^2}{1+h^2}$, we have also

$$P\left(\frac{1}{2}\right) = \frac{1}{4} - \frac{h^2}{4} - x(1-x) \ge \frac{1}{4} - \frac{h^2}{4} - \frac{h^2}{(1+h^2)^2} \ge \frac{1}{4} - \frac{1}{16} - \frac{4}{25} > 0.$$

It follows that $x < y < \frac{1}{2}$. Hence $\varphi(x) < \varphi(y)$ and $z < 1$.

Case b: $x - h\varphi(x) < 0$ and $y < t$. The function $g(t) := \Delta(f; x, t, z) = \Delta(f; y - h\varphi(y), t, y + h\varphi(y))$, $t \in [0, 1]$ satisfies the conditions $g(y - h\varphi(y)) = 0 = g(y + h\varphi(y))$ and $\omega_2^\varphi(g, h) = \omega_2^\varphi(f, h)$. Consequently, from Lemma 2.5.10 we deduce

$$|\Delta(f; x, t, z)| \le \left(\frac{1}{2} + \frac{3}{2}\left(\frac{t-y}{h\varphi(y)}\right)^2\right)\omega_2^\varphi(f, h).$$

Using Lemma 2.5.6 (ii), we get $\left(\frac{t-y}{h\varphi(y)}\right)^2 < \left(\frac{t-x}{h\varphi(x)}\right)^2$. Now we have

$$|f(t) - f(x)| = \left|\frac{t-x}{z-x}(f(z) - f(x)) - \Delta(f; x, t, z)\right|$$
$$\le \frac{|t-x|}{2h\varphi(y)} \cdot \omega_1^\varphi(f, 2h) + \left(\frac{1}{2} + \frac{3}{2}\left(\frac{t-y}{h\varphi(y)}\right)^2\right)\omega_2^\varphi(f, h).$$

The inequality (2.151) follows.

Case c: $x - h\varphi(x) < 0$ and $t \le y$. From Lemma 2.5.8 we obtain

$$|\Delta(f; x, t, z)| \le \omega_2^\varphi(f, h).$$

Hence we have

$$
\begin{aligned}
|f(t) - f(x)| &= \left| \frac{t-x}{z-x} (f(z) - f(x)) - \Delta(f; x, t, z) \right| \\
&\le \frac{|t-x|}{2h\varphi(y)} \cdot \omega_1^\varphi(f, 2h) + \omega_2^\varphi(f, h).
\end{aligned}
$$

Consequently, (2.151) follows. The theorem is completely proved.   □

3

Absolute Optimal Constants

3.1 Introduction

We point out the optimality of the estimate given in Theorem 2.2.2, for $s = 2$, in a stronger sense than the optimality of the constants. Let x be an interior point of I and denote by $\mathcal{U}_x(I)$ the family of all linear positive functionals F defined on a subspace $V_F \subset \mathcal{F}_b(I)$, with the property $F(e_0) = 1$, $F(e_1) = x$ and which are admissible related to the point x.

Theorem 3.1.1. *For any $\lambda \in \left[0, \frac{1}{2}\right]$ and any $h > 0$, we have*

$$\sup_{F \in \mathcal{U}_x(I)} \sup_{f \in V_F \backslash \Pi_1} \frac{|F(f) - f(x)|}{\left[\frac{2}{1+2\lambda} + h^{-2}F((e_1 - xe_0)^2)\right] \cdot \omega_2^\lambda(f, h)} = 1. \tag{3.1}$$

Proof. Let M be the supremum in (3.1). From Theorem 2.2.2 it follows that $M \leq 1$. We prove the converse inequality. Choose $\varepsilon > 0$ arbitrarily and choose $a, b \in I$ such that $a < x < b, b - a \leq 2h$ and $\frac{x-a}{b-a} < \varepsilon$. Consider the functional $F : \mathcal{F}_b(I) \to \mathcal{F}(I)$ defined by

$$F(f) := \frac{b-x}{b-a} f(a) + \frac{x-a}{b-a} f(b), \quad f \in \mathcal{F}_b(I). \tag{3.2}$$

Consider also the function

$$f_0(t) := \begin{cases} 0, & t \in I \cap (-\infty, a), \\ \frac{t-a}{x-a}, & t \in [a, x], \\ 1, & t \in I \cap (x, \infty). \end{cases} \tag{3.3}$$

We have $\omega_2^\lambda(f, h) = \frac{1}{2} + \lambda$. Indeed, let $u, w \in I$, such that $w - u \leq 2h$ and let $v = (1-t)u + tw$, with $\left|t - \frac{1}{2}\right| \leq \lambda$. Since $(f(w) - f(v)) \cdot (f(u) - f(v)) \leq 0$, we obtain

$$|\Delta(f; u, v, w)| = |t(f(u) - f(v)) + (1-t)(f(w) - f(v))|$$
$$\leq \max\{t|f(u) - f(v)|, (1-t)|f(w) - f(v)|\} \leq \frac{1}{2} + \lambda.$$

Also it is easy to see that

$$|F(f_0) - f_0(x)| = \frac{b - x}{b - a} > 1 - \varepsilon, \text{ and}$$

$$\frac{2}{1 + 2\lambda} + h^{-2}F((e_1 - xe_0)^2) = \frac{2}{1 + 2\lambda} + h^{-2}(b - x)(x - a) < \frac{2}{1 + 2\lambda} + 4\varepsilon.$$

Consequently, $M > \frac{1-\varepsilon}{1+2\varepsilon(1+2\lambda)}$. Since $\varepsilon > 0$ was arbitrarily chosen, it follows that $M \geq 1$. $\qquad\qquad\qquad\qquad\qquad\qquad\qquad\qquad\qquad\qquad\qquad\qquad\qquad\square$

Corollary 3.1.1. *Let* $\lambda \in \left[0, \frac{1}{2}\right]$.

a) If I is an arbitrary interval, then for any $h > 0$ we have

$$\sup_{F \in \mathcal{U}_x(I)} \sup_{f \in V_F \backslash \Pi_1} \frac{|F(f) - f(x)|}{\omega_2^\lambda(f, h)} \geq \frac{2}{1 + 2\lambda}. \tag{3.4}$$

b) If I is a finite interval, then for any $h \geq \frac{1}{2} \cdot \text{length}(I)$, we have

$$\sup_{F \in \mathcal{U}_x(I)} \sup_{f \in V_F \backslash \Pi_1} \frac{|F(f) - f(x)|}{\omega_2^\lambda(f, h)} = \frac{2}{1 + 2\lambda}. \tag{3.5}$$

Proof. The point a) follows from Theorem 3.1.1. The point b) follows from the point a) and from Theorem 2.2.2. $\qquad\qquad\qquad\qquad\qquad\qquad\qquad\qquad\qquad\qquad\qquad\square$

We can note that in obtaining the inverse inequalities in Theorem 3.1.1 we use the simplest type of functionals in the class $\mathcal{U}_x(I)$, excepting the Dirac functional. Based on the previous corollary we conclude that $C = \frac{2}{1+2\lambda}$ is the **absolute optimal constant** that can appear in a general estimate of the form $|F(f) - f(x)| \leq C\omega_2^\lambda(f, h)$, $F \in \mathcal{U}_x(I)$ $f \in V_F$, $h > 0$. In the case of the classical second order modulus, the value $C = 1$ is the absolute optimal constant that can appear in a general estimate of the form $|F(f) - f(x)| \leq C\omega_2(f, h)$, $F \in \mathcal{U}_x(I)$ $f \in V_F$, $h > 0$.

For the modulus ω_2 and the discrete functionals with equidistant knots we have the following stronger result, given in [105].

Theorem 3.1.2. *Let $F : \mathcal{F}[0, 1] \to \mathbb{R}$ be a functional with equidistant knots of the form $F(f) := \sum_{k=0}^{n} f\left(\frac{k}{n}\right) v_k$, $f \in \mathcal{F}[0, 1]$, where $v_k \in \mathbb{R}$, $0 \leq k \leq n$. Suppose $F(e_0) = 1$ and $F(e_1) = x$, $x \in (0, 1)$. If the number x is irrational, then for any $h > 0$ we have*

$$\sup_{f \in C[0,1] \backslash \Pi_1} \frac{|F(f) - f(x)|}{\omega_2(f, h)} \geq 1. \tag{3.6}$$

Proof. Let $0 \leq k \leq n - 1$ be such that $\frac{k}{n} < x < \frac{k+1}{n}$. Denote $y := nx - k$. Let $0 < \varepsilon < \frac{1}{2}$ arbitrarily. We can choose $p \in \mathbb{N}$ such that $\{py\} < \varepsilon$. Denote $l := [py]$, $q := np$, $m := kp + l$, $\delta := \{py\}/q$. Then we have $x = \frac{m}{q} + \delta$, and $0 < \delta q < \varepsilon$. Let us construct the function $g \in C[0, 1]$ in the following way. Firstly, set

$$g(r) := \begin{cases} \frac{r}{\delta}, & 0 \le r \le \delta, \\ \frac{1-rq}{1-\delta q}, & \delta \le r \le \frac{1}{q}. \end{cases}$$

Next define g on the interval $[0, 1]$ by periodicity, with the period $\frac{1}{q}$.

We have $|F(g) - g(x)| = |-g(x)| = 1$. In order to estimate $\omega_2(g, h)$, let $0 \le x_1 < x_2 < x_3 \le 1$, be such that $2x_2 = x_1 + x_3$ and $x_2 - x_1 \le h$. Let us represent $x_j = \frac{s_j}{q} + r_j$, where $s_j \in \mathbb{N} \cup \{0\}$ and $0 \le r_j < \frac{1}{q}$, $j = 1, 2, 3$. Since g is periodic with the period $\frac{1}{q}$ we can write $|g(x_1) - 2g(x_2) + g(x_3)| = |g(r_1) - 2g(r_2) + g(r_3)|$. From the condition $2x_2 = x_1 + x_3$ it follows that there are only these possible cases: $2r_2 = r_1 + r_3$, $2r_2 = r_1 + r_3 + \frac{1}{q}$ and $2r_2 = r_1 + r_3 - \frac{1}{q}$. We may suppose without any loss of generality that $r_1 \le r_3$. Denote $R := |g(r_1) - 2g(r_2) + g(r_3)|$. We have

Case 1. $2r_2 = r_1 + r_3$. Hence $r_1 \le r_2 \le r_3$. If $r_1 \ge \delta$ or $r_3 \le \delta$, clearly we have $R = 0$. If $r_2 \le \delta < r_3$, then $R = \frac{r_3 - \delta}{\delta(1 - q\delta)} < \frac{1}{1 - q\delta}$, since $0 < r_3 - \delta < r_3 - r_2 = r_2 - r_1 < \delta$. If $r_1 < \delta < r_2$, we have $R = \frac{\delta - r_1}{\delta(1 - q\delta)} < \frac{1}{1 - q\delta}$.

Case 2. $2r_2 = r_1 + r_3 + 1/q$. Then $r_1 \le r_3 < r_2$. We have $r_2 > \delta$, since $\delta < \frac{1}{2q}$. If $r_3 \le \delta$, we have $R = \left| \frac{r_1 + r_3 - \delta}{\delta(1 - q\delta)} \right| \le \frac{1}{1 - q\delta}$. If $r_1 \le \delta < r_3$, we have $R = \frac{r_1}{\delta(1 - q\delta)} \le \frac{1}{1 - q\delta}$. If $\delta < r_1$, then $R = \frac{1}{1 - q\delta}$.

Case 3. $2r_2 = r_1 + r_3 - 1/q$. Then $r_2 \le r_1 \le r_3$. Since $\delta < \frac{1}{2q}$, we have $r_3 > \delta$. If $\delta < r_2$, then $R = \frac{1}{1 - q\delta}$. If $r_2 \le \delta < r_1$, we have $R = \left| \frac{\delta - 2r_2}{\delta(1 - q\delta)} \right| \le \frac{1}{1 - q\delta}$. If $r_1 \le \delta < r_3$, we have $R = \left| \frac{r_1 - 2r_2}{\delta(1 - q\delta)} \right| \le \frac{r_1}{\delta(1 - q\delta)} \le \frac{1}{1 - q\delta}$.

Therefore in all cases, one obtains $R \le \frac{1}{1 - q\delta} \le \frac{1}{1 - \varepsilon}$. Hence $\omega_2(g, h) \le \frac{1}{1 - \varepsilon}$. Since $0 < \varepsilon < \frac{1}{2}$ was arbitrarily chosen, the theorem is proved. $\square$

An application of this theorem will be presented for the Bernstein operators, see Corollary 4.2.1. The results above justifies considering the following definition.

Definition 3.1.1. *Let $F : V \to \mathbb{R}$, $V \subset \mathcal{F}_b(I)$, be a positive linear functional such that $\Pi_1 \subset V$ and $F(e_0) = 1$, $F(e_1) = x$, where $x \in I$. Define the function*

$$\varphi_{F,x}(h) := \sup_{\substack{f \in V \\ f \ne linear}} \frac{|F(f) - f(x)|}{\omega_2(f, h)}, \quad h > 0. \tag{3.7}$$

*We define the **critical value** of F, related to the point x, to be the number*

$$h_{F,x} := \inf\{h > 0 \mid \varphi_{F,x}(h) \le 1\}. \tag{3.8}$$

(In the case where this set is empty, we take $h_{F,x} = \infty$.)

Remark 3.1.1. There exist functionals as above, see for instance the results given for Bernstein operators, in Chapter 4, for which the critical value is not trivial. By taking into account relation (2.52), this means that the critical value is smaller, (possibly very small) than $\frac{1}{2}$ length (I). This fact points out an important difference between the estimates in terms of the first and the second moduli of continuity. Indeed, if

$F : C(I) \to \mathbb{R}$ is a linear positive functional induced by a Borel positive measure μ, if it is such that $F(e_0) = 1$ and if there is $x \in I$ such that $\mu(\{x\}) = 0$ and $\mu(I \setminus [x - h, x + h]\} > 0$, then

$$\sup_{\substack{f \in C(I) \\ f \neq \text{linear}}} \frac{|F(f) - f(x)|}{\omega_1(f, h)} > 1.$$

Also, note that if $F : \mathcal{F}[0, 1] \to \mathbb{R}$ is a discrete positive functional with equidistant knots, such that $F(e_0) = 1$ and $F(e_1) = x$, $x \in (0, 1)$, irrational, then $\varphi_{F,x}(h) = 1$, for all $h \geq h_{F,x}$, and $\varphi_{F,x}(h) > 1$, for $0 < h < h_{F,x}$.

In the next sections we obtain sufficient conditions for estimates with absolute optimal constants.

3.2 Discrete functionals and the classical second order modulus ω_2

In this section we consider estimates with the modulus ω_2 for discrete linear positive functionals. From Section 3.1 it follows that the value 1 is the absolute optimal constant that can appear in estimates with the usual second order modulus. We give a general sufficient condition that assures estimates with the absolute optimal constant 1, for discrete linear positive functionals.

From Lemma 2.2.1 we deduce:

Lemma 3.2.1. We have

$$|\Delta(f; a, t, b)| \leq \omega_2(f, h), \tag{3.9}$$

for all functions $f \in \mathcal{F}_b(I)$ and all $h > 0$, $a, b, t \in I$, $a < b \leq a + 2h$, $a \leq t \leq b$.

Consider a sequence of points $y_0 < y_1 < \cdots < y_m$ of I. For any $t \in (y_0, y_m)$ define:

$$\sigma(t) := \max\{i \in \mathbb{N} \cup \{0\}, \; y_i < t\}, \quad \overline{\sigma}(t) := \max\{i \in \mathbb{N} \cup \{0\}, \; y_i \leq t\}, \tag{3.10}$$

$$\tau(t) := \min\{i \in \mathbb{N}, \; y_i > t\}, \quad \overline{\tau}(t) := \min\{\in \mathbb{N}, \; y_i \geq t\}. \tag{3.11}$$

Let also $h > 0$. If $u, v - h \in (y_0, y_m)$, write

$$\pi(u, v) := \max\{\tau(u), \overline{\tau}(v - h)\}. \tag{3.12}$$

For any numbers $t < z \leq t + h$, denote

$$\alpha(t, z, h) := t + 2^s(z - t), \text{ where } s := \min\{r \in \mathbb{N}, \; 2^r(z - t) > h\}. \tag{3.13}$$

If $x \in (y_0, y_m)$, $h > 0$ and $x - y_0 \leq h$, denote by $M_{x,h}(y_0, y_1, \ldots, y_m)$, the set of all the pairs (u, v), for which there are $k \in \mathbb{N_0}$, the indices $0 \leq j_0 < \cdots < j_k < m$ and the numbers u_i, $(0 \leq i \leq k + 1)$ such that the following four relations are satisfied:

$$y_{j_0} < u_0 = x < y_{j_1} < u_1 < \cdots < y_{j_k} < u_k = u < u_{k+1} = v < y_m, \quad (3.14)$$

$$u_i - h \le y_{j_i}, \; (0 \le i \le k), \quad (3.15)$$

$$u_{i+1} = \alpha(y_{j_i}, u_i, h), \; (0 \le i \le k), \quad (3.16)$$

$$y_m - y_{j_k} > 2h. \quad (3.17)$$

Finally, if $(u, v) \in M_{x,h}(y_0, y_1, \ldots, y_m)$, then denote by $ord(u, v)$ the smallest index k for which the indices j_i and the numbers u_i satisfying (3.14)–(3.17) exist.

Lemma 3.2.2. *If $(u, v) \in M_{x,h}(y_0, y_1, \ldots, y_m)$, then,*

$$v > u + \frac{h}{2} \text{ and } \pi(u, v) \le \tau\left(v - \frac{h}{2}\right). \quad (3.18)$$

Proof. Let $v = \alpha(y_j, u, h)$, where $y_j \in [u - h, u)$. If $u - y_j \le \frac{h}{2}$, then we have $v - u = v - y_j - (u - y_j) > h - \frac{h}{2} = \frac{h}{2}$. If $u - y_j > \frac{h}{2}$, then we have $v - u = u - y_j$. Consequently one obtains $\pi(u, v) \le \tau\left(v - \frac{h}{2}\right)$. □

The main result of this section is the following, (see [91] and [109]).

Theorem 3.2.1. *Consider the sequence of points of I $y_0 < y_1 < \cdots < y_m$, $m \ge 1$ and let the positive linear functional $F : B(I) \to \mathbf{R}$ be defined by*

$$F(f) := \sum_{i=0}^{m} f(y_i)\gamma_i, \quad f \in B(I), \quad (3.19)$$

where $\gamma_i > 0$. Let $h > 0$ and $x \in (a, b)$, $x \le a + h$. Suppose that

$$F(e_1 - xe_0) = 0 \quad (3.20)$$

and in the case $b - a > 2h$ suppose also that

$$\sum_{i=\pi(u,v)}^{m} (v - y_i)\gamma_i \ge 0 \text{ for all } (u, v) \in M_{x,h}(y_0, y_1, \ldots, y_m). \quad (3.21)$$

Then,

$$|F(f) - F(e_0)f(x)| \le F(e_0)\omega_2(f, h) \quad (3.22)$$

holds true for any $f \in B(I)$.

Proof. Define

$$M := \sum_{i=0}^{\sigma(x)} (x - y_i)\gamma_i = \sum_{i=\tau(x)}^{m} (y_i - x)\gamma_i \quad (3.23)$$

and

$$\lambda_{i,j} := \frac{y_j - y_i}{M} \cdot \gamma_i\gamma_j, \quad 0 \le i \le \sigma(x), \; \tau(x) \le j \le m. \quad (3.24)$$

First we prove the particular case of the theorem where $b - a \leq 2h$. We use the following decomposition:

$$F(f) - F(e_0)f(x) = \sum_{i=0}^{\sigma(x)} \sum_{j=\tau(x)}^{m} \lambda_{i,j}\Delta(f; y_i, x, y_j). \tag{3.25}$$

We have

$$\sum_{i=0}^{\sigma(x)} \sum_{j=\tau(x)}^{m} \lambda_{i,j} = \sum_{0 \leq i \leq m, y_i \neq x} \gamma_i \leq F(e_0). \tag{3.26}$$

Then the particular case of the theorem follows from Lemma 3.2.1.

We consider now the general case and prove the theorem by induction with respect to m. If $m = 1$ we must have $b - a \leq 2h$. Otherwise, there exists $(x, v) \in M_{x,h}(y_0, y_1)$ such that $v = \alpha(y_0, x, h)$. We have $\pi(x, v) = 1$. From (3.21) we deduce that $(v - y_1)\gamma_1 \geq 0$, in contradiction with $v < y_1$. Therefore the particular case, already proved, of the theorem includes the case $m = 1$.

Suppose now the statement is true until $m-1$, $m \geq 2$ and prove it for m. Consider the functional F defined in (3.19) with $b - a > 2h$. For any $0 \leq i \leq \sigma(x)$, denote

$$z_i := \alpha(y_i, x, h), \quad p_i := \pi(x, z_i) \quad \text{and} \quad \eta_i := (x - y_i)\frac{\gamma_i}{M}. \tag{3.27}$$

Let $f \in B(I)$ be fixed. It is then sufficient to consider only the case where $F(f) - F(e_0)f(x) \geq 0$. There exists the decomposition

$$F(f) - F(e_0)f(x)$$
$$= \sum_{i=0}^{\sigma(x)} \left[\sum_{j=\tau(x)}^{p_i-1} \lambda_{i,j}\Delta(f; y_i, x, y_j) + G_i(f) - G_i(e_0)f(x) \right], \tag{3.28}$$

where the functionals G_i are defined by

$$G_i(\varphi) := \frac{\gamma_i}{M}\left(\sum_{j=p_i}^{m} \gamma_j(y_j - x) \right)\varphi(y_i) + \eta_i \sum_{j=p_i}^{m} \gamma_j\varphi(y_j), \quad \text{for all } \varphi \in B(I). \tag{3.29}$$

Here, the sums $\sum_{j=\tau(x)}^{p_i-1}$ are considered empty if $\tau(x) = p_i$. One can immediately obtain that

$$G_i(e_0) = \sum_{j=p_i}^{m} \lambda_{i,j}. \tag{3.30}$$

From Lemma (3.2.1) we have $\Delta(f; y_i, x, y_j) \leq \omega_2(f, h)$, $0 \leq i \leq \sigma(x)$, $\tau(x) \leq j \leq p_i - 1$. Then, by taking into account (3.28), (3.30) and (3.26), it follows that in order to prove (3.22) it suffices to show for all $0 \leq i \leq \sigma(x)$ that

$$G_i(f) - G_i(e_0)f(x) \leq G_i(e_0)\omega_2(f, h). \tag{3.31}$$

Let $0 \leq i \leq \sigma(x)$ be fixed. If $y_m - y_i \leq 2h$, then (3.31) follows from the particular, already proved, case of the theorem. Consider now the case $y_m - y_i > 2h$. One can choose $l \in \Pi_1$ such that the function $g := f + l$ satisfies

$$g(y_i) = 0 \quad \text{and} \quad g(x) = -\omega_2(f, h). \tag{3.32}$$

By denoting $u_k := y_i + 2^k(x - y_i)$, $k \geq 0$ and by taking $s \in \mathbf{N}$ such that $u_s = z_i$, one has the identity

$$g(z_i) = \sum_{k=1}^{s} 2^{s-k}(g(u_k) - 2g(u_{k-1}) + g(y_i)) + 2^s g(x) - (2^s - 1)g(y_i).$$

Since $\omega_2(g, h) = \omega_2(f, h)$, one obtains

$$g(z_i) \leq -\omega_2(g, h). \tag{3.33}$$

From (3.29) we get $G_i(e_1 - xe_0) = 0$ and, hence, inequality (3.31) is equivalent to the following one:

$$G_i(g) - G_i(e_0)g(x) \leq G_i(e_0)\omega_2(g, h). \tag{3.34}$$

Set

$$T := \sum_{j=\tau(z_i)}^{m} (y_j - z_i)\gamma_j \Big/ \sum_{j=p_i}^{\overline{\sigma}(z_i)} (z_i - y_j)\gamma_j.$$

Since $(x, z_i) \in M_{x,h}(y_0, y_1, \ldots, y_m)$, from condition (3.21) for the pair (x, z_i), we deduce that T is well defined and also that $0 < T \leq 1$. Consider the functional

$$H_i(\varphi) := T \cdot \sum_{j=p_i}^{\overline{\sigma}(z_i)} \eta_i \gamma_j \varphi(y_j) + \sum_{j=\tau(z_i)}^{m} \eta_i \gamma_j \varphi(y_j), \quad \varphi \in B([y_{p_i}, \infty) \cap I)).$$

By using the hypothesis of induction we can apply the theorem for $m - p_i$ to the functional H_i and to the point $z_i \in (y_{p_i}, y_{p_i} + h]$. Indeed, by the definition of T one has $H_i(e_1 - z_i e_0) = 0$. Let us assume that $b - y_{p_i} > 2h$, since otherwise the theorem is clearly applicable. If $(u, v) \in M_{z_i, h}(y_{p_i}, y_{p_i+1}, \ldots, y_m)$, then $(u, v) \in M_{x,h}(y_0, y_1, \ldots, y_m)$. Consequently, since $\eta_i > 0$, the condition in (3.21) for the functional H_i, the point z_i and the pair (u, v) coincides with the same condition for the functional F, the point x and the same pair. Therefore, we conclude that

$$H_i(g) \leq g(z_i)H_i(e_0) + H_i(e_0)\omega_2(g, h) \leq 0.$$

By taking into account that $g(y_i) = 0$ and that $\Delta(g; y_i, x, y_j) \leq \omega_2(g, h)$, $p_i \leq j \leq \overline{\sigma}(z_i)$, we get

$$G_i(g) - G_i(e_0)g(x) = (1 - T) \sum_{j=p_i}^{\overline{\sigma}(z_i)} \lambda_{i,j} \left(\Delta(g; y_i, x, y_j) + g(x)\right)$$
$$+ H_i(g) - G_i(e_0)g(x)$$
$$\leq G_i(e_0)\omega_2(g, h).$$

Then (3.34) is valid and the proof is completed.   □

3.3 General functionals and the second order modulus with parameter ω_2^λ

The results in this sections are given in [101] and some variants of them in [105] and [98]. We shall obtain sufficient conditions for a linear positive integral functional F under which we have

$$|F(f) - f(x)| \leq \frac{2}{1 + 2\lambda}\, \omega_2^\lambda(f, h), \ (f \in \mathcal{L}_\sigma(I)),$$

where $F(f) = \int_I f\, d\sigma, \lambda \in \left[0, \frac{1}{2}\right), x \in I$ and $h > 0$.

3.3.1 A particular case

We establish the following notation. If $y \in I$, $I := [a, b]$, $\rho_1 > 0$ and $\rho_2 > 0$, then set

$$A(y, \rho_1, \rho_2) := \big([y - \rho_1, y) \cup (y + \rho_2, \infty)\big) \cap I, \qquad (3.35)$$

$$B(y, \rho_1, \rho_2) := \big((-\infty, y - \rho_2) \cup (y, y + \rho_1]\big) \cap I. \qquad (3.36)$$

The basic result is the following.

Theorem 3.3.1. *Let σ be a positive regular finite Borel measure on the interval $[a, b]$. Let $F : \mathcal{L}_\sigma[a, b] \to \mathbb{R}$ be the positive linear functional represented by σ. Let $\lambda \in \left[0, \frac{1}{2}\right)$ and denote $\mu := \frac{1}{2} - \lambda$. Let $h > 0, k \geq 0$ and $\eta > 0$ be such that*

$$\eta < \eta_{\lambda,k} := \min \left\{ \frac{2\mu(1 - \mu^d)}{1 + k}, 2\mu^d, \frac{2}{2 + 2k + \max\{1, k\}} \right\}, \qquad (3.37)$$

where $d := \left[\log_\mu \frac{\mu}{2+k}\right]$. Let $x \in (a, b)$ and suppose that

$$F(e_1 - xe_0) = 0. \qquad (3.38)$$

In the case $b - a > 2h$, suppose also that $x \leq a + \eta h$ and

$$F\big((ye_0 - e_1)\chi_{A(y, \eta h, k\eta h)}\big) \geq 0, \quad \text{for all } y \in (x + (1 + k)\eta h, b - k\eta h). \qquad (3.39)$$

Then we have

$$|F(f) - F(e_0)f(x)| \leq \frac{2}{1 + 2\lambda}\, F(e_0)\, \omega_2^\lambda(f, h), \quad \text{for all } f \in \mathcal{L}_\sigma[a, b]. \qquad (3.40)$$

Proof. Let λ, h, k, η, f be fixed as in the hypothesis. Consider only the case where $F(f) - F(e_0)f(x) \geq 0$.

Let us represent $F(f) = \int_{[a,b]} f(t)\, d\sigma(t)$, where σ is the regular positive Borel measure associated to F. Define:

$$M := \int_{[a,x)} (x - t)\, d\sigma(t) = \int_{(x,b]} (t - x)\, d\sigma(t).$$

We restrict ourselves only to the case $M > 0$, since if $M = 0$, then $F(f) = F(e_0) f(x)$. Using the notation (2.12), we have the following decomposition:

$$F(f) - F(e_0) f(x) = \int_{[a,x)} d\sigma(s) \int_{(x,b]} \frac{t - s}{M} \Delta(f; s, x, t)\, d\sigma(t), \quad f \in C[a, b].$$

Also we have

$$\int_{[a,x)} d\sigma(s) \int_{(x,b]} \frac{t - s}{M}\, d\sigma(t) \le F(e_0).$$

Therefore, in order to prove (3.40) it is sufficient to show that for any $s \in [a, x)$ we have

$$\int_{(x,b]} \frac{t - s}{M} \Delta(f; s, x, t)\, d\sigma(t) \le \frac{2}{1 + 2\lambda} \omega_2^\lambda(f, h) \int_{(x,b]} \frac{t - s}{M}\, d\sigma(t). \tag{3.41}$$

In the particular case where $b - a \le 2h$, the inequality in (3.41) follows immediately, for any $s \in [a, x)$, from Lemma 2.2.1 and consequently, (3.40) holds.

Consider now the general case. Denote $\rho := 2\mu - \eta > 0$ and $m := \left[\frac{b-a}{\rho h}\right]$. We prove the theorem by induction with regard to m. If $m \le 1$, then the theorem follows from the particular, already proved case, $b - a \le 2h$. Consider the theorem true until $m \le m_0$ and prove it for $m_0 + 1$. Consider that $\left[\frac{b-a}{\rho h}\right] = m_0 + 1$. We shall prove (3.41) for any $s \in [a, x)$.

Let $s \in [a, x)$ be fixed. If $b - s \le 2h$, then inequality (3.41) follows from Lemma 2.2.1.

Consider now that $b - s > 2h$. Set

$$z := s + (x - s)\mu^{-r},$$

where

$$r = \min\{j \in \mathbb{N} \mid (x - s)\mu^{-j} > 2\mu h\}.$$

Hence $z - s \le 2h$ and $z < b$. We have $z > x + (1 + k)\eta h$. Indeed, set

$$J_j := [x - 2\mu^j h, x - 2\mu^{j+1} h), \quad j \ge 1.$$

We have $s \in [a, x) \subset (x - \eta_{\lambda,k} h, x) \subset [x - 2\mu^d h, x) = \bigcup_{j \ge d} J_j$. Note that $s \in J_r$.

Then $r \ge d$. It follows that

$$z - x = z - s - (x - s) = (x - s)\left(\mu^{-r} - 1\right) \ge (x - \sup J_r)\left(\mu^{-r} - 1\right)$$

$$= 2\mu^{r+1}\left(\mu^{-r} - 1\right)h = 2\mu\left(1 - \mu^r\right)h \ge 2\mu\left(1 - \mu^d\right)h$$

$$\ge (1 + k)\eta_{\lambda,k} h > (1 + k)\eta h.$$

Then put $p := z - \eta h$. We have $p \geq a + z - s - \eta h > a + 2\mu h - \eta h = a + \rho h$ and also $p \in (x, b)$.

We can choose $l \in \Pi_1$ such that the function $g := f + l$ satisfies the conditions

$$g(s) = 0 \quad \text{and} \quad g(x) = -\frac{2}{1 + 2\lambda} \omega_2^\lambda(f, h). \tag{3.42}$$

We have $\omega_2^\lambda(f, h) = \omega_2^\lambda(g, h)$ and $\Delta(f; u, y, v) = \Delta(g; u, y, v)$, for any points $a \leq u < y < v \leq b$. Therefore in order to prove (3.41) it suffices to prove the following inequality, on the hypothesis that the theorem is true for $m \leq m_0$.

$$\int_{(x,b]} \frac{t - s}{M} \Delta(g; s, x, t) \, d\sigma(t) \leq \frac{2}{1 + 2\lambda} \omega_2^\lambda(g, h) \int_{(x,b]} \frac{t - s}{M} \, d\sigma(t). \tag{3.43}$$

We have

$$g(z) \leq -\frac{2}{1 + 2\lambda} \omega_2^\lambda(g, h). \tag{3.44}$$

Indeed, denote $u_j := s + (x - s)\mu^{-j}$, $j \geq 0$. Consequently $z = u_r$. One has

$$g(z) = \sum_{j=1}^r \mu^{-j} \big[\mu g(u_{r-j+1}) - g(u_{r-j}) + (1 - \mu)g(s) \big]$$
$$+ \mu^{-r} g(x) - \big(\mu^{-r} - 1 \big) g(s)$$
$$= \sum_{j=1}^r \mu^{-j} \Delta(g; s, u_{r-j}, u_{r-j+1}) + \mu^{-r} g(x)$$
$$\leq \left[\sum_{j=1}^r \mu^{-j} - \mu^{-r} \frac{2}{1 + 2\lambda} \right] \omega_2^\lambda(g, h)$$
$$= -\frac{2}{1 + 2\lambda} \omega_2^\lambda(g, h).$$

Let $q := z + k\eta h$. In the proof of relation (3.43) we distinguish between two cases.

Case 1: $z \leq x + (1 + k + \max\{1, k\})\eta h$. We have $q \leq s + 2h$. Indeed, it follows that $q - s = z - x + x - s + k\eta h \leq (2 + 2k + \max\{1, k\})\eta h \leq 2h$. Since $b > s + 2h$, it follows that $q < b$. Define

$$T := \begin{cases} \int_{(q,b]}(t - z) \, d\sigma(t) \big/ \int_{[p,z]}(z - t) \, d\sigma(t), & \text{if } \sigma([p, z]) > 0, \\ 0, & \text{if } \sigma([p, z]) = 0. \end{cases} \tag{3.45}$$

We can apply condition (3.39) for the choice $y = z$, that is

$$\int_{[p,z]}(z - t) \, d\sigma(t) = \int_{[p,z)}(z - t) \, d\sigma(t) \geq \int_{(q,b]}(t - z) \, d\sigma(t).$$

Therefore $T \in [0, 1]$. Also, if $\sigma([p, z]) = 0$, then $\sigma((q, b]) = 0$. Consider the functional.

$$H(\varphi) := T \int_{[p,\,z]} \frac{x-s}{M}\, \varphi(t)\, \mathrm{d}\sigma(t) + \int_{(q,\,b]} \frac{x-s}{M}\, \varphi(t)\, \mathrm{d}\sigma(t), \quad \text{for } \varphi \in C[p,\, b].$$

$$(3.46)$$

Using the relations in (3.42) we have

$$\int_{(x,\,b]} \frac{t-s}{M}\, \Delta(g;s,x,t)\, \mathrm{d}\sigma(t)$$

$$= \int_{(x,\,p)\cup(z,\,q]} \frac{t-s}{M}\, \Delta(g;s,x,t)\, \mathrm{d}\sigma(t)$$

$$+ (1-T) \int_{[p,\,z]} \frac{t-s}{M}\, \Delta(g;s,x,t)\, \mathrm{d}\sigma(t)$$

$$+ H(g) + \frac{2}{1+2\lambda}\, \omega_2^\lambda(g,h) \left[T \int_{[p,\,z]} \frac{t-s}{M}\, \mathrm{d}\sigma(t) + \int_{(q,\,b]} \frac{t-s}{M}\, \mathrm{d}\sigma(t) \right].$$

We have $\Delta(g;s,x,t) \le \frac{2}{1+2\lambda}\, \omega_2^\lambda(g,h)$, for any $t \in (x,q]$. Therefore, in order to prove (3.43) it suffices to show that $H(g) \le 0$. But the functional H satisfies the conditions in the hypothesis of the theorem, with regard to the point $z \in (p,\,b)$, $z = p + \eta h$. Indeed, the equality $H(e_1 - z e_0) = 0$ is immediate from the definition of H and, in the case $b - p > 2h$, we have for any $y \in (z + (1+k)\eta h,\ b - k\eta h)$ that

$$H\left((y e_0 - e_1)\chi_{A(y,\eta h, k\eta h)}\right) = \frac{x-s}{M} \cdot F\left((y e_0 - e_1)\chi_{A(y,\eta h, k\eta h)}\right) \ge 0.$$

Because $p \ge a + \rho h$, it follows that $\left[\frac{b-p}{\rho h}\right] \le m_0$. Hence, by using the hypothesis of induction, we can apply the theorem to the functional H. One obtains

$$H(g) \le g(z) H(e_0) + \frac{2}{1+2\lambda} H(e_0)\omega_2^\lambda(g,h) \le 0.$$

Case 2: $z > x + (1 + k + \max\{1,k\})\eta h$. Put $v := z - \max\{1,k\}\eta h$ and $u := v - \eta h$. Define:

$$U = \begin{cases} \int_{(z,\,\min\{q,b\}]}(t-z)\, \mathrm{d}\sigma(t) \big/ \int_{[u,\,v)}(z-t)\, \mathrm{d}\sigma(t), & \text{if } \sigma([u,\,v)) > 0, \\ 0, & \text{if } \sigma([u,\,v)) = 0, \end{cases} \quad (3.47)$$

and consider the functional

$$K(\varphi) := U \int_{[u,\,v)} \frac{x-s}{M}\, \varphi(t)\, \mathrm{d}\sigma(t) + \int_{(z,\,\min\{q,b\}]} \frac{x-s}{M}\, \varphi(t)\, \mathrm{d}\sigma(t), \quad (3.48)$$

$\varphi \in C[u,\ \min\{q,\, b\}]$.

From the condition given in Case 2 it follows that $v \in (x + (1+k)\eta h,\ b - k\eta h)$. We apply condition (3.39) for the choice $y = v$, i.e.,

$$\int_{[u,\,v)} (v-t)\, \mathrm{d}\mu(t) \ge \int_{(v+k\eta h,\,b]} (t-v)\, \mathrm{d}\mu(t).$$

Consequently we have

$$\int_{[u,v)} (v - t)\, d\mu(t) \geq \int_{(z,\,\min\{q,b\}]} (t - v)\, d\mu(t)$$

and then

$$\int_{[u,v)} (z - t)\, d\mu(t) \geq \int_{(z,\,\min\{q,b\}]} (t - z)\, d\mu(t).$$

Thus one has $U \in [0, 1]$. Also, if $\sigma([u, v)) = 0$, then $\sigma((z, \min\{q, b\}]) = 0$. From (3.47) and (3.48) it follows that $K(e_1 - ze_0) = 0$. Also $\min\{q, b\} - u \leq (1 + k + \max\{1, k\})\eta h < 2h$. Therefore the functional K satisfies the particular case of the theorem, with regard to the point z. Hence

$$K(g) \leq g(z)K(e_0) + \frac{2}{1 + 2\lambda}\, \omega_2^\lambda(g, h)K(e_0) \leq 0.$$

Now we distinguish between two subcases.
Subcase 2.1: $b \leq q$. We have the following decomposition:

$$\int_{(x,b]} \frac{t - s}{M}\, \Delta(g; s, x, t)\, d\sigma(t)$$

$$= \int_{(x,u)\cup[v,z]} \frac{t - s}{M}\, \Delta(g; s, x, t)\, d\sigma(t)$$

$$+ (1 - U)\int_{[u,v)} \frac{t - s}{M}\, \Delta(g; s, x, t)\, d\sigma(t)$$

$$+ K(g) + \frac{2}{1 + 2\lambda}\, \omega_2^\lambda(g, h) \left[U\int_{[u,v)} \frac{t - s}{M}\, d\sigma(t) + \int_{(z,b]} \frac{t - s}{M}\, d\sigma(t) \right].$$

Since $\Delta(g; s, x, t) \leq \frac{2}{1+2\lambda}\omega_2^\lambda(g, h)$, for $t \in (x, z]$ and $K(g) \leq 0$, we obtain (3.43).
Subcase 2.2: $b > q$. Consider the number T and the functional H defined in (3.45) and (3.46). As in Case 1, we have $H(g) \leq 0$. Consider the decomposition

$$\int_{(x,b]} \frac{t - s}{M}\, \Delta(g; s, x, t)\, d\sigma(t)$$

$$= \int_{(x,u)\cup[v,p)} \frac{t - s}{M}\, \Delta(g; s, x, t)\, d\sigma(t)$$

$$+ (1 - U)\int_{[u,v)} \frac{t - s}{M}\, \Delta(g; s, x, t)\, d\sigma(t)$$

$$+ (1 - T)\int_{[p,z]} \frac{t - s}{M}\, \Delta(g; s, x, t)\, d\sigma(t) + H(g) + K(g)$$

$$+ \frac{2}{1 + 2\lambda}\, \omega_2^\lambda(g, h) \left[U\int_{[u,v)} \frac{t - s}{M}\, d\sigma(t) \right.$$

$$+ T\int_{[p,z]} \frac{t - s}{M}\, d\sigma(t) + \left. \int_{(z,b]} \frac{t - s}{M}\, d\sigma(t) \right].$$

Since $\Delta(g; s, x, t) \leq \frac{2}{1+2\lambda} \omega_2^\lambda(g, h)$, $t \in (x, z]$ and $K(g) \leq 0$, $H(g) \leq 0$ we obtain (3.43). $\qquad\qquad\qquad\qquad\qquad\qquad\qquad\qquad\qquad\qquad\qquad\qquad\qquad\square$

Remark 3.3.1. For a fixed linear positive functional F on $C[a, b]$, $x \in (a, b)$ that satisfies (3.38), for $\lambda \in \left[0, \frac{1}{2}\right)$, $h > 0$ and $k \geq 0$, denote by $P(\eta)$, $\eta > 0$ the condition given in (3.39). The implication

$$P(\eta_1) \Rightarrow P(\eta_2), \quad \text{if } 0 < \eta_1 < \eta_2, \tag{3.49}$$

holds. Indeed, if $P(\eta_1)$ is true, since $(a + (1+k)\eta_2 h, b - k\eta_2 h) \subset (a + (1+k)\eta_1 h, b - k\eta_1 h)$ and

$$\int_{[y-\eta_2 h, y)} (y - t)\, d\sigma(t) - \int_{(y+k\eta_2 h, b]} (t - y)\, d\sigma(t) \geq \int_{[y-\eta_1 h, y)} (y - t)\, d\sigma(t) - \int_{(y+k\eta_1 h, b]} (t - y)\, d\sigma(t),$$

it follows that $P(\eta_2)$.

Consequently if in Theorem 3.3.1 we replace the value $\eta_{\lambda,k}$ by a smaller value, the modified statement remains true.

One can see that $\eta_{\lambda,k}$ is the supremum of the values η that satisfies the conditions appearing in the proof of Theorem 3.3.1, i.e., $z > x + (1 + k)\eta h$, (at the beginning of the proof) and $q \leq s + 2h$, (in Case 1).

Corollary 3.3.1. *If in the hypothesis of Theorem 3.3.1 we replace the value $\eta_{\lambda,k}$ by one of the two values*

$$\eta_{\lambda,k}^1 := \min\left\{\frac{2\mu}{2+k}, \frac{2}{2+3k}\right\}, \quad \text{or}$$

$$\eta_{\lambda,k}^2 := \frac{2\mu}{2+k+k\mu},$$

then condition (3.40) holds true.

Proof. By taking into account Remark 3.3.1, it suffices to prove that $\eta_{\lambda,k}^1 \leq \eta_{\lambda,k}$ and $\eta_{\lambda,k}^2 \leq \eta_{\lambda,k}^1$.

i) We have: $\frac{2\mu(1-\mu^d)}{1+k} > \frac{2\mu}{1+k}\left(1 - \mu^{\log_\mu \frac{1}{2+k}}\right) = \frac{2\mu}{2+k} \geq \eta_{\lambda,k}^1$ and $2\mu^d \geq \frac{2\mu}{2+k} \geq \eta_{\lambda,k}^1$. If $k \geq 1$, then we have $\frac{2}{2+2k+\max\{1,k\}} = \frac{2}{2+3k} \geq \eta_{\lambda,k}^1$ and if $k \leq 1$, then we have $\frac{2}{2+2k+\max\{1,k\}} = \frac{2}{3+2k} > \frac{1}{2+2k} \geq \eta_{\lambda,k}^1$. Therefore $\eta_{\lambda,k}^1 \leq \eta_{\lambda,k}$.

ii) The inequality $\eta_{\lambda,k}^1 \geq \eta_{\lambda,k}^2$ is immediate. $\qquad\qquad\qquad\qquad\qquad\square$

3.3.2 The main results

The main result is the following.

Theorem 3.3.2. *Let I be an arbitrary interval of the real axis and let σ be a positive regular Borel measure on I. Suppose that e_0, $e_1 \in \mathcal{L}_\sigma(I)$. Let $F : \mathcal{L}_\sigma(I) \to \mathbb{R}$ be the positive linear functional represented by σ. Let $k \geq 0$, $\lambda \in \left[0, \frac{1}{2}\right)$ and denote $\mu := \frac{1}{2} - \lambda$. Set $d := \left[\log_\mu \frac{\mu}{2+k}\right]$ and*

$$\eta_{\lambda,k} := \min\left\{\frac{2\mu(1-\mu^d)}{1+k}, 2\mu^d, \frac{2}{2+2k+\max\{1,k\}}\right\}. \qquad (3.50)$$

Let $\eta \in (0, \eta_{\lambda,k})$. Set $c := \left[\log_\mu \frac{2\mu-\eta}{4}\right]$ and

$$\Theta_{\lambda,k,\eta} := \min\{2\mu(1-\mu^c) - \eta, \ 2\mu^c, \ 1 - \frac{\eta}{2}(1+k+\max\{1,k\})\}. \qquad (3.51)$$

We have $\Theta_{\lambda,k,\eta} > 0$. Let $\Theta \in (0, \Theta_{\lambda,k,\eta})$, x be an interior point of I and $h > 0$ be such that $length(I) > 2h$. Suppose also, that the following conditions are satisfied:

$$F\left((xe_0 - e_1)\chi_{A(x,\Theta h,\Theta h)}\right) \geq 0, \ and \ F\left((e_1 - xe_0)\chi_{B(x,\Theta h,\Theta h)}\right) \geq 0, \qquad (3.52)$$

$$F\left((ye_0 - e_1)\chi_{A(y,\eta h,k\eta h)}\right) \geq 0, \ for \ all \ y \in (x+(\eta+\Theta)h, \ b - k\eta h), \qquad (3.53)$$

and

$$F\left((e_1 - ye_0)\chi_{B(y,\eta h,k\eta h)}\right) \geq 0, \ for \ all \ y \in (a+k\eta h, \ x-(\eta+\Theta)h). \qquad (3.54)$$

Then we have

$$|F(f) - f(x)| \leq |f(x)|\,|F(e_0) - 1| + h^{-1}|F(e_1 - xe_0)|\,\omega_1(f,h) \qquad (3.55)$$
$$+\frac{2}{1+2\lambda}\,F(e_0)\,\omega_2^\lambda(f,h),$$

for all $f \in \mathcal{L}_\sigma(I)$.

Proof.

Step I. We may consider only the case $\omega_1(f,h) < \infty$ and $\omega_2^\lambda(f,h) < \infty$.

Firstly, show that $\Theta_{\lambda,k,\eta} > 0$. Indeed, we have $d \geq 1$ and hence $\eta < 2\mu$. It follows that

$$2\mu(1-\mu^c) - \eta > 2\mu(1 - \mu^{\log_\mu \frac{2\mu-\eta}{4} - 1}) - \eta = \mu - \frac{\eta}{2} > 0.$$

Also

$$1 - \frac{\eta}{2}(1+k+\max\{1,k\}) > \frac{1+k}{2+2k+\max\{1,k\}}.$$

Consequently, $\Theta_{\lambda,k,\eta} > 0$.

Let a sequence of intervals $([a_n, b_n])_n$, be such that $x \in (a_n, b_n)$, $[a_n, b_n] \subset [a_{n+1}, b_{n+1}]$ for $n \in \mathbb{N}$ and $I = \bigcup_{n\in\mathbb{N}}[a_n, b_n]$. Fix indices $n \in \mathbb{N}$. Define the functional

$$F_n(\varphi) := \int_{[a_n,b_n]} \varphi(t)\,d\sigma(t), \quad \varphi \in \mathcal{L}_\sigma[a_n, b_n]. \qquad (3.56)$$

Note that, for any $\varphi \in \mathcal{L}_\sigma(I)$, we have $F_n(\varphi) = F(\varphi \cdot \chi_{[a_n,b_n]})$. By using the Lebesgue convergence theorem, it follows that

$$F(\varphi) = \lim_{n\to\infty} F_n(\varphi), \quad \varphi \in \mathcal{L}_\sigma(I).$$

Suppose that the inequality

$$|F_n(f) - f(x)| \leq |f(x)| \, |F_n(e_0) - 1| + h^{-1}|F_n(e_1 - xe_0)| \, \omega_1(f, h) \qquad (3.57)$$
$$+ \frac{2}{1 + 2\lambda} \, F_n(e_0) \, \omega_2^\lambda(f, h), \quad (f \in C[a_n, b_n])$$

holds for any $n \in \mathbb{N}$. Taking $f \in \mathcal{L}_\sigma(I) \cap C(I)$ we obtain

$$|F(f \cdot \chi_{[a_n, b_n]}) - f(x)| \leq |f(x)| \, |F(e_0 \cdot \chi_{[a_n, b_n]}) - 1|$$
$$+ h^{-1}|F((e_1 - xe_0) \cdot \chi_{[a_n, b_n]})| \, \omega_1(f, h)$$
$$+ \frac{2}{1 + 2\lambda} \, F(e_0 \cdot \chi_{[a_n, b_n]}) \, \omega_2^\lambda(f, h).$$

Then by passing to the limit $n \to \infty$, (3.55) follows. Therefore it remains to prove condition (3.57).

Step II. Let $n \in \mathbb{N}$ and $f \in C[a_n, b_n]$ be fixed. Firstly, we have

$$|F_n(f) - f(x)| \leq |f(x)| \, |F_n(e_0) - 1| + |F_n(f) - F_n(e_0)f(x)|.$$

We can consider only the case $F_n(f) - F_n(e_0)f(x) \geq 0$. Denote $\alpha := \max\{x - \Theta h, a_n\}$ and $\beta := \min\{x + \Theta h, b_n\}$. By using the condition (3.52), it follows that there are the numbers $\rho_1, \rho_2 \in [0, 1]$, such that $\rho_1 F_n\big((xe_0 - e_1) \cdot \chi_{[\alpha, x)}\big) = F_n\big((e_1 - xe_0) \cdot \chi_{(\beta, b_n]}\big)$ and $\rho_2 F_n\big((e_1 - xe_0) \cdot \chi_{(x, \beta]}\big) = F_n\big((xe_0 - e_1) \cdot \chi_{[a_n, \alpha)}\big)$. We decompose the measure $\sigma|_{[a_n, .b_n]}$ on the form $\sigma|_{[a_n, b_n]} = \sigma_1 + \sigma_2 + \sigma_3$, where

$$\sigma_1 := \rho_1 \sigma|_{[\alpha, x)} + \sigma|_{(\beta, b_n]}, \quad \sigma_2 := \sigma|_{[a_n, \alpha)} + \rho_2 \sigma|_{(x, \beta]},$$
$$\sigma_3 := (1 - \rho_1)\sigma|_{[\alpha, x)} + (1 - \rho_2)\sigma|_{(x, \beta]} + \sigma|_{\{x\}}.$$

We may consider that the measures σ_1, σ_2 and σ_3 are supported, respectively, on the intervals: $[\alpha, b_n]$, $[a_n, \beta]$ and $[\alpha, \beta]$.

Let now the positive linear functionals $G_1 : C[\alpha, b_n] \to \mathbb{R}$, $G_2 : C[a_n, \beta] \to \mathbb{R}$ and $G_3 : C[\alpha, \beta] \to \mathbb{R}$, be represented by the measures σ_1, σ_2, σ_3, respectively. From above it follows that $G_i(e_1 - xe_0) = 0$, $(i = 1, 2)$.

Step III. Let us show that

$$G_1(f) - G_1(e_0)f(x) \leq \frac{2}{1 + 2\lambda} \, G_1(e_0)\omega_2^\lambda(f, h). \qquad (3.58)$$

Denote $M_1 := \int_{[\alpha, x)}(x - t) \, d\sigma_1(t) = \int_{(\beta, b]}(t - x) \, d\sigma_1(t)$. Suppose that $M_1 > 0$, since if $M_1 = 0$, the relation (3.58) is obvious. We repeat for the functional G_1 the reasoning made for the functional F in the proof of Theorem 3.3.1. Since $\sigma_1|_{(x, \beta]} = 0$ and $\sigma_1|_{(\beta, b]} = \sigma|_{(\beta, b]}$, we obtain that it is sufficient to show for any $s \in [\alpha, x)$, that

$$\int_{(\beta, b_n]} \frac{t - s}{M_1} \Delta(f; s, x, t) \, d\sigma(t) \leq \frac{2}{1 + 2\lambda} \, \omega_2^\lambda(f, h) \int_{(\beta, b_n]} \frac{t - s}{M_1} \, d\sigma(t). \qquad (3.59)$$

By taking into account Theorem 3.3.1, we have to consider only the case $b_n - s > 2h$. For fixed $s \in [\alpha, x)$, define the numbers r, z, p, and q as in the proof of Theorem 3.3.1. We have $z \leq s + 2h$. Also, it follows that $z > x + (\eta + \Theta)h$. Indeed, define the intervals J_j, $j \geq 1$, as in the proof of Theorem 3.3.1. We have

$$s \in [\alpha, x) \subset [x - \Theta h, x) \subset [x - 2\mu^c h, x) = \bigcup_{j \geq c} J_j.$$

From the definition of z it follows that $s \in J_r$. Hence $r \geq c$. We have

$$z - x = z - s - (x - s) = (x - s)\left(\mu^{-r} - 1\right) \geq (x - \max J_r)\left(\mu^{-r} - 1\right)$$

$$= 2\mu^{r+1}\left(\mu^{-r} - 1\right)h = 2\mu\left(1 - \mu^r\right)h \geq 2\mu\left(1 - \mu^c\right)h > (\eta + \Theta)h.$$

As in the proof of Theorem 3.3.1, by using the function g defined in the same mode, one can reduce the proof of the inequality in (3.59) to the following one:

$$\int_{(\beta, b_n]} \frac{t - s}{M_1} \Delta(g; s, x, t) \, d\sigma(t) \leq \frac{2}{1 + 2\lambda} \omega_2^\lambda(g, h) \int_{(\beta, b_n]} \frac{t - s}{M_1} \, d\sigma(t). \qquad (3.60)$$

We distinguish two cases.

Case 1: $z \leq x + ((1 + \max\{1, k\})\eta + \Theta)h$. We have $q < s + 2h$. Indeed,

$$q - s = z - x + x - s + k\eta h \leq ((1 + k + \max\{1, k\})\eta + 2\Theta)h < 2h.$$

Since $b_n > s + 2h$, it follows that $q < b_n$. Define the number T and the functional H as in (3.45) and (3.46). Since $z > x + (\eta + \Theta)h$ we can apply relation (3.53) for the choice $y := z$. Consequently $T \in [0, 1]$. The functional H satisfies the condition in the hypothesis of Theorem 3.3.1 and from this we derive $H(g) \leq 0$. We can use a decomposition of the term $\int_{(\beta, b_n]} \frac{t-s}{M_1} \Delta(g; s, x, t) \, d\sigma(t)$, similarly as in Case 1 of the proof of Theorem 3.3.1 for the term $\int_{(x, b]} \frac{t-s}{M} \Delta(g; s, x, t) \, d\sigma(t)$, but where M is replaced by M_1, x by β and b by b_n. By using the same argument we obtain (3.59).

Case 2: $z \geq x + ((1 + \max\{1, k\})\eta + \Theta)h$. Let the numbers u, v, U and the functional K be as in Case 2 of the proof of Theorem 3.3.1. Since $v \in (x + (\eta + \Theta)h, b_n - k\eta h)$, we can apply relation (3.53) with the choice $y := v$ and we obtain $K(g) \leq 0$. The subcases 2.1 and 2.2 from the proof of Theorem 3.3.1 can be adapted here with the only modification consisting in the replacement of M, x and b by M_1, β and b_n, respectively. Then (3.59) follows and consequently (3.58).

By using the symmetry, a similar inequality to (3.58) holds for the functional G_2.

Step IV. For the functional G_3, we suppose firstly that $G_3 \neq 0$. In order to estimate the difference, $G_3(f) - G_3(e_0)f(x)$ we shall apply Theorem 2.1.2, with the following choices: $V := \mathcal{L}_{\sigma_3}[\alpha, \beta]$, $I := [\alpha, \beta]$, $F := (G_3(e_0))^{-1} G_3$, $\Omega_1 := \omega_1$, $\Omega_2 := \omega_2^\lambda$ and the function $\Psi(y) := \frac{2}{1+2\lambda}$, $y \in [0, \infty)$.

The corresponding condition (2.15) in Theorem 2.1.2 is equivalent to the inequality

$$|\Delta(f; t_1, x, t_2)| \leq \frac{2}{1 + 2\lambda} \omega_2^\lambda(f, h), \quad t_1 \in [\alpha, x), \ t_2 \in (x, \beta], \qquad (3.61)$$

which is true from Lemma 2.2.1.

The condition (2.17) in Theorem 2.1.2 can be rewritten in the form

$$|f(t) - f(x)| \le |t - x|h^{-1}\omega_1(f, h) + \frac{2}{1 + 2\lambda}\omega_2^\lambda(f, h), \quad (t \in [\alpha, \beta]). \qquad (3.62)$$

We can consider, for a choice, that $x \le t$. Since length$(I) > 2h$, it follows that either $x + h \in I$, or $t - h \in I$. By the symmetry we may consider only the case $x + h \in I$. Note that, since $\Theta < 1$, we have $t < x + h$. We have

$$f(t) - f(x) = \frac{t - x}{h}(f(x + h) - f(x)) - \Delta(f; x, t, x + h).$$

By Lemma 2.2.1 we deduce (3.62). Therefore, we can apply Theorem 2.1.2 and we obtain

$$|G_3(f) - G_3(e_0)f(x)| \le |G_3(e_1 - xe_0)|h^{-1}\omega_1(f, h) + \frac{2}{1 + 2\lambda}G_3(e_0)\omega_2^\lambda(f, h).$$

A similar inequality remains true for the functional G_3 in the case $G_3 = 0$.

Finally, since $F_n(e_0) = G_1(e_0) + G_2(e_0) + G_3(e_0)$ and $F_n(e_1 - xe_0) = G_3(e_1 - xe_0)$, we obtain (3.55). The proof is finished. $\square$

Remark 3.3.2. Similarly as in Remark 3.3.1 we can note that in Theorem 3.3.2 it is possible to replace the values $\eta_{\lambda,k}$ and $\Theta_{\lambda,k,\eta}$ by any positive smaller values and the theorem remains true. Moreover one can see that the values of $\eta_{\lambda,k}$ and $\Theta_{\lambda,k,\eta}$ are the best possible, in order that the conditions used in the proof hold true.

Corollary 3.3.2. *If in the hypothesis of Theorem 3.3.2 we replace the conditions $\eta < \eta_{\lambda,k}$ and $\Theta < \Theta_{\lambda,k,\eta}$, by the two conditions*

$$\eta < \eta_{\lambda,k}^1 := \min\left\{\frac{2\mu}{2 + k}, \frac{2}{2 + 3k}\right\}, \qquad (3.63)$$

$$\Theta < \Theta_{\lambda,k,\eta}^1 := \min\{\mu - \frac{\eta}{2}, 1 - \frac{\eta}{2}(1 + k + \max\{1, k\})\}, \qquad (3.64)$$

then the estimate (3.55) holds true.

Proof. We saw in Corollary 3.3.1 that $\eta_{\lambda,k}^1 \le \eta_{\lambda,k}$. Let $\eta < \eta_{\lambda,k}^1$ and define the numbers c and d as in Theorem 3.3.2. The inequality $2\mu(1 - \mu^c) - \eta > \mu - \frac{\eta}{2}$ is true, since it was shown at the beginning of the proof of Theorem 3.3.2 for any $\eta < \eta_{\lambda,k}$. Also the inequality $2\mu^c \ge \mu - \frac{\eta}{2}$ is immediate. It follows that $\Theta_{\lambda,k,\eta} \ge \Theta_{\lambda,k,\eta}^1$. By taking into account Remark 3.3.2, we can apply Theorem 3.3.2. $\square$

Corollary 3.3.3. *If in the hypothesis of Theorem 3.3.2 we replace the conditions $\eta < \eta_{\lambda,k}$ and $\Theta < \Theta_{\lambda,k,\eta}$, by the two conditions*

$$\eta < \eta_{\lambda,k}^2 := \frac{2\mu}{2 + k + k\mu}, \qquad (3.65)$$

$$\Theta < \Theta_{\lambda,k,\eta}^2 := \mu - \frac{\eta}{2}(1 + k\mu), \qquad (3.66)$$

then the estimate (3.55) holds true.

Proof. Condition (3.65) implies immediately condition (3.63). It suffices to show that $\Theta^2_{\lambda,k,\eta} \leq \Theta_{\lambda,k,\eta}$. Since $\Theta^2_{\lambda,k,\eta} \leq \mu - \frac{\eta}{2}$ it remains to show that $\Theta^2_{\lambda,k,\eta} \leq 1 - \frac{\eta}{2}(1 + k + \max\{1, k\})$. Because $1 - \frac{\eta k}{2} > 0$, it suffices to check this inequality for $\mu = \frac{1}{2}$. In this case this inequality is equivalent to $\eta < \frac{2}{k+2\max\{1,k\}}$, which follows from (3.65). $\square$

In the case $\Theta = \eta$ and $k = 1$, the conditions in the hypothesis of Theorem 3.3.2 can be simplified in the following way:

Corollary 3.3.4. *Let I be an arbitrary interval of the real axis and let σ be a positive regular Borel measure on I. Suppose that e_0, $e_1 \in \mathcal{L}_\sigma(I)$. Let $F : \mathcal{L}_\sigma(I) \to \mathbb{R}$ be the positive linear functional represented by σ. Let $\lambda \in \left[0, \frac{1}{2}\right)$. Denote $\mu := \frac{1}{2} - \lambda$ and*

$$
\overline{\eta}_\lambda := \begin{cases} \mu(1 - \mu), & 0 < \mu < \frac{1}{3}, \\ 2\mu^2, & \frac{1}{3} \leq \mu < \sqrt{2} - 1, \\ \mu(1 - \mu^2), & \sqrt{2} - 1 \leq \mu \leq \frac{1}{2}. \end{cases} \tag{3.67}
$$

Let $\eta \in (0, \overline{\eta}_\lambda)$, x be an interior point of I and $h > 0$ be such that $length(I) > 2h$. Suppose also that the following conditions are satisfied:

$$
F\left((ye_0 - e_1)\chi_{A(y,\eta h,k\eta h)}\right) \geq 0, \text{ for all } y \in \{x\} \cup (x + (\eta + \Theta)h, b - k\eta h), \tag{3.68}
$$

and

$$
F\left((e_1 - ye_0)\chi_{B(y,\eta h,k\eta h)}\right) \geq 0, \text{ for all } y \in \{x\} \cup (a + k\eta h, x - (\eta + \Theta)h). \tag{3.69}
$$

Then we have

$$
|F(f) - f(x)| \leq |f(x)| \, |F(e_0) - 1| + h^{-1}|F(e_1 - xe_0)| \, \omega_1(f, h) \tag{3.70}
$$
$$
+ \frac{2}{1 + 2\lambda} F(e_0) \, \omega_2^\lambda(f, h),
$$

for all $f \in \mathcal{L}_\sigma(I)$.

Proof. We apply Theorem 3.3.2 for $k = 1$ and $\Theta = \eta$. Note that conditions (3.52), (3.53) and (3.54) in Theorem 3.3.2 are equivalent to conditions (3.68) and (3.69) in the present corollary. Define $d := \left\lceil \log_\mu \frac{\mu}{3} \right\rceil$. With the notation (3.50), we have

$$
\eta_{\lambda,1} = \min\left\{\mu(1 - \mu^d), 2\mu^d, \frac{2}{5}\right\}.
$$

For any $\eta \in (0, \eta_{\lambda,1})$, define $c := \left\lceil \log_\mu \frac{2\mu - \eta}{4} \right\rceil$. With the notation (3.51) we have

$$
\Theta_{\lambda,1,\eta} = \min\left\{\mu(1 - \mu^c) - \eta, 2\mu^d, 1 - \frac{3}{2} \cdot \eta\right\}.
$$

The inequality $\eta < \Theta_{\lambda,1,\eta}$ is equivalent to $\eta < T_\lambda$, where

$$T_\lambda = \min\left\{\mu(1 - \mu^c),\ 2\mu^c,\ \frac{2}{5}\right\}. \tag{3.71}$$

It remains to show that if $\eta \in (0, \bar{\eta}_\lambda)$, then

$$\eta < \min\{\eta_{\lambda,1},\ T_\lambda\}. \tag{3.72}$$

Firstly observe that we have

$$d = \begin{cases} 1. & 0 < \mu < \frac{1}{3}, \\ 2. & \frac{1}{3} \le \mu \le \frac{1}{2}. \end{cases}$$

For any $\mu \in \left(0, \frac{1}{2}\right]$ we have $c = 1$, if $0 < \eta < 2\mu - 4\mu^2$, and $c = 2$, if $2\mu - 4\mu^2 \le \eta < 2\mu - 4\mu^3$. Also we have $\min\{\mu(1 - \mu),\ 2\mu\} = \mu(1 - \mu)$ and

$$\min\{\mu(1 - \mu^2),\ 2\mu^2\} = \begin{cases} 2\mu^2, & 0 < \mu < \sqrt{2} - 1, \\ \mu(1 - \mu^2), & \sqrt{2} - 1 \le \mu \le \frac{1}{2}. \end{cases}$$

We shall prove the condition given in (3.72) in several cases.

Case 1: $0 < \mu < \frac{1}{3}$. We have $d = 1$ and $\eta_{\lambda,1} = \mu(1 - \mu) = \bar{\eta}_\lambda$. There is the inequality $\mu(1 - \mu) < 2\mu - 4\mu^2$. If $\eta \in (0, \bar{\eta}_\lambda)$, we get $c = 1$. Hence $T_\lambda = \eta_{\lambda,1}$. Consequently, inequality (3.72) follows.

Case 2: $\frac{1}{3} \le \mu < \sqrt{2} - 1$. We have $d = 2$. and $\eta_{\lambda,1} = \min\left\{2\mu^2, \frac{2}{5}\right\} = 2\mu^2 = \bar{\eta}_\lambda$. The inequalities

$$2\mu - 4\mu^2 < 2\mu^2 < 2\mu - 4\mu^3$$

are immediate.

Consider firstly that $\eta \in (0,\ 2\mu - 4\mu^2)$. Then $c = 1$. Hence $T_\lambda = \mu(1 - \mu)$. But $2\mu - 4\mu^2 \le \mu(1 - \mu)$. Consequently, inequality (3.72) is true.

Let now $\eta \in [2\mu - 4\mu^2,\ 2\mu^2)$. Then $c = 2$. We obtain $T_\lambda = \eta_{\lambda,1}$ and relation (3.72) is true.

Case 3: $\sqrt{2} - 1 \le \mu \le \frac{1}{2}$. Then $d = 2$ and $\eta_{\lambda,1} = \min\left\{\mu(1 - \mu^2), \frac{2}{5}\right\} = \mu(1 - \mu^2)$. Here the following inequalities hold:

$$2\mu - 4\mu^2 < \mu(1 - \mu^2) < 2\mu - 4\mu^2.$$

If $\eta \in (0,\ 2\mu - 4\mu^2)$, then $c = 1$. Hence $T_\lambda = \mu(1 - \mu)$. We have the inequality $2\mu - 4\mu^2 < \mu(1 - \mu)$. Relation (3.72) is proved.

If $\eta \in [2\mu - 4\mu^2,\ \mu(1 - \mu^2))$, then $c = 2$ and hence $T_\lambda = \eta_{\lambda,1}$. Condition (3.72) follows directly from the above.

The proof is finished. $\qquad\square$

4

Estimates for the Bernstein Operators

4.1 Various types of estimates

4.1.1 Introduction

The Bernstein operators B_n, $n \in \mathbb{N}$ assign to each function $f \in \mathcal{F}[0, 1]$, the polynomials

$$B_n(f, x) := \sum_{k=0}^{n} p_{n,k}(x) \cdot f\left(\frac{k}{n}\right), \quad x \in [0, 1],$$

$$\text{where } p_{n,k}(x) := \binom{n}{k} x^k (1 - x)^{n-k}. \tag{4.1}$$

These operators are, very probably, the most studied linear positive operators. They were generalized and modified in a great number of variants, like the well-known Bernstein-type operators of Kantorovich [53], Durrmeyer [31], Stancu [131], Brass [17], Lupaş [62], Bleimann, Butzer and Hahn [15] and so on. The bibliographical lists drawn up by E. Stark [133] and H. Gonska and J. Meier [43] contain about 1500 titles of papers, until 1986, dedicated to Bernstein operators and to their generalizations and modifications. The advantages of the Bernstein operators consist in their simplicity, and on their sharp properties of approximation. From certain points of view the Bernstein operators play an extremal position in some classes of operators.

We mention here only some basic properties of these operators. For a more detailed discussion see the monographs by G.G. Lorentz [59], or by R. DeVore and G.G. Lorentz [28].

A first important property is the preservation of the linear functions. Their moments,

$$m_{n,s}(x) := B_n((e_1 - xe_0)^s, x), \quad n \in \mathbb{N}, \ s = 0, 1, \dots, \ x \in [0, 1], \tag{4.2}$$

satisfy the recursion formula

$$m_{n,s+1}(x) = \frac{x(1 - x)}{n} \left[m'_{n,s}(x) + s \cdot m_{n,s-1}(x) \right], \ s \geq 1, \ n \in \mathbb{N}, \ x \in [0, 1], \tag{4.3}$$

see G.G. Lorentz [59]. We obtain

$$m_{n,0}(x) = 1, \quad m_{n,1}(x) = 0, \quad m_{n,2}(x) = \frac{x(1-x)}{n}, \tag{4.4}$$

$$m_{n,3}(x) = \frac{x(1-x)(1-2x)}{n^2},$$

$$m_{n,4}(x) = \frac{x(1-x)}{n^2}\left[\left(3 - \frac{6}{n}\right)x(1-x) + \frac{1}{n}\right].$$

Applying, for instance, formula (1.39), for the functionals $F := B_n(\cdot, x)$, and $h = \frac{1}{\sqrt{n}}$, it follows that the sequence $(B_n(f))_n$ is uniformly convergent to f, for any $f \in C[0, 1]$.

The following asymptotic formula, given by Voronovskaja [138],

$$\lim_{n \to \infty} n(B_n(f, x) - f(x)) = -\frac{x(1-x)}{2} \cdot f''(x), \tag{4.5}$$

holds for $f \in B[0, 1]$, if $f''(x)$ exists at a certain point $x \in [0, 1]$. If $f \in C^2[0, 1]$, then the convergence is uniform.

Since the operator B_n preserves linear functions there is the inequality

$$f \le B_n(f), \quad n \in \mathbb{N} \tag{4.6}$$

for any convex function f. Moreover, as it was noted first by O. Aramă [7], for a convex function $f \in \mathcal{F}[0, 1]$ we have the inequality

$$B_{n+1}(f, x) \le B_n(f, x), \quad x \in [0, 1], \ n \in \mathbb{N}. \tag{4.7}$$

The derivatives of order $r \ge 1$ of the Bernstein polynomials can be expressed by

$$B_n^{(r)}(f, x) = \frac{r!n!}{(n-r)!n^r} \sum_{k=0}^{n-r}\left[f; \frac{k}{n}, \frac{k+1}{n}, \dots, \frac{k+r}{n}\right] p_{n-r,k}(x). \tag{4.8}$$

This formula, obtained by T. Popoviciu [115], shows that if f is convex of order $k \ge -1$, (see Definition 1.1.1), then $B_n(f)$ is also convex of order k. For formula (4.8) one obtains that

$$B_n(f, x) = \sum_{k=0}^{n}\binom{n}{k}\Delta_{\frac{1}{n}}^k f(0) \cdot x^k. \tag{4.9}$$

Consequently, if $f \in \Pi_r$, then $B_n(f) \in \Pi_r$. Also, using formula (4.8) it is proved in [90] that if f is quasiconvex of order $k \ge 0$, then $B_n(f)$ is itself quasiconvex of order k. A function f on an interval I is quasiconvex of order $k \ge 0$, in the sense of E. Popoviciu [114], if the inequality

$$[f; x_2, \dots, x_{k+2}] \le \max\{[f; x_1, \dots, x_{k+1}], \ [f; x_3, \dots, x_{k+3}]\} \tag{4.10}$$

holds for every system of points $x_1 < \cdots < x_{k+3}$ from I.

The preservation of the convexity of higher order is essential for the possibility of approximation of derivatives of higher order, as we can see from a theorem of Sendov and Popov, see Theorem 1.2.3. The fact that

$$\lim_{n\to\infty} \|B_n^{(r)}(f) - f^{(r)}\| = 0, \quad \text{if } f \in C^r[0, 1], \ r \geq 1, \tag{4.11}$$

was pointed out by I. Chlodovsky, see Lorentz [59] and S. Wigert [140].

The operators B_n have also the variational diminishing property, i.e.,

$$\text{Var } B_n(f) \leq \text{Var } f, \tag{4.12}$$

where Var g is the total variation of a function g. Moreover, the number of zeros of the polynomial $B_n(f)$ does not exceed the number of sign changes of f. (Polya, in Schonberg [121]).

The first estimate of the degree of approximation by Bernstein operators was given by T. Popoviciu [116] using the first order modulus

$$\|B_n(f) - f\| \leq \frac{3}{2} \cdot \omega_1\left(f, \frac{1}{\sqrt{n}}\right), \quad f \in C[0, 1], \ n \in \mathbb{N}. \tag{4.13}$$

The optimal estimate of this type was obtained by Sikkema [128]:

$$\sup_{f \in C[0,1]\setminus\Pi_0} \sup_{n \in \mathbb{N}} \frac{\|B_n(f) - f\|}{\omega_1\left(f, \frac{1}{\sqrt{n}}\right)} = \frac{4306 + 837\sqrt{6}}{5832} = 1.08988\ldots. \tag{4.14}$$

A careful look at Sikkema's proof shows that the same value of this supremum is obtained if we replace the space $C[0, 1]$ by the space $B[0, 1]$.

Also the best asymptotical constant, given by Esseen [24], is known to be

$$\sup_{f \in C[0,1]\setminus\Pi_0} \limsup_{n \to \infty} \frac{\|B_n(f) - f\|}{\omega_1\left(f, \frac{1}{\sqrt{n}}\right)}$$

$$= 2\sum_{k=0}^{\infty}(k + 1)(\lambda(2k + 2) - \lambda(2k)) = 1.04556\ldots, \tag{4.15}$$

where $\lambda(x) = \frac{1}{\sqrt{2\pi}} \int_{-\infty}^{x} e^{-t^2/2}\, dt$.

From the estimate of Mond, see (1.39), we obtain

$$|B_n(f, x) - f(x)| \leq 2 \cdot \omega_1\left(f, \sqrt{\frac{x(1 - x)}{n}}\right), \quad f \in B[0, 1], \ n \in \mathbb{N}, \ x \in (0, 1). \tag{4.16}$$

In the case of continuously differentiable functions, the first estimate with the first order modulus of the derivative, was given by T. Popoviciu [115]. The pointwise version of it is:

$$|B_n(f, x) - f(x)| \le C\sqrt{\frac{x(1-x)}{n}} \cdot \omega_1\left(f', \sqrt{\frac{x(1-x)}{n}}\right), \quad f \in C^1[0, 1],$$

$$(4.17)$$

where $x \in (0, 1)$, $n \in \mathbb{N}$ and $C > 0$ is an absolute constant. The best value of the constant C appearing in (4.17) will be given in the next subsection. For the global approximation the best constant was obtained by F. Schurer and W. Steutel [122]:

$$\sup_{f \in C^1[0,1] \setminus \Pi_1} \sup_{n \in \mathbb{N}} \frac{\|B_n(f) - f\|}{\frac{1}{\sqrt{n}} \cdot \omega_1\left(f', \frac{1}{\sqrt{n}}\right)} = \frac{1}{4}. \tag{4.18}$$

As regards the second order modulus of continuity, Y.A.Brudnyi [18] showed that there exists a constant $C > 0$, such that

$$\|B_n(f) - f\| \le C \cdot \omega_2\left(f, \frac{1}{\sqrt{n}}\right), \quad n \in \mathbb{N}, \ f \in C[0, 1]. \tag{4.19}$$

The pointwise version of this estimate, namely there exists $C > 0$ such that

$$|B_n(f, x) - f(x)| \le C \cdot \omega_2\left(f, \sqrt{\frac{x(1-x)}{n}}\right), \quad x \in (0, 1), \ n \in \mathbb{N}, \ f \in C[0, 1],$$

$$(4.20)$$

was first obtained by Jia-ding Cao [20]

The first precise constants which appear in both estimates (4.19) and (4.20) were given by H. Gonska [36], see also [41], as an application of some estimates for general linear positive operators. From relation (2.20) one obtains for the Bernstein operators the estimate

$$|B_n(f, x) - f(x)| \le \left(\frac{3}{2} + \frac{3}{4}h^{-2}B_n((e_1 - xe_0)^2, x)\right) \cdot \omega_2(f, h),$$

for $f \in C[0, 1]$, $h > 0$ and $x \in [0, 1]$. Consequently we can take $C = 1.6875$ in estimate (4.19) and $C = 2.25$ in estimate (4.20).

The first estimates, in norm, using the Ditzian–Totik modulus with undetermined constants were given by X.-l. Zhou [143] and by Ditzian and Totik [30]. A first estimate with explicit constants was obtained by Gonska and Tachev [46]

For the global approximation, the second order Ditzian–Totik modulus and some equivalent moduli of it are more appropriate, see the works by Ditzian [29] and Totik [134], [135], Ditzian and Totik [30], Lorentz and Schumaker [61], K.G. Ivanov [49], D-X. Zhou [141].

Finally, we mention briefly some inverse and saturation theorems. We shall use the Landau symbols "O" and "o". For the pointwise approximation these results can be expressed with the aid of the usual second order modulus. H. Berens and G.G. Lorentz [12] showed that for any $0 < \alpha < 2$ the following conditions are equivalent for a function $f \in C[0, 1]$:

$$a) \ |B_n(f, x) - f(x)| = O\left(\left(\frac{x(1-x)}{n}\right)^{\alpha/2}\right), \quad x \in [0, 1], \ n \in \mathbb{N}, \quad (4.21)$$

$$b) \ \omega_2(f, h) = O\left(h^\alpha\right), \quad h > 0.$$

For the saturation case $\alpha = 2$, Lorentz [60] showed that the following conditions are equivalent for a function f on $[0, 1]$:

$$a) \ |B_n(f, x) - f(x)| \leq M \cdot \frac{x(1-x)}{2n} + o_x\left(\frac{1}{n}\right), \quad x \in [0, 1], \ n \in \mathbb{N},$$

$$b) \ |B_n(f, x) - f(x)| \leq M \cdot \frac{x(1-x)}{2n}, \quad x \in [0, 1], \ n \in \mathbb{N}, \quad (4.22)$$

$$c) \ f \in W_\infty^2, \quad \text{and} \quad \omega_2(f, h) \leq Mh^2,$$

where W_∞^2 is the space of all functions f which have an absolutely continuous derivative f' on $[0, 1]$ and $|f''| \leq M$ a.e. for some $M > 0$. In particular, $B_n(f, x) - f(x) = o_x\left(\frac{1}{n}\right)$ if and only if f is linear, see Bajsanski and Bojanic [10].

4.1.2 Applications of general estimates

In the sequel we point out some estimates with the second order moduli that can be derived for Bernstein operators, from the general results given in Chapter 2.

We start with the modulus ω_2^λ:

Corollary 4.1.1. *For any $f \in B[0, 1]$, $n \in \mathbb{N}$, $\lambda \in \left[0, \frac{1}{2}\right]$, $x \in (0, 1)$ and $k > 0$ we have*

$$|B_n(f, x) - f(x)| \leq \begin{cases} \left(\frac{2}{1+2\lambda} + \frac{1}{k^2}\right) \omega_2^\lambda\left(f, k\sqrt{\frac{x(1-x)}{n}}\right), & 0 < k \leq \frac{\sqrt{3}}{2}, \\ \left(\frac{2}{1+2\lambda} + \frac{3}{4k^4}\right) \omega_2^\lambda\left(f, k\sqrt{\frac{x(1-x)}{n}}\right), & k > \frac{\sqrt{3}}{2}. \end{cases} \quad (4.23)$$

For $k = 1$ we have

$$|B_n(f, x) - f(x)| \leq \frac{11 + 6\lambda}{4 + 8\lambda} \cdot \omega_2^\lambda\left(f, \sqrt{\frac{x(1-x)}{n}}\right). \quad (4.24)$$

Proof. We apply Theorem 2.2.1 for $b = 0$ and $p = 1$, in the case $0 < k \leq \frac{\sqrt{3}}{2}$ and for $b = 0$ and $p = 2$, in the case $\frac{\sqrt{3}}{2} < k$ and we take into account relations (4.4). □

For the particular case of the modulus ω_2, we obtain:

Corollary 4.1.2. *For any $f \in B[0, 1]$, $n \in \mathbb{N}$, $x \in (0, 1)$ and $k > 0$ we have*

$$|B_n(f, x) - f(x)| \leq \begin{cases} \left(1 + \frac{1}{2k^2}\right) \omega_2\left(f, k\sqrt{\frac{x(1-x)}{n}}\right), & 0 < k \leq \frac{\sqrt{3}}{2}, \\ \left(1 + \frac{3}{8k^4}\right) \omega_2\left(f, k\sqrt{\frac{x(1-x)}{n}}\right), & k > \frac{\sqrt{3}}{2}. \end{cases} \quad (4.25)$$

For $k = 1$ we have

$$|B_n(f, x) - f(x)| \leq \frac{11}{8} \cdot \omega_2\left(f, \sqrt{\frac{x(1-x)}{n}}\right).$$ (4.26)

In a similar mode we obtain from Theorem 2.2.3:

Corollary 4.1.3. *For any $f \in B[0, 1]$, $n \in \mathbb{N}$, $x \in (0, 1)$ and $k > 0$ we have*

$$|B_n(f, x) - f(x)| \leq \begin{cases} \left(1 + \frac{1}{k^2}\right) \omega_2^\star\left(f, k\sqrt{\frac{x(1-x)}{n}}\right), & 0 < k \leq \frac{\sqrt{3}}{2}, \\ \left(1 + \frac{3}{4k^4}\right) \omega_2^\star\left(f, k\sqrt{\frac{x(1-x)}{n}}\right), & k > \frac{\sqrt{3}}{2}. \end{cases}$$ (4.27)

For $k = 1$ we have

$$|B_n(f, x) - f(x)| \leq \frac{7}{4} \cdot \omega_2^\star\left(f, \sqrt{\frac{x(1-x)}{n}}\right).$$ (4.28)

For the global approximation, the estimates below follow.

Corollary 4.1.4. *For any $f \in B[0, 1]$, $n \in \mathbb{N}$, $\lambda \in \left[0, \frac{1}{2}\right]$, and $k > 0$ we have*

$$\|B_n(f) - f\| \leq \begin{cases} \left(\frac{2}{1+2\lambda} + \frac{1}{4k^2}\right) \omega_2^\lambda\left(f, \frac{k}{\sqrt{n}}\right), & 0 < k \leq \frac{\sqrt{3}}{2}, \\ \left(\frac{2}{1+2\lambda} + \frac{3}{16k^4}\right) \omega_2^\lambda\left(f, \frac{k}{\sqrt{n}}\right), & k > \frac{\sqrt{3}}{2}. \end{cases}$$ (4.29)

For $k = 1$ we have

$$\|B_n(f) - f\| \leq \frac{35 + 6\lambda}{16 + 32\lambda} \cdot \omega_2^\lambda\left(f, \frac{1}{\sqrt{n}}\right).$$ (4.30)

Corollary 4.1.5. *For any $f \in B[0, 1]$, $n \in \mathbb{N}$ and $k > 0$ we have*

$$\|B_n(f) - f\| \leq \begin{cases} \left(1 + \frac{1}{4k^2}\right) \omega_2^\star\left(f, \frac{k}{\sqrt{n}}\right), & 0 < k \leq \frac{\sqrt{3}}{2}, \\ \left(1 + \frac{3}{16k^4}\right) \omega_2^\star\left(f, \frac{k}{\sqrt{n}}\right), & k > \frac{\sqrt{3}}{2}. \end{cases}$$ (4.31)

For $k = 1$ we have

$$\|B_n(f) - f\| \leq \frac{19}{16} \cdot \omega_2^\star\left(f, \frac{1}{\sqrt{n}}\right).$$ (4.32)

In the case of the classical second order modulus we can give a more refined estimate

Corollary 4.1.6. *For any $f \in B[0, 1]$, $n \in \mathbb{N}$ and $k > 0$ we have*

$$\|B_n(f) - f\| \leq \begin{cases} \left(1 + \frac{1}{8k^2}\right) \omega_2\left(f, \frac{k}{\sqrt{n}}\right), & 0 < k \leq \frac{\sqrt{3}}{2}, \\ \left(1 + \frac{1}{16k^4 - 8k^2 + 3}\right) \omega_2\left(f, \frac{k}{\sqrt{n}}\right), & k > \frac{\sqrt{3}}{2}. \end{cases}$$ (4.33)

For $k = 1$ we have

$$\|B_n(f) - f\| \leq \frac{12}{11} \cdot \omega_2\left(f, \frac{1}{\sqrt{n}}\right).$$ (4.34)

Proof. For $k \leq \frac{\sqrt{3}}{2}$ we apply Corollary 2.2.1. Now let $k > \frac{\sqrt{3}}{2}$. We apply Theorem 2.2.1 with the choice $p = 2$ and $b = b_k$, where

$$b_k := \frac{4k^2 - 3}{4k^2(4k^2 - 1)}.$$

We get

$$|B_n(f, x) - f(x)| \leq \left(1 + \frac{1}{2}P(x(1-x))\right) \omega_2 \left(f, \frac{k}{\sqrt{n}}\right),$$

where

$$P(t) := \frac{1}{(1 - b_k)^2} \left[\frac{1}{k^4}\left(\left(3 - \frac{6}{n}\right)t^2 + \frac{1}{n} \cdot t\right) - \frac{2b_k}{k^2} \cdot t + (b_k)^2\right], \ t \in \mathbb{R}.$$

For $n \geq 1$, after short computations we get

$$P\left(\frac{1}{4}\right) \leq \frac{1}{(1 - b_k)^2} \left[\frac{3}{16k^4} - \frac{b_k}{2k^2} + (b_k)^2\right] = \frac{2}{16k^4 - 8k^3 + 3}.$$

Then,

$$P(x(1-x)) \leq \max\left\{P(0), P\left(\frac{1}{4}\right)\right\}$$

$$\leq \max\left\{\left(\frac{4k^2 - 3}{16k^4 - 8k^2 + 3}\right)^2, \frac{2}{16k^4 - 8k^2 + 3}\right\}$$

$$= \frac{2}{16k^4 - 8k^2 + 3}.$$

From this we obtain (4.33). □

For the particular case $k = 1$ the optimal estimate will be obtained in the next section, by using the method given in Chapter 3.

From Corollaries 2.3.1 and 2.3.2 we obtain

Corollary 4.1.7. *We have*

$$|B_n(f, x) - f(x)| \leq \left(\frac{1}{8} + \frac{1}{2k^2}\right) \omega_2^d\left(f, k\sqrt{\frac{x(1-x)}{n}}\right), \tag{4.35}$$

for $f \in \mathcal{F}[0, 1]$, $n \in \mathbb{N}$, $x \in (0, 1)$, $0 < k \leq 2$, and

$$|B_n(f, x) - f(x)| \leq \left(\frac{k}{8} + \frac{1}{2k}\right) \sqrt{\frac{x(1-x)}{n}} \cdot \omega_1\left(f', k\sqrt{\frac{x(1-x)}{n}}\right), \tag{4.36}$$

for $f \in \mathcal{D}[0, 1]$, $n \in \mathbb{N}$, $x \in (0, 1)$, $0 < k \leq 2$. For $k = 1$ we have

$$|B_n(f, x) - f(x)| \leq \frac{5}{8} \cdot \sqrt{\frac{x(1-x)}{n}} \cdot \omega_1\left(f', \sqrt{\frac{x(1-x)}{n}}\right). \tag{4.37}$$

Remark 4.1.1. For $k = 2$, estimate (4.36), namely

$$|B_n(f, x) - f(x)| \leq \frac{1}{2} \cdot \sqrt{\frac{x(1-x)}{n}} \cdot \omega_1 \left(f', 2\sqrt{\frac{x(1-x)}{n}} \right), \qquad (4.38)$$

for $f \in \mathcal{D}[0, 1]$, $n \in \mathbb{N}$, $x \in (0, 1)$, is the pointwise version of the optimal global estimate of F. Schurer and F. W. Steutel (4.18). The situation differs from those given in the case of the modulus ω_2, where, from the estimate with the best constants in front, the moments m_0 and m_2 do not follow the estimate with the global best constant in front of the modulus.

From Corollaries 2.4.3 and 2.4.5 we obtain the following corollaries:

Corollary 4.1.8. *We have*

$$|B_n(f, x) - f(x)| \leq \frac{x(1-x)}{n} \cdot M_2(f), \qquad (4.39)$$

for any $f \in \mathcal{F}[0, 1]$, $x \in [0, 1]$ and $n \in \mathbb{N}$.

Corollary 4.1.9. *We have*

$$|B_n(f, x) - f(x)| \leq \frac{x(1-x)}{2n} \cdot \|f''\|, \qquad (4.40)$$

for any $f \in C^2[0, 1]$, $x \in [0, 1]$ and $n \in \mathbb{N}$.

Finally, from Theorem 2.5.1 we arrive at the following corollary.

Corollary 4.1.10. *We have*

$$\|B_n(f) - f\| \leq \left(1 + \frac{3}{2k^2} \right) \omega_2^\varphi \left(f, \frac{k}{\sqrt{n}} \right), \qquad (4.41)$$

for any $f \in B[0, 1]$, $n \in \mathbb{N}$ and $k > 0$. For $k = 1$ we have

$$\|B_n(f) - f\| \leq \frac{5}{2} \cdot \omega_2^\varphi \left(f, \frac{1}{\sqrt{n}} \right). \qquad (4.42)$$

Estimates for simultaneous approximation by Bernstein operators will be obtained in the next chapter, see Corollary 5.1.3, Remarks 5.1.3 and Corollary 5.1.4.

4.2 Best constant in the estimate with modulus ω_2

4.2.1 Introduction. Main result

In this section we point out the optimal constant C that can appear in the estimate (4.19). This problem is analogous, but not similar, to the extremal problem solved by P.C.Sikkema [128], see (4.14).

We define

$$C_2 := \sup_{n \in \mathbb{N}} \sup_{f \in C[0,1] \setminus \Pi_1} \frac{\|B_n(f) - f\|}{\omega_2\left(f, \frac{1}{\sqrt{n}}\right)}. \tag{4.43}$$

In [39], Gonska mentioned that $C_2 \geq 1$ and made the conjecture that $C_2 = 1$.

Note that from Corollary 4.1.6, we obtain $C_2 \leq \frac{12}{11} = 1.0909\ldots$.

In connection with the constant C_2 we mention also the following result of Gonska and D.X.Zhou [45]: there exists a constant $0 < c < 1$ such that for any $\frac{1}{2} \leq a < 1$ there exists $N(a) \in \mathbb{N}$ such that for all $n \geq N(a)$, there holds:

$$\sup_{1-a \leq \frac{k}{n} \leq a} \left| B_n\left(f, \frac{k}{n}\right) - f\left(\frac{k}{n}\right) \right| \leq c \cdot \omega_2\left(f, \frac{1}{\sqrt{n}}\right).$$

Also, in a paper of D.Kacsó [52] the following inequality for a convex function f is given:

$$B_n\left(f, \frac{k}{[\sqrt{n}]+1}\right) - f\left(\frac{k}{[\sqrt{n}]+1}\right) \leq \frac{5}{8} \cdot \omega_2\left(f, \frac{1}{\sqrt{n}}\right),$$

for all $0 \leq k \leq [\sqrt{n}] + 1$, where $[\cdot]$ denotes the integer part of a number.

As we noted for Sikkema's result, also in the case of the second modulus we may extend the problem to the space of bounded functions. The main result of the present section is the following theorem, proved in [109], that implies $C_2 = 1$.

Theorem 4.2.1. *For any $n \in \mathbb{N}$ we have*

$$\sup_{f \in B[0,1] \setminus \Pi_1} \frac{\|B_n(f) - f\|}{\omega_2\left(f, \frac{1}{\sqrt{n}}\right)} = \sup_{f \in C[0,1] \setminus \Pi_1} \frac{\|B_n(f) - f\|}{\omega_2\left(f, \frac{1}{\sqrt{n}}\right)} = 1. \tag{4.44}$$

The direct part of the theorem consists in the proof of the inequality

$$|B_n(f, x) - f(x)| \leq \omega_2\left(f, \frac{1}{\sqrt{n}}\right), \quad f \in B[0,1], \ x \in (0,1), \ n \in \mathbb{N}. \tag{4.45}$$

For $x = 0, 1$ we have $B_n(f, 0) = f(0)$ and $B_n(f, 1) = f(1)$.

The proof of (4.45) will be given in Subsections 4.2.2 and 4.2.3 and is based on the result given in Section 3.2.

The inverse part of the theorem, namely the fact that for any n the two suprema in (4.44) are not less than 1, can be easily derived. So, if we choose, for $0 < \varepsilon < \frac{1}{n}$ the functions f_ε, given by $f_\varepsilon(t) := \frac{\varepsilon - t}{\varepsilon}$, $(0 \leq t \leq \varepsilon)$ and $f_\varepsilon(t) := 0$, $(\varepsilon \leq t \leq 1)$, then we have $\omega_2\left(f_\varepsilon, \frac{1}{\sqrt{n}}\right) = 1$ and on the other hand $B_n(f_\varepsilon, \varepsilon) - f_\varepsilon(\varepsilon) = p_{n,0}(\varepsilon)$. But $\lim_{\varepsilon \to 0+} p_{n,0}(\varepsilon) = 1$. This method was used in obtaining the result mentioned in [39]. A similar construction was used in [82].

By combining the results given in Theorem 3.1.2 and Theorem 4.2.1 we arrive at

Corollary 4.2.1. *For any $n \in \mathbb{N}$ any $h \geq \frac{1}{\sqrt{n}}$ and any irrational point $x \in (0, 1)$ we have*

$$\sup_{f \in C[0,1] \setminus \Pi_1} \frac{|B_n(f, x) - f(x)|}{\omega_2(f, h)} = 1. \tag{4.46}$$

4.2.2 Proof of the direct part of the theorem for $n \geq 60$

In the next two sections, we shall denote for $t > 0$:

$$\sigma(t) := \max\{i \in \mathbb{N} \cup \{0\}, \; \frac{i}{n} < t\}, \quad \bar{\sigma}(t) := \max\{i \in \mathbb{N} \cup \{0\}, \; \frac{i}{n} \leq t\},$$

$$\tau(t) := \min\{i \in \mathbb{N}, \; \frac{i}{n} > t\}, \quad \bar{\tau}(t) := \min\{i \in \mathbb{N}, \; \frac{i}{n} \geq t\}.$$

Therefore, if $t \in (0, 1)$, these notations coincide with the definitions (3.10) and (3.11) for the corresponding sequence $y_i = \frac{i}{n}$, $(0 \leq i \leq n)$.

For $x \in [0, 1]$ and $0 \leq s \leq n$ set

$$\varphi_{n,s}(x) := \sum_{k=s}^{n} p_{n,k}(x), \quad \Psi_{n,s}(x) := \sum_{k=s}^{n} \left(\frac{k}{n} - x\right) p_{n,k}(x), \tag{4.47}$$

where $p_{n,k}(x) := \binom{n}{k} x^k (1 - x)^{n-k}$. Next, for $x \in [0, 1)$ and $1 \leq s \leq n$, denote

$$\Theta_{n,s}(x) := \frac{n}{s(1 - x)} \sum_{j=s}^{n} \frac{s!(n - s)!}{j!(n - j)!} \left(\frac{x}{1 - x}\right)^{j-s}. \tag{4.48}$$

In Sikkema's paper [127] it is shown, for $1 \leq s \leq n$, that

$$\Psi_{n,s}(x) = \binom{n-1}{s-1} x^s (1 - x)^{n-s+1}, \tag{4.49}$$

implying

$$\Theta_{n,s}(x) = \frac{\varphi_{n,s}(x)}{\Psi_{n,s}(x)}, \quad \text{for } x \in (0, 1). \tag{4.50}$$

Take $h := \frac{1}{\sqrt{n}}$. For $q \geq 0$ and $n \in \mathbb{N}$ such that $qh < 1$, put

$$c_n(q) := h \cdot \min_{x \in [0, 1-qh]} \Theta_{n, \tau(x+qh)}(x). \tag{4.51}$$

The function $\Theta_{n,s}$ is increasing and, consequently, we have

$$c_n(q) = \min_{\tau(qh) \leq s \leq n} h \cdot \Theta_{n,s}\left(\max\left\{0, \frac{s-1}{n} - qh\right\}\right). \tag{4.52}$$

We use the following elementary lemma.

Lemma 4.2.1. *If the strictly positive numbers* λ_j, μ_j, d_j, $(1 \leq j \leq m)$, $m \geq 2$, *satisfy the following conditions:*

$$d_{j+1} \leq d_j, \quad (1 \leq j \leq m - 1) \tag{4.53}$$

$$\frac{\lambda_{j+1}}{\lambda_j} \leq \frac{\mu_{j+1}}{\mu_j}, \quad (1 \leq j \leq m - 1), \tag{4.54}$$

then

$$\frac{\mu_1 d_1 + \cdots + \mu_m d_m}{\mu_1 + \cdots + \mu_m} \leq \frac{\lambda_1 d_1 + \cdots + \lambda_m d_m}{\lambda_1 + \cdots + \lambda_m}. \tag{4.55}$$

We omit the proof of this lemma, since it can be easily obtained by induction.

Lemma 4.2.2. *For all $n \geq 60$, one has*

$$c_n(1) \geq \frac{4}{5}. \tag{4.56}$$

Proof. By numerical computations, using (4.52) and (4.48) one obtains that the numbers $c_n(1)$ are strictly increasing for $60 \leq n \leq 99$ and:

$$c_{60}(1) = 0.80021\ldots < c_{61}(1) = 0.80060\cdots < \cdots < c_{99}(1) = 0.80995\ldots .$$

It remains to prove (4.56) for $n \geq 100$. Assume that $c_{n-1}(1) \geq \frac{4}{5}$, $(n \geq 100)$ and prove by induction that $c_n(1) \geq \frac{4}{5}$. Fix $s \in \mathbf{N}$ such that $\tau\left(\frac{1}{\sqrt{n}}\right) \leq s \leq n$. Denote

$$y := \frac{s-1}{n} - \frac{1}{\sqrt{n}}, \qquad v := \frac{s-1}{n-1} - \frac{1}{\sqrt{n-1}}.$$

If $y \leq 0$, one has $\frac{1}{\sqrt{n}} \cdot \Theta_{n,s}(0) = \frac{\sqrt{n}}{s} \geq \frac{\sqrt{n}}{\sqrt{n}+1} \geq \frac{4}{5}$. If $s = n$, then $y > 0$ and $\frac{1}{\sqrt{n}} \cdot \Theta_{n,s}(y) = \frac{1}{\sqrt{n}(1-y)} = \frac{\sqrt{n}}{1+\sqrt{n}} \geq \frac{4}{5}$. Therefore we can consider that $y > 0$ and $s \leq n-1$. Since $y > 0$ and $n \geq \sqrt{n} + \sqrt{n-1}$, we have $v > 0$. Also $\frac{s}{n-1} > \frac{1}{\sqrt{n-1}}$. Then, the induction yields

$$\frac{1}{\sqrt{n-1}} \cdot \Theta_{n-1,s}(v) \geq \frac{4}{5}. \tag{4.57}$$

Now we prove the inequality

$$\frac{\sum_{k=s}^{n-1} p_{n,k+1}(y)\left(\frac{k}{n-1} - v\right)}{\sum_{k=s}^{n-1} p_{n,k+1}(y)} \leq \frac{\sum_{k=s}^{n-1} p_{n-1,k}(v)\left(\frac{k}{n-1} - v\right)}{\sum_{k=s}^{n-1} p_{n-1,k}(v)}. \tag{4.58}$$

This inequality, (for $s \leq n-2$) follows from Lemma 4.2.1, by choosing: $m := n - s$, $\lambda_j := p_{n-1,n-j}(v)$, $\mu_j := p_{n,n-j+1}(y)$ and $d_j := \frac{n-j}{n-1} - v > 0$, $(1 \leq j \leq m)$. Indeed, in the case $m \geq 2$, condition (4.53) is immediate and it remains to show (4.54). This is equivalent to

$$\frac{p_{n-1,k-1}(v)}{p_{n-1,k}(v)} \leq \frac{p_{n,k}(y)}{p_{n,k+1}(y)}, \quad \text{i.e.,} \quad \frac{k}{k+1} \leq \frac{v(1-y)}{y(1-v)}, \quad \text{for } s+1 \leq k \leq n-1.$$

It suffices to check it for $k = n - 1$. In this case it is equivalent to the obvious inequality

$$\frac{n-1}{n} \leq \frac{s-1-\sqrt{n-1}}{s-1-\sqrt{n}} \cdot \frac{n-s+1+\sqrt{n}}{n-s+\sqrt{n-1}}.$$

Denote now

$$T := \frac{1}{\sqrt{n-1}} \cdot \sum_{k=s}^{n-1} p_{n,k+1}(y) \Big/ \sum_{k=s}^{n-1} p_{n,k+1}(y)\left(\frac{k}{n-1} - v\right).$$

We have

$$T = \frac{1}{\sqrt{n-1}} \cdot \frac{\varphi_{n,s}(y) - p_{n,s}(y)}{\frac{n}{n-1}\Psi_{n,s}(y) + \left(\frac{n}{n-1}y - \frac{1}{n-1} - v\right)\varphi_{n,s}(y) + \left(v - \frac{s-1}{n-1}\right)p_{n,s}(y)}$$

$$= \frac{\sqrt{n-1}(\varphi_{n,s}(y) - p_{n,s}(y))}{n\Psi_{n,s}(y) + (-1 - \sqrt{n} + \sqrt{n-1})\varphi_{n,s}(y) - \sqrt{n-1}\cdot p_{n,s}(y)}.$$

Since

$$\frac{p_{n,s}(y)}{\Psi_{n,s}(y)} = \frac{n}{s(1-y)} = \frac{n^2}{s(n-s+1+\sqrt{n})},$$

we obtain

$$T = \frac{\sqrt{n-1}\left(\frac{1}{\sqrt{n}}\cdot\Theta_{n,s}(y) - \frac{n\sqrt{n}}{s(n-s+1+\sqrt{n})}\right)}{\sqrt{n} + (-1 - \sqrt{n} + \sqrt{n-1})\cdot\frac{1}{\sqrt{n}}\cdot\Theta_{n,s}(y) - \frac{n\sqrt{n(n-1)}}{s(n-s+1+\sqrt{n})}}.$$

Consider the polynomial of degree 2,

$$P(t) := (-1 - \sqrt{n} + \sqrt{n-1})t^2 + \left(\sqrt{n} - \sqrt{n-1} - \frac{n\sqrt{n(n-1)}}{s(n-s+1+\sqrt{n})}\right)t$$

$$+ \frac{n\sqrt{n(n-1)}}{s(n-s+1+\sqrt{n})}.$$

The inequality

$$T \leq \frac{1}{\sqrt{n}}\cdot\Theta_{n,s}(y) \tag{4.59}$$

is equivalent to

$$P\left(\frac{1}{\sqrt{n}}\cdot\Theta_{n,s}(y)\right) \geq 0. \tag{4.60}$$

We distinguish between the following two cases:

Case 1: The relation (4.60) is true. Since (4.58) can be rewritten in the form $T \geq \frac{1}{\sqrt{n-1}}\cdot\Theta_{n-1,s}(v)$, by combining (4.57),(4.58) and (4.59) we get

$$\frac{1}{\sqrt{n}}\cdot\Theta_{n,s}(y) \geq \frac{4}{5}. \tag{4.61}$$

Case 2: The relation (4.60) is false. In this case we shall prove (4.61) by using a different argument. Note that the polynomial P admits the real roots $t_1 = t_1(n,s)$ and $t_2 = t_2(n,s)$ such that $t_1 < 0 < t_2$. The condition in Case 2 implies that $t_2 < \frac{1}{\sqrt{n}}\cdot\Theta_{n,s}(y)$. Then in order to prove (4.61) it suffices to show that

$$P\left(\frac{4}{5}\right) > 0, \quad \text{for } n \geq 100, \ 1 \leq s \leq n-1.$$

We have

$$P\left(\frac{4}{5}\right) = \frac{16}{25}(-1 - \sqrt{n} + \sqrt{n-1}) + \frac{4}{5}(\sqrt{n} - \sqrt{n-1})$$

$$+ \frac{1}{5} \cdot \frac{n\sqrt{n(n-1)}}{s(n-s+1+\sqrt{n})}$$

$$\geq -\frac{16}{25} + \frac{4}{25}\frac{1}{\sqrt{n}+\sqrt{n-1}} + \frac{4}{5} \cdot \frac{n\sqrt{n(n-1)}}{(n+\sqrt{n}+1)^2}$$

$$> -\frac{16}{25} + \frac{2}{25}\frac{1}{\sqrt{n}} + \frac{4}{5}\frac{n^2-n}{(n+\sqrt{n}+1)^2}.$$

The positivity of this expression is equivalent to $2n^2\sqrt{n} - 15n^2 - 32n\sqrt{n} - 13n - 6\sqrt{n} + 1 > 0$. But this last inequality holds for $n = 100$ and consequently for any $n \geq 100$.

Therefore (4.61) holds in both cases.   □

Now we introduce some new notation. For fixed $n \geq 1$, $x \in (0, 1)$ and for $\alpha \geq 0$ let

$$M_r(\alpha, \infty) := (M_r(\alpha, \infty))(x) := \begin{cases} \Psi_{n,\tau(x+\alpha h)}(x), & \text{if } x + \alpha h < 1 \\ 0, & \text{if } x + \alpha h \geq 1 \end{cases}, \quad (4.62)$$

$$T_r(\alpha, \infty) := (T_r(\alpha, \infty))(x) := \begin{cases} \varphi_{n,\tau(x+\alpha h)}(x), & \text{if } x + \alpha h < 1 \\ 0, & \text{if } x + \alpha h \geq 1 \end{cases}, \quad (4.63)$$

$$M_l(\alpha, \infty) := (M_l(\alpha, \infty))(x) := \begin{cases} \Psi_{n,\tau(1-x+\alpha h)}(1-x), & \text{if } x > \alpha h \\ 0, & \text{if } x \leq \alpha h \end{cases}, \quad (4.64)$$

$$T_l(\alpha, \infty) := (T_l(\alpha, \infty))(x) := \begin{cases} \varphi_{n,\tau(1-x+\alpha h)}(1-x), & \text{if } x > \alpha h \\ 0, & \text{if } x \leq \alpha h \end{cases}. \quad (4.65)$$

By using the symmetrical relation $p_{n,k}(1-x) = p_{n,n-k}(x)$, $0 \leq k \leq n$, it follows, in the case $x > \alpha h$, that

$$M_l(\alpha, \infty) = \sum_{k=0}^{\sigma(x-\alpha h)} p_{n,k}(x)\left(x - \frac{k}{n}\right), \quad T_l(\alpha, \infty) = \sum_{k=0}^{\sigma(x-\alpha h)} p_{n,k}(x).$$

Moreover, for $0 \leq \alpha < \beta$, denote

$$M_r(\alpha, \beta] := M_r(\alpha, \infty) - M_r(\beta, \infty), \quad T_r(\alpha, \beta] := T_r(\alpha, \infty) - T_r(\beta, \infty), \quad (4.66)$$

$$M_l(\alpha, \beta] := M_l(\alpha, \infty) - M_l(\beta, \infty), \quad T_l(\alpha, \beta] := T_l(\alpha, \infty) - T_l(\beta, \infty). \quad (4.67)$$

Since $B_n(e_i) = e_i$, $i = 0, 1$, it follows that $M_r(0, \infty) = M_l(0, \infty)$. Set

$$M(0, \infty) := M_r(0, \infty) = M_l(0, \infty). \quad (4.68)$$

If E is an expression constructed with terms of the types defined in (4.62)–(4.68) and we need to specify the point x, we denote $(E)(x)$. If E and E^* are two such expressions, we say that E^* is the symmetrical expression of E if it is obtained from E by a permutation of the indices r and l but with the same x.

In the following part of this section, we consider $n \in \mathbf{N}$ and $x \in (0, 1)$ to be fixed.

Lemma 4.2.3. *Let* $0 \le \alpha < \beta$, $\beta - \alpha \ge h$. *We have*

$$h \cdot \frac{T_r(\alpha, \beta]}{M_r(\alpha, \beta]} \ge \frac{2}{\alpha + \beta + h}, \quad if \ x + \alpha h < 1, \tag{4.69}$$

and

$$h \cdot \frac{T_l(\alpha, \beta]}{M_l(\alpha, \beta]} \ge \frac{2}{\alpha + \beta + h}, \quad if \ x > \alpha h. \tag{4.70}$$

Proof. First suppose that $x + \beta h < 1$. Denote $s := \tau(x + \alpha h)$ and $t := \tau(x + \beta h)$. The condition $\beta - \alpha \ge h$ guarantees that $t > s$. We have $p_{n,k+1}(x) < p_{n,k}(x)$, if $s \le k \le n - 1$. By using the Chebychev inequality it follows that

$$\frac{M_r(\alpha, \beta]}{T_r(\alpha, \beta]} \le \frac{1}{t - s} \cdot \sum_{k=s}^{t-1} \left(\frac{k}{n} - x \right) = \frac{1}{2} \left(\left(\frac{s}{n} - x \right) + \left(\frac{t-1}{n} - x \right) \right)$$

$$\le \frac{h}{2}(\alpha + \beta + h).$$

Now, suppose that $x + \beta h \ge 1$. By applying once again the Chebychev inequality we get

$$\frac{M_r(\alpha, \beta]}{T_r(\alpha, \beta]} = \frac{\sum_{k=s}^{n} p_{n,k}(x) \left(\frac{k}{n} - x \right)}{\sum_{k=s}^{n} p_{n,k}(x)} \le \frac{1}{n - s + 1} \sum_{k=s}^{n} \left(\frac{k}{n} - x \right)$$

$$= \frac{1}{2} \left(\left(\frac{s}{n} - x \right) + (1 - x) \right) \le \frac{h}{2}(\alpha + \beta + h).$$

The inequality (4.70) is similar. $\qquad\square$

Lemma 4.2.4. *Let* $0 \le \alpha < \beta$. *If* $x + \alpha h < 1$, *then*

$$\frac{M_r(\beta, \infty)}{M_r(\alpha, \infty)} \le \exp(-2(\beta^2 - \alpha^2) + 2h(\beta + \alpha)), \tag{4.71}$$

and, if $x > \alpha h$, *then*

$$\frac{M_l(\beta, \infty)}{M_l(\alpha, \infty)} \le \exp(-2(\beta^2 - \alpha^2) + 2h(\beta + \alpha)). \tag{4.72}$$

Proof. By taking into account the symmetry we prove only the first inequality. Denote $s := \tau(x + \alpha h)$ and $t := \tau(x + \beta h)$. The inequality is obvious if $x + \beta h \ge 1$ or $\beta \le \alpha + h$. In the contrary case we have $s < t \le n$. From (4.49) it follows that

$$M_r(\alpha, \infty) = \binom{n-1}{s-1} x^s (1-x)^{n-s+1}, \quad M_r(\beta, \infty) = \binom{n-1}{t-1} x^t (1-x)^{n-t+1}.$$

Denote $T := \ln \frac{M_r(\beta,\infty)}{M_r(\alpha,\infty)}$. We have:

$$T = \ln \left(\frac{x}{1-x} \right)^{t-s} \frac{(s-1)!(n-s)!}{(t-1)!(n-t)!} = \sum_{i=s}^{t-1} \ln \frac{n-i}{i} \cdot \frac{x}{1-x}$$

$$= \sum_{j=0}^{t-s-1} \ln \frac{n-s-j}{s+j} \cdot \frac{x}{1-x}$$

$$= \sum_{j=0}^{t-s-1} F\left(\frac{s+j}{n} - x \right),$$

where $F(u) := \ln \frac{1-u-x}{u+x} \cdot \frac{x}{1-x}$, $u \in (-x, 1-x)$.

Consider also the function $G(u) := -4u - F(u)$, $u \in (-x, 1-x)$. We have $G(0) = 0$ and $G'(u) = -4 + \frac{1}{(1-x-u)(x+u)} \geq 0$. Consequently, $G(u) > 0$ for $0 < u < 1-x$ and hence $F(u) < -4u$ for such u. It follows that

$$T < -4 \cdot \sum_{j=0}^{t-s-1} \left(\frac{s+j}{n} - x \right) = -4(t-s)\left(\frac{t+s-1}{2n} - x \right).$$

Let $\beta h = \frac{p+\varepsilon}{n}$, $p \in \mathbb{N} \cup \{0\}$, $\varepsilon \in (0,1]$, $\frac{s}{n} = x + \frac{q+\mu}{n}$, $q \in \mathbb{N} \cup \{0\}$, $\mu \in (0,1]$ and $\alpha h = \frac{q+\mu-v}{n}$, $v \in (0,1]$, with q, μ given above. From the definition of t it follows that $t - 1 \leq p - q + s + \varepsilon - \mu < t$. We distinguish between two cases.

Case 1: $\mu \leq \varepsilon$. Then $t = p - q + s + 1$. We have successively

$$(t-s)\left(\frac{t+s-1}{2n} - x \right) = (p-q+1)\left(\frac{p-q+2s}{2n} - \frac{s-q-\mu}{n} \right)$$

$$= \frac{1}{2n}(p-q+1)(p+q+2\mu)$$

$$= \frac{1}{2n}(\beta\sqrt{n} - \varepsilon - \alpha\sqrt{n} + \mu - v + 1)(\beta\sqrt{n} - \varepsilon + \alpha\sqrt{n} + \mu + v)$$

$$= \frac{1}{2}(\beta^2 - \alpha^2) - \frac{h}{2}(\alpha + \beta) + R,$$

where $R := \alpha h(1-v) + \beta h(1-\varepsilon+\mu) + \frac{h^2}{2}((\varepsilon-\mu)^2 - (\varepsilon-\mu) + v - v^2) \geq \beta h(1-\varepsilon+\mu) + \frac{h^2}{2}((\varepsilon-\mu)^2 - (\varepsilon-\mu))$. Consider the polynomial $g(u) := \frac{1}{2}u^2 - (\beta + \frac{h}{2})u + \beta h$. From above we have $R \geq g(h(\varepsilon-\mu))$. Since g is decreasing on the interval $[0,h]$, one obtains $R \geq g(h) = 0$. Therefore we have

$$(t-s)\left(\frac{t+s-1}{2n} - x \right) \geq \frac{1}{2}(\beta^2 - \alpha^2) - \frac{h}{2}(\alpha + \beta). \qquad (4.73)$$

Case 2: $\varepsilon < \mu$. Then $t = p - q + s$. We have

$$(t-s)\left(\frac{t+s-1}{2n}-x\right) = (p-q)\left(\frac{2s+p-q-1}{2n}-\frac{s-q-\mu}{n}\right)$$

$$= \frac{1}{2n}(p-q)(p+q-1+2\mu)$$

$$= \frac{1}{2n}(\beta\sqrt{n}-\varepsilon-\alpha\sqrt{n}+\mu-\nu)(\beta\sqrt{n}-\varepsilon+\alpha\sqrt{n}+\mu+\nu-1)$$

$$= \frac{1}{2}(\beta^2-\alpha^2)-\frac{h}{2}(\alpha+\beta)+R_1,$$

where

$$R_1 := \alpha h(1-\nu) + \beta h(\mu-\varepsilon) + \frac{h^2}{2}((\mu-\varepsilon)^2-(\mu-\varepsilon)+\nu-\nu^2)$$

$$\geq (\mu-\varepsilon)(\beta-\frac{h}{2})h \geq 0.$$

Hence the inequality (4.73) holds in both cases and, consequently, we have (4.71). □

Lemma 4.2.5. *Let $n \in \mathbb{N}$, $d > 0$ and $\alpha \geq 0$ such that $\alpha h < 1$. If $(d+\alpha)c_n(\alpha) > 1$, $x \in (0,1)$ and $\rho \geq \alpha$ are such that $x + \rho h < 1$, then*

$$(d+\rho)hT_r(\rho,\infty) - M_r(\rho,\infty) > 0. \tag{4.74}$$

Proof. Denote $y = x+(\rho-\alpha)h$, and $v = x+(\rho+d)h$. As we defined $\tau(v) \in \mathbb{N}$ at the beginning of this section, $\tau(v)$ exists even in the case $v \geq 1$. We have successively:

$$((d+\rho)hT_r(\rho,\infty) - M_r(\rho,\infty))(x) = \sum_{j=\tau(v-dh)}^{n}\left(v-\frac{j}{n}\right)p_{n,j}(x)$$

$$= (1-x)^{n-\tau(v)}x^{\tau(v)}\sum_{j=\tau(v-dh)}^{n}\left(v-\frac{j}{n}\right)\binom{n}{j}\left(\frac{x}{1-x}\right)^{j-\tau(v)}$$

$$\geq (1-x)^{n-\tau(v)}x^{\tau(v)}\sum_{j=\tau(v-dh)}^{n}\left(v-\frac{j}{n}\right)\binom{n}{j}\left(\frac{y}{1-y}\right)^{j-\tau(v)}$$

$$= \left(\frac{1-x}{1-y}\right)^{n-\tau(v)}\left(\frac{x}{y}\right)^{\tau(v)}\sum_{j=\tau(y+\alpha h)}^{n}\left(y+(d+\alpha)h-\frac{j}{n}\right)p_{n,j}(y)$$

$$= \left(\frac{1-x}{1-y}\right)^{n-\tau(v)}\left(\frac{x}{y}\right)^{\tau(v)}((d+\alpha)hT_r(\alpha,\infty) - M_r(\alpha,\infty))(y)$$

$$\geq \left(\frac{1-x}{1-y}\right)^{n-\tau(v)}\left(\frac{x}{y}\right)^{\tau(v)}\{M_r(\alpha,\infty)(y)\}[(d+\alpha)c_n(\alpha)-1] > 0. \quad □$$

Lemma 4.2.6. *If $n \geq 60$ and $x \in (0,1)$, then there exist $\lambda_r \in [0,1]$ and $\lambda_l \in [0,1]$ such that the inequalities*

$$\lambda_r\left\{\frac{3}{4}hT_r\left(0,\frac{1}{2}\right] - M_r\left(0,\frac{1}{2}\right]\right\} + \frac{3}{4}hT_r(1,\infty) - M_r(1,\infty) \geq 0, \tag{4.75}$$

$$\lambda_r\left\{(1+\rho)hT_r\left(\rho,\frac{1}{2}\right] - M_r\left(\rho,\frac{1}{2}\right]\right\} + (1+\rho)hT_r(1,\infty) - M_r(1,\infty) \geq 0, \tag{4.76}$$

for all $\rho \in [0, \frac{1}{4}]$, and

$$(1 - \lambda_r) M_r \left(0, \frac{1}{2}\right] \geq \lambda_l M_l \left(0, \frac{1}{2}\right] + M_l(1, \infty), \qquad (4.77)$$

as well as their symmetrical relations hold.

Proof. Define:

$$k_1(h) := \frac{1 + 2h}{5 - 5h} \cdot \exp(-2 + 2h), \quad k_2(h) := \frac{8(1 + 2h)}{5(1 - 2h)} \cdot \frac{\exp(-2 + 2h)}{9 - 4h^2}.$$

Then define

$$\lambda_r := \max \left\{ k_1(h) \frac{M(0, \infty)}{M_r \left(0, \frac{1}{2}\right]}, \ k_2(h) \right\},$$

and λ_l by symmetry. Clearly, $\lambda_r > 0$ and $\lambda_l > 0$ and if (4.77) and its symmetrical relation are true, then it follows that $\lambda_r \leq 1$ and $\lambda_l \leq 1$.

By using the symmetry we can omit proof of the symmetrical relations of (4.75)–(4.77). If $x \geq 1 - h$, then $T_r(1, \infty) = 0 = M_r(1, \infty)$ and (4.75), (4.76) are obvious. Consider now that $x < 1 - h$. It is easy to see that $\frac{3}{4} h T_r \left(0, \frac{1}{2}\right] > M_r \left(0, \frac{1}{2}\right]$ and $(1 + \rho) h T_r \left(\rho, \frac{1}{2}\right] > M_r \left(\rho, \frac{1}{2}\right]$, for all $\rho \in \left[0, \frac{1}{4}\right]$. By applying Lemmas 4.2.2, 4.49 and then 4.2.4, one obtains

$$\frac{M_r(1, \infty) - \frac{3}{4} h T_r(1, \infty)}{\frac{3}{4} h T_r \left(0, \frac{1}{2}\right] - M_r \left(0, \frac{1}{2}\right]} \leq \frac{\left(1 - \frac{3}{5}\right) M_r(1, \infty)}{\left(\frac{3}{4} \cdot \frac{4}{1 + 2h} - 1\right) M_r \left(0, \frac{1}{2}\right]}$$

$$= \frac{1 + 2h}{5(1 - h)} \cdot \frac{M_r(1, \infty)}{M_r \left(0, \frac{1}{2}\right]} \leq \lambda_r.$$

Consequently, (4.75) is true. The relation (4.76), for $\rho = \frac{1}{4}$, follows immediately from Lemma 4.2.2. Suppose now that $\rho \in \left[0, \frac{1}{4}\right)$. We have

$$\frac{M_r(1, \infty) - (1 + \rho) h T_r(1, \infty)}{(1 + \rho) h T_r \left(\rho, \frac{1}{2}\right] - M_r \left(\rho, \frac{1}{2}\right]} \leq \frac{\left(1 - \frac{4}{5}(1 + \rho)\right) M_r(1, \infty)}{\left(\frac{4(1 + \rho)}{1 + 2\rho + 2h} - 1\right) M_r \left(\rho, \frac{1}{2}\right]}$$

$$= \frac{(1 - 4\rho)(1 + 2\rho + 2h)}{5(3 + 2\rho - 2h)} \cdot \frac{M_r(1, \infty)}{M_r(\rho, \infty)} \cdot \frac{1}{1 - M_r \left(\frac{1}{2}, \infty\right) / M_r(\rho, \infty)}$$

$$\leq \frac{(1 - 4\rho)(1 + 2\rho + 2h)}{5(3 + 2\rho - 2h)} \cdot \frac{\exp(-2 + 2\rho^2 + 2h + 2\rho h)}{1 - \exp\left(-\frac{1}{2} + 2\rho^2 + h + 2h\rho\right)}$$

$$\leq \exp(f(\rho)),$$

where

$$f(\rho) := \ln \frac{(1 - 4\rho)(1 + 2\rho + 2h)\exp(-2 + 2\rho^2 + 2h + 2\rho h)}{5(3 + 2\rho - 2h)\left(\frac{1}{2} - 2\rho^2 - h - 2\rho h\right)\left(\frac{3}{4} + \rho^2 + \frac{1}{2}h + \rho h\right)}.$$

The last inequality follows from the inequality $1 - \exp(-u) > u - \frac{1}{2}u^2$, $(u > 0)$, since $-\frac{1}{2} + 2\rho^2 + h + 2\rho h < 0$. We obtain successively

$$f'(\rho) = -\frac{4}{1 - 4\rho} + \frac{2}{1 + 2\rho + 2h} - \frac{2}{3 + 2\rho - 2h} + \frac{4\rho + 2h}{\frac{1}{2} - 2\rho^2 - h - 2\rho h}$$

$$-\frac{2\rho + h}{\frac{3}{4} + \rho^2 + \frac{1}{2}h + \rho h} + 4\rho + 2h$$

$$\leq -\frac{4}{1 - 4\rho} + \frac{2}{1 + 2h} - \frac{4}{7 - 4h} + \frac{4\rho + 2h}{\frac{1}{2} - 2\rho^2 - h - 2\rho h} + 4\rho + 2h$$

$$= \frac{-8\rho^2 + 4\rho + 6h - 2}{(1 - 4\rho)\left(\frac{1}{2} - 2\rho^2 - h - 2\rho h\right)} + \frac{10 - 2h + 20h^2 - 16h^3}{7 + 10h - 8h^2} + 4\rho$$

$$\leq \frac{-3 + 12h}{(1 - 4\rho)(1 - 2h)} + \frac{10}{7} + 4\rho.$$

If we denote by $g(\rho)$ the latter expression, we have

$$g'(\rho) = \frac{4(-3 + 12h)}{(1 - 4\rho)^2(1 - 2h)} + 4 < \frac{4(-3 + 12h)}{1 - 2h} + 4 < 0.$$

Consequently, $f'(\rho) \leq g(0) = \frac{-11 + 64h}{7(1 - 2h)} < 0$. Therefore $f(\rho) \leq f(0)$, $\rho \in \left[0, \frac{1}{4}\right)$ and hence

$$\frac{M_r(1, \infty) - (1 + \rho)hT_r(1, \infty)}{(1 + \rho)hT_r\left(\rho, \frac{1}{2}\right] - M_r\left(\rho, \frac{1}{2}\right]} \leq k_2(h) \leq \lambda_r.$$

Thus (4.75) and (4.76) are completely proved for $x \in (0, 1)$.

In order to prove (4.77), note that it can be rewritten in the form

$$\lambda_r M_r\left(0, \frac{1}{2}\right] + \lambda_l M_l\left(0, \frac{1}{2}\right] + M_l(1, \infty) + M_r\left(\frac{1}{2}, \infty\right) \leq M(0, \infty). \qquad (4.78)$$

In the proof of (4.78) we shall use the following immediate properties of the functions $k_1(h)$ and $k_2(h)$:

a) $k_1(h)$ and $k_2(h)$ are increasing.

b) $k_2(h)(1 - 2h)$ is increasing.

c) $k_2(h) < 1$, $(n \geq 60)$.

d) $k_2(h) < \exp\left(-\frac{3}{2} + 3h\right)$, $(n \geq 60)$.

Also, we shall use the inequality $1 - \exp(-u) < u$, for $u > 0$. For $\mu \geq 0$, $\gamma \geq 0$ let

$$E(\mu, \gamma) := (M(0, \infty))^{-1}\left\{\mu M_r\left(0, \frac{1}{2}\right] + \gamma M_l\left(0, \frac{1}{2}\right] + M_l(1, \infty) + M_r\left(\frac{1}{2}, \infty\right)\right\}.$$

By using Lemma 4.2.4 and the above relations one obtains the following estimates:

$$\text{i)} \quad E\left(k_1(h) \cdot \frac{M(0, \infty)}{M_r\left(0, \frac{1}{2}\right]}, \; k_1(h) \frac{M(0, \infty)}{M_l\left(0, \frac{1}{2}\right]}\right)$$

$$\leq 2k_1(h) + \exp(-2 + 2h) + \exp\left(-\frac{1}{2} + h\right)$$

$$=: T_1(n).$$

We have $T_1(n) \leq T_1(60) = 0.96656\ldots < 1$.

$$\text{ii)} \quad E\left(k_1(h) \cdot \frac{M(0, \infty)}{M_r\left(0, \frac{1}{2}\right]}, \; k_2(h)\right)$$

$$= k_1(h) + k_2(h) + \frac{M_l\left(\frac{1}{2}, \infty\right)}{M(0, \infty)}\left[-k_2(h) + \frac{M_l(1, \infty)}{M_l\left(\frac{1}{2}, \infty\right)}\right] + \frac{M_r\left(\frac{1}{2}, \infty\right)}{M(0, \infty)}$$

$$\leq k_1(h) + k_2(h) + \exp\left(-\frac{1}{2} + h\right)\left[-k_2(h) + \exp\left(-\frac{3}{2} + 3h\right) + 1\right]$$

$$\leq k_1(h) + k_2(h)\left(\frac{1}{2} - h\right) + \exp(-2 + 4h) + \exp\left(-\frac{1}{2} + h\right) =: T_2(n).$$

We have $T_2(n) \leq T_2(60) = 0.98729\ldots < 1$.

$$\text{iii)} \quad E\left(k_2(h), \; k_1(h) \cdot \frac{M(0, \infty)}{M_l\left(0, \frac{1}{2}\right]}\right)$$

$$= k_1(h) + k_2(h) + \frac{M_r\left(\frac{1}{2}, \infty\right)}{M(0, \infty)}(1 - k_2(h)) + \frac{M_l(1, \infty)}{M(0, \infty)}$$

$$\leq k_1(h) + k_2(h) + \exp\left(-\frac{1}{2} + h\right)(1 - k_2(h)) + \exp(-2 + 2h)$$

$$\leq k_1(h) + k_2(h)\left(\frac{1}{2} - h\right) + \exp\left(-\frac{1}{2} + h\right)$$

$$+ \exp(-2 + 2h) < T_2(n) < 1.$$

$$\text{iv)} E(k_2(h), \; k_2(h)) = 2k_2(h) + \frac{M_r\left(\frac{1}{2}, \infty\right)}{M(0, \infty)}(1 - k_2(h))$$

$$+ \frac{M_l\left(\frac{1}{2}, \infty\right)}{M(0, \infty)}\left(\frac{M_l(1, \infty)}{M_l\left(\frac{1}{2}, \infty\right)} - k_2(h)\right)$$

$$\leq 2k_2(h) + \exp\left(-\frac{1}{2} + h\right)(1 - k_2(h))$$

$$+ \exp\left(-\frac{1}{2} + h\right)\left[\exp\left(-\frac{3}{2} + 3h\right) - k_2(h)\right]$$

$$\leq 2k_2(h)\left(\frac{1}{2} - h\right) + \exp\left(-\frac{1}{2} + h\right) + \exp(-2 + 4h) =: T_3(n).$$

We have $T_3(n) \leq T_3(60) = 0.95641\ldots < 1$.

It follows from above that $E(\lambda_r, \lambda_l) < 1$, for $n \geq 60$, which implies (4.78). $\qquad\square$

Now we introduce new notation. If $x \in (0, 1)$ is fixed and $\alpha \geq 0$ is such that $x + \alpha h < 1$, define the functional $L_r(\alpha, \infty)$ on $B[0, 1]$ by:

$$L_r(\alpha, \infty) := \sum_{j=\tau(x+\alpha h)}^{n} p_{n,j}(x) f\left(\frac{j}{n}\right), \quad f \in B[0, 1], \qquad (4.79)$$

and then extend the definition by putting $L_r(\alpha, \infty) = 0$ in the case $x + \alpha h \geq 1$. Symmetrically, we define in the case $x > \alpha h$:

$$L_l(\alpha, \infty) := \sum_{j=0}^{\sigma(x-\alpha h)} p_{n,j}(x) f\left(\frac{j}{n}\right), \quad f \in B[0, 1], \qquad (4.80)$$

and in the case $x \leq \alpha h$, put $L_l(\alpha, \infty) = 0$. Also for $0 \leq \alpha < \beta$ define the functionals:

$$L_r(\alpha, \beta] := L_r(\alpha, \infty) - L_r(\beta, \infty), \quad L_l(\alpha, \beta] := L_l(\alpha, \infty) - L_l(\beta, \infty). \quad (4.81)$$

We give now the main result of this subsection.

Proposition 4.2.1. *The inequality (4.45) holds for any $n \geq 60$.*

Proof. Let $x \in (0, 1)$ and let $n \in \mathbb{N}$, $n \geq 60$ be fixed. Let λ_r and λ_l be a pair of numbers satisfying the relations given in Lemma 4.2.6. Define the following functionals on $B[0, 1]$:

$$F_r := \lambda_r L_r\left(0, \frac{1}{2}\right] + L_r(1, \infty) + \frac{\lambda_r M_r\left(0, \frac{1}{2}\right] + M_r(1, \infty)}{M_l\left(0, \frac{1}{2}\right]} L_l\left(0, \frac{1}{2}\right],$$

$$F_l := \lambda_l L_l\left(0, \frac{1}{2}\right] + L_l(1, \infty) + \frac{\lambda_l M_l\left(0, \frac{1}{2}\right] + M_l(1, \infty)}{M_r\left(0, \frac{1}{2}\right]} L_r\left(0, \frac{1}{2}\right],$$

$$F_m(\cdot) := B_n(\cdot, x) - F_r(\cdot) - F_l(\cdot).$$

The functionals F_r and F_l are positive and satisfy the conditions $F_r(e_1 - xe_0) = 0$, $F_l(e_1 - xe_0) = 0$. Consequently, we have also $F_m(e_1 - xe_0) = 0$. Let $i_1 := \overline{\tau}(x - h)$, if $x \geq h$ and $i_1 := 0$, if $x < h$ and let $i_2 := \overline{\sigma}(x + h)$, if $x \leq 1 - h$ and $i_2 := n$, if $x > 1 - h$. The functional F_m has the form

$$F_m(f) = \sum_{i=i_1}^{i_2} \mu_i \, p_{n,i}(x) \, f\left(\frac{i}{n}\right),$$

where

$$\mu_i := \begin{cases} 1 - \lambda_r - \dfrac{\lambda_l M_l\left(0, \frac{1}{2}\right] + M_l(1, \infty)}{M_r\left(0, \frac{1}{2}\right]}, & \text{if } \tau(x) \le i \le \overline{\sigma}\left(x + \frac{h}{2}\right), \\ 1, & \text{if } \tau\left(x + \frac{h}{2}\right) \le i \le i_2 \end{cases}$$

and, for $i \le \sigma(x)$, the coefficients μ_i are defined in a symmetrical manner. Also, if there exists $j \in \mathbb{N}$ such that $x = \frac{j}{n}$, then $\mu_j := 1$. From (4.77) and its symmetrical relation it follows that F_m is positive. Since $\frac{i_2 - i_1}{n} \le 2h$, we can apply Theorem 3.2.1 and therefore it follows that (3.22) holds for $F := F_m$.

Let us represent by $\{t_0 < t_1 < \cdots < t_m\}$ the set

$$\left\{ \frac{j}{n} \,\middle|\, j \in \mathbb{N} \text{ and } \frac{j}{n} \in \left[x - \frac{h}{2}, x\right) \cup \left(x, x + \frac{h}{2}\right] \cup (x + h, 1] \right\}.$$

The functional F_r admits a representation of the form

$$F_r(f) = \sum_{i=0}^{m} \gamma_i \, f(t_i), \quad f \in B[0, 1],$$

where $\gamma_i > 0$. Denote by $\tau^*(t), \overline{\tau}^*(t), \sigma^*(t), \overline{\sigma}^*(t)$, and $\pi^*(u, v)$, when $t, u, v \in (t_0, t_m)$, the indices defined as in (3.10),(3.11) and (3.12), corresponding to the sequence $t_0 < t_1 < \cdots < t_m$. We have $x \le t_0 + h$. In order to apply Theorem 3.2.1 to the functional F_r, it remains to prove

$$\sum_{i=\pi^*(u,v)}^{m} (v - t_i)\gamma_i \ge 0, \text{ for any } (u, v) \in M_{x,h}(t_0, t_1, \ldots, t_m).$$

Let $(u, v) \in M_{x,h}(t_0, t_1, \ldots, t_m)$. Denote by $ord^*(u, v)$ the order of the pair (u, v) corresponding to the sequence $t_0 < t_1 < \cdots < t_m$. If $ord^*(u, v) = 0$, i.e., $u = x$, we have

$$v > x + \frac{3}{4}h. \tag{4.82}$$

Indeed, let $t_i \in \left[x - \frac{h}{2}, x\right)$ be such that $v = \alpha(t_i, x, h)$ and let $s \in \mathbb{N}$ be such that $v = t_i + 2^s(x - t_i)$. If $x - t_i \le \frac{h}{4}$, since $v - t_i > h$, it follows that $v - x > \frac{3}{4}h$. If $\frac{h}{4} < x - t_i \le \frac{h}{2}$, it follows that $s = 2$ and hence $v - x = v - t_i - (x - t_i) = 3(x - t_i) > \frac{3}{4}h$. Consequently, by using Lemma 3.2.2 we obtain that the condition $v \le x + \frac{5}{4}h$ implies that $ord^*(u, v) = 0$.

We consider several cases.

Case a): $v \in (x, x + h]$. Then $u = x$ and $\pi^*(u, v) = \tau^*(x)$. Using (4.75) one obtains

$$\sum_{j=\pi^*(u,v)}^{m} (v - t_j)\gamma_j$$

$$= \lambda_r \left((v - x)T_r\left(0, \frac{1}{2}\right] - M_r\left(0, \frac{1}{2}\right] \right) + (v - x)T_r(1, \infty) - M_r(1, \infty)$$

$$\geq \lambda_r \left(\frac{3}{4}hT_r\left(0, \frac{1}{2}\right] - M_r\left(0, \frac{1}{2}\right] \right) + \frac{3}{4}hT_r(1, \infty) - M_r(1, \infty) \geq 0.$$

Case b): $v \in \left(x + h, x + \frac{5}{4}h\right]$. Let $v = x + (1 + \rho)h$, with $\rho \in \left(0, \frac{1}{4}\right]$. We have $u = x$ and $\pi^*(u, v) = \overline{\tau}^*(x + \rho h) \leq \tau^*(x + \rho h)$. Using (4.76) one gets

$$\sum_{j=\pi^*(u,v)}^{m} (v - t_j)\gamma_j \geq \sum_{j=\tau^*(x+\rho h)}^{m} (v - t_j)\gamma_j$$

$$= \lambda_r \left((1 + \rho)hT_r\left(\rho, \frac{1}{2}\right] - M_r\left(\rho, \frac{1}{2}\right] \right) + (1 + \rho)hT_r(1, \infty) - M_r(1, \infty) \geq 0.$$

Case c): $v \in \left(x + \frac{5}{4}h, x + \frac{3}{2}h\right]$. From Lemma 3.2.2 it follows that $u < x + h$. Hence $\pi^*(u, v) \leq \tau^*(x + h)$. An application of Lemma 4.2.2 yields

$$\sum_{j=\pi^*(u,v)}^{m} (v - t_j)\gamma_j \geq \sum_{j=\tau^*(x+h)}^{m} (v - t_j)\gamma_j = (v - x)T_r(1, \infty) - M_r(1, \infty)$$

$$\geq \frac{5}{4}hT_r(1, \infty) - M_r(1, \infty) \geq 0.$$

Case d): $v \in \left(x + \frac{3}{2}h, 1\right)$. Then, from Lemma 3.2.2 we have $\pi^*(u, v) \leq \tau^*\left(v - \frac{h}{2}\right)$. We write $v = x + \left(\frac{1}{2} + \rho\right)h$, where $\rho > 1$ is such that $x + \rho h < 1$. One obtains

$$\sum_{j=\pi^*(u,v)}^{m} (v - t_j)\gamma_j \geq \sum_{j=\tau^*\left(v-\frac{h}{2}\right)}^{m} (v - t_j)\gamma_j \geq \left(\frac{1}{2} + \rho\right)hT_r(\rho, \infty) - M_r(\rho, \infty).$$

From Lemma 4.2.2 it follows that $\frac{3}{2}c_n(1) > 1$, and then, using Lemma 4.2.5 it follows that $\left(\frac{1}{2} + \rho\right)hT_r(\rho, \infty) - M_r(\rho, \infty) > 0$.

Therefore we can apply Theorem 3.2.1 to the functional F_r and we obtain that (3.22) is satisfied for $F := F_r$. A similar condition holds for the functional F_l. Since $F_r(e_0) + F_l(e_0) + F_m(e_0) = 1$, we obtain (4.45). $\qquad\square$

4.2.3 Proof of the direct part of the theorem for $1 \leq n \leq 59$

In the sequel we use the notation of (4.47),(4.48),(4.62)–(4.68),(4.79)–(4.81) from the previous subsection.

Lemma 4.2.7. *Let* $5 \leq n \leq 59$ *and* $x \in (0, 1)$. *We have*

$$\left(\frac{3}{4} + \rho\right) h T_r(\rho, \infty) > M_r(\rho, \infty), \quad \text{if } \rho \geq 0 \text{ and } x + \rho h < 1, \qquad (4.83)$$

$$\left(\frac{1}{2} + \rho\right) h T_r(\rho, \infty) > M_r(\rho, \infty), \quad \text{if } \rho \geq \frac{3}{4} \text{ and } x + \rho h < 1. \qquad (4.84)$$

Proof. Using formula (4.52) we obtain that the sequences $c_n(0)$ and $c_n\left(\frac{3}{4}\right)$, $5 \leq n \leq 59$ are increasing and, moreover,

$$c_5(0) = 1.71155\ldots < c_6(0) = 1.78353\ldots < \cdots < c_{59}(0) = 2.25515\ldots$$

and

$$c_5\left(\frac{3}{4}\right) = 0.81484\ldots < c_6\left(\frac{3}{4}\right) = 0.83556\ldots < \cdots < c_{59}\left(\frac{3}{4}\right) = 0.97104\ldots.$$

Then, from Lemma 4.2.5 one obtains (4.83) and (4.84). $\qquad \square$

Let $n \in \mathbb{N}$ and $x \in (0, 1)$ be fixed. If $d > 0$ and $\rho \geq 0$ are such that $x + \rho h < 1$ and $(d + \rho) h T_r(\rho, \infty) - M_r(\rho, \infty) > 0$, denote

$$\Omega_r(d, \rho) := \frac{M_r(1, \infty) - (d + \rho) h T_r(1, \infty)}{(\rho + d) h T_r(\rho, \infty) - M_r(\rho, \infty)}. \qquad (4.85)$$

In a symmetrical manner we define $\Omega_l(d, \rho)$.

Based on Lemma 4.2.7 we can consider, for fixed $7 \leq n \leq 59$ and $x \in (0, 1)$, the number γ_r defined by:

$$\gamma_r := \begin{cases} \max\{\tilde{a}, \tilde{b}, \tilde{c}\}, & \text{if } x + h < 1, \\ 0, & \text{if } x + h \geq 1, \end{cases} \qquad (4.86)$$

where

$$\tilde{a}_r := \Omega_r\left(\frac{3}{4}, 0\right), \quad \tilde{b}_r := \sup_{\rho \in \left(0, \frac{1}{4}\right]} \Omega_r(1, \rho), \quad \tilde{c}_r := \sup_{\rho \in \left(\frac{3}{4}, 1\right]} \Omega_r\left(\frac{1}{2}, \rho\right).$$

$$(4.87)$$

Furthermore, we define the numbers $\tilde{a}_l, \tilde{b}_l, \tilde{c}_l, \gamma_l$ by symmetry.

One can see that, in the case $x + h < 1$ we have $\tilde{a} > 0$ and hence $\gamma_r > 0$. Symmetrically, we have $\gamma_l > 0$, if $x > h$.

Lemma 4.2.8. *Let* $7 \leq n \leq 59$ *and* $x \in (0, 1)$. *If the conditions*

$$\frac{1}{1 + \gamma_r} M_r(0, 1] + \frac{1}{1 + \gamma_l}\left(M_l\left(0, \frac{1}{2}\right] + M_l\left(\frac{3}{4}, 1\right]\right) \geq M(0, \infty), \qquad (4.88)$$

and

$$\frac{1}{1 + \gamma_l} M_l(0, 1] + \frac{1}{1 + \gamma_r}\left(M_r\left(0, \frac{1}{2}\right] + M_r\left(\frac{3}{4}, 1\right]\right) \geq M(0, \infty) \qquad (4.89)$$

are satisfied, then (4.45) holds.

Proof. Set $\lambda_r := \frac{\gamma_r}{1+\gamma_r}$, $\lambda_l := \frac{\gamma_l}{1+\gamma_l}$. Consider the functional

$$G_r := \beta_r \left(L_l\left(0, \frac{1}{2}\right] + L_l\left(\frac{3}{4}, 1\right] \right) + \lambda_r L_r(0, 1] + L_r(1, \infty),$$

where

$$\beta_r := \frac{\lambda_r M_r(0, 1] + M_r(1, \infty)}{M_l\left(0, \frac{1}{2}\right] + M_l\left(\frac{3}{4}, 1\right]}.$$

Symmetrically, one can define the functional G_l. Then define the functional G_m by

$$G_m(\cdot) := B_n(\cdot, x) - G_r(\cdot) - G_l(\cdot).$$

One obtains immediately that $G_r(e_1 - xe_0) = 0$ and $G_l(e_1 - xe_0) = 0$. Consequently, $G_m(e_1 - xe_0) = 0$.

Let $i_1 := \overline{\tau}(x - h)$ if $x \geq h$ and $i_1 := 0$ if $x < h$ and let $i_2 := \overline{\sigma}(x + h)$ if $x \leq 1 - h$ and $i_2 := n$ if $x > 1 - h$. The functional G_m is of the form

$$G_m(f) := \sum_{i=i_1}^{i_2} \mu_i \, p_{n,i}(x) f\left(\frac{i}{n}\right), \quad f \in B[0, 1],$$

where

$$\mu_i := \begin{cases} 1 - \lambda_r - \beta_l, & \text{if } \frac{i}{n} \in \left(x, x + \frac{h}{2}\right] \cup \left(x + \frac{3}{4}h, x + h\right], \\ 1 - \lambda_r, & \text{if } \frac{i}{n} \in \left(x + \frac{h}{2}, x + \frac{3}{4}h\right], \end{cases}$$

and μ_i are defined in a symmetrical manner for $\overline{\tau}(x - h) \leq i \leq \overline{\sigma}(x)$. Also if there exists $j \in \mathbb{N}$ such that $x = \frac{j}{n}$, then $\mu_j = 1$.

Conditions (4.88) and (4.89) imply that $\mu_i \geq 0$, for $\overline{\tau}(x - h) \leq i \leq \overline{\sigma}(x + h)$. Since $\frac{1}{n}(i_2 - i_1) \leq 2h$, we can apply Theorem 3.2.1 and obtain that (3.22) holds for $F := G_m$.

We denote by $\{t_0, t_1, \dots, t_m\}$ the set

$$\left\{ \frac{i}{n} \,\Big|\, i \in \mathbb{N} \text{ and } \frac{i}{n} \in \left[x - h, x - \frac{3}{4}h\right) \cup \left[x - \frac{h}{2}, x\right) \cup (x, 1] \right\}.$$

The functional G_r admits a representation of the form

$$G_r(f) := \sum_{i=0}^{m} \nu_i f(t_i), \quad f \in B[0, 1],$$

where $\nu_i > 0$.

We have $x \leq t_0 + h$. Denote by $\tau^\star(t)$, $\overline{\tau}^\star(t)$, $\sigma^\star(t)$, $\overline{\sigma}^\star(t)$, for $t \in (t_0, t_m)$ and by $\pi^\star(u, v)$, for $u, v - h \in (t_0, t_m)$ the indices defined in (3.10)–(3.12), corresponding to the sequence $t_0 < t_1 < \cdots < t_m$. In order to apply Theorem 3.2.1 to the functional G_r it remains to prove the corresponding condition to (3.21), namely

$$\sum_{j=\pi^\star(u,v)}^{m} (v - t_j)\nu_j \geq 0, \quad \text{for any } (u, v) \in M_{x,h}(t_0, t_1, \dots, t_m). \tag{4.90}$$

Let $(u, v) \in M_{x,h}(t_0, t_1, \ldots, t_m)$. Denote by $ord^\star(u, v)$ the order of the pair (u, v) corresponding to the sequence $t_0 < t_1 < \cdots < t_m$. If $ord^\star(u, v) = 0$, that is, $u = x$, then we have

$$v > x + \frac{3}{4}h. \tag{4.91}$$

Indeed, let $t_i \in \left[x - h, x - \frac{3}{4}h\right) \cup \left[x - \frac{h}{2}, x\right)$ be such that $v = \alpha(t_i, x, h)$. If $t_i \in \left[x - \frac{h}{2}, x\right)$, this inequality was proved in Proposition 4.2.1, see (4.82). If $t_i \in \left[x - h, x - \frac{3}{4}h\right)$, then $v - x = x - t_i > \frac{3}{4}h$. '

By using Lemma 3.2.2 one obtains that if $v \leq x + \frac{5}{4}h$, then $ord^\star(u, v) = 0$. Consider the following cases.

Case a): $v \in (x, x + h]$. Then $u = x$ and $\pi^\star(u, v) = \tau^\star(x)$. From (4.83) for $\rho = 0$ and the inequality $\frac{3}{4}hT_r(1, \infty) \leq M_r(1, \infty)$ we infer $\frac{3}{4}hT_r(0, 1] > M_r(0, 1]$. Then, using (4.91) and (4.87), one obtains

$$\sum_{j=\pi^\star(u,v)}^{m} (v - t_j)v_j = \lambda_r((v - x)T_r(0, 1] - M_r(0, 1])$$

$$+(v - x)T_r(1, \infty) - M_r(1, \infty)$$

$$\geq \frac{\tilde{a}_r}{1 + \tilde{a}_r}\left(\frac{3}{4}hT_r(0, 1] - M_r(0, 1]\right)$$

$$+\frac{3}{4}hT_r(1, \infty) - M_r(1, \infty) = 0.$$

Case b): $v \in \left(x + h, x + \frac{5}{4}h\right]$. Let $v = x + (1 + \rho)h$, $\rho \in \left(0, \frac{1}{4}\right]$. We have $u = x$, $\pi^\star(u, v) = \overline{\tau}^\star(x + \rho h) \leq \tau^\star(x + \rho h)$ and $x < 1 - h$. Since $M_r(\rho, 1] < (1 + \rho)hT_r(\rho, 1]$, we have $\Omega_r(1, \rho) > -1$, $\rho \in \left(0, \frac{1}{4}\right]$. We get successively

$$\sum_{j=\pi^\star(u,v)}^{m} (v - t_j)v_j \geq \sum_{j=\tau^\star(x+\rho h)}^{m} (v - t_j)v_j$$

$$= \lambda_r((1 + \rho)hT_r(\rho, 1] - M_r(\rho, 1]) + (1 + \rho)hT_r(1, \infty) - M_r(1, \infty)$$

$$\geq \frac{\Omega_r(1, \rho)}{1 + \Omega_r(1, \rho)}((1 + \rho)hT_r(\rho, 1] - M_r(\rho, 1])$$

$$+(1 + \rho)hT_r(1, \infty) - M_r(1, \infty) = 0.$$

Case c): $v \in \left(x + \frac{5}{4}h, x + \frac{3}{2}h\right]$. Let $v = x + \left(\frac{1}{2} + \rho\right)h$, with $\rho \in \left(\frac{3}{4}, 1\right]$. We have $x < 1 - h$. Using Lemma 3.2.2, it follows that $\pi^\star(u, v) \leq \tau^\star\left(v - \frac{h}{2}\right)$. We have

$$\sum_{j=\pi^\star(u,v)}^{m} (v - t_j)v_j \geq \sum_{j=\tau^\star\left(v-\frac{h}{2}\right)}^{m} (v - t_j)v_j$$

$$= \lambda_r\left(\left(\frac{1}{2} + \rho\right)hT_r(\rho, 1] - M_r(\rho, 1]\right)$$

$$+ \left(\frac{1}{2} + \rho\right) hT_r(1, \infty) - M_r(1, \infty) =: U.$$

Note that $\left(\frac{1}{2} + \rho\right) hT_r(\rho, 1] \geq M_r(\rho, 1]$. If $\left(\frac{1}{2} + \rho\right) hT_r(1, \infty) \geq M_r(1, \infty)$, then obviously $U \geq 0$. Otherwise it follows that $\Omega_r\left(\frac{1}{2}, 0\right) > 0$ and hence we have

$$U \geq \frac{\Omega_r\left(\frac{1}{2}, \rho\right)}{1 + \Omega_r\left(\frac{1}{2}, \rho\right)} \left(\left(\frac{1}{2} + \rho\right) hT_r(\rho, 1] - M_r(\rho, 1]\right)$$

$$+ \left(\frac{1}{2} + \rho\right) hT_r(1, \infty) - M_r(1, \infty) = 0.$$

Case d): $v \in \left(x + \frac{3}{2}h, 1\right)$. From Lemma 3.2.2 we have $\pi^*(u, v) \leq \tau^*\left(v - \frac{h}{2}\right)$. We write $v = x + \left(\frac{1}{2} + \rho\right)h, \ \rho > 1$ with $x + \rho h < 1$. From (4.84) one obtains

$$\sum_{j=\pi^*(u,v)}^{m} (v - t_j)v_j \geq \sum_{j=\tau^*\left(v-\frac{h}{2}\right)}^{m} (v - t_j)v_j$$

$$= \left(\frac{1}{2} + \rho\right) hT_r(\rho, \infty) - M_r(\rho, \infty) > 0.$$

Thus, relation (4.90) is valid in all cases and from Theorem 3.2.1 we derive that (3.22) holds for $F := G_r$. A similar relation holds for G_l. Since $G_r(e_0) + G_l(e_0) + G_m(e_0) = 1$, (4.45) follows. $\qquad \square$

Lemma 4.2.9. *Let $n \in \mathbb{N}$, $0 < x_1 < x_2 < 1$, $d > 0$ and $0 < \rho_2 < \rho_1$ be such that:*
i) $x_1 + \rho_1 h = x_2 + \rho_2 h < 1$,
ii) $\tau(x_1 + h) = \tau(x_2 + h)$,
iii) $((d + \rho_i)hT_r(\rho_i, \infty) - M_r(\rho_i, \infty))(x_i) > 0, \ i = 1, 2$.
Then

$$(\Omega_r(d, \rho_1))(x_1) \leq (\Omega_r(d, \rho_2))(x_2). \tag{4.92}$$

Proof. Let $T \geq (\Omega_r(d, \rho_2))(x_2)$, be arbitrarily chosen. Denote $v := x_1 + (d + \rho_1)h$ and

$$U_1 := (1 - x_1)^{n-\tau(v)} x_1^{\tau(v)}, \quad U := \left(\frac{1 - x_1}{1 - x_2}\right)^{n-\tau(v)} \left(\frac{x_1}{x_2}\right)^{\tau(v)},$$

$$P(t) := \sum_{j=\tau(v-dh)}^{n} \left(v - \frac{j}{n}\right) \binom{n}{j} \left(\frac{t}{1-t}\right)^{j-\tau(v)}, \quad (t \in (0, 1)),$$

$$Q(t) := \sum_{j=\tau(x_1+h)}^{n} \left(v - \frac{j}{n}\right) \binom{n}{j} \left(\frac{t}{1-t}\right)^{j-\tau(v)}, \quad (t \in (0, 1)).$$

We write successively

$$
(T \cdot [(d + \rho_1)hT_r(\rho_1, \infty) - M_r(\rho_1, \infty)] + (d + \rho_1)hT_r(1, \infty) - M_r(1, \infty)) (x_1)
$$

$$
= T \sum_{j=\tau(v-dh)}^{n} \left(v - \frac{j}{n} \right) p_{n,j}(x_1) + \sum_{j=\tau(x_1+h)}^{n} \left(v - \frac{j}{n} \right) p_{n,j}(x_1)
$$

$$
= U_1 \cdot \{T \cdot P(x_1) + Q(x_1)\} \geq U_1 \cdot \{T \cdot P(x_2) + Q(x_2)\}
$$

$$
= U \left\{ T \sum_{j=\tau(v-dh)}^{n} \left(v - \frac{j}{n} \right) p_{n,j}(x_2) + \sum_{j=\tau(x_1+h)}^{n} \left(v - \frac{j}{n} \right) p_{n,j}(x_2) \right\}
$$

$$
= U (T \cdot [(d + \rho_2)hT_r(\rho_2, \infty) - M_r(\rho_2, \infty)]
$$

$$
+ (d + \rho_2)hT_r(1, \infty) - M_r(1, \infty)) (x_2) \geq 0.
$$

Consequently, $T \geq (\Omega_r(d, \rho_1))(x_1)$. Since $T \geq (\Omega_r(d, \rho_1))(x_2)$ was arbitrarily chosen, (4.92) follows.   □

For $x \in (0, 1)$ and $1 \leq l \leq k$ denote

$$
\eta_{n,l,k}(x) := \prod_{i=l}^{k-1} \left(\frac{n - i}{i} \cdot \frac{x}{1 - x} \right). \tag{4.93}
$$

Here $\eta_{n,l,k}(x) = 0$, if $k > n$ and also $\eta_{n,l,k}(x) = 1$ if $l = k$. From (4.49) it follows that, if $l = \tau(x + \alpha h)$ and $k = \tau(x + \beta h)$, $0 \leq \alpha \leq \beta$, $x + \alpha h < 1$, then

$$
\left(\frac{M_r(\beta, \infty)}{M_r(\alpha, \infty)} \right) (x) = \eta_{n,l,k}(x). \tag{4.94}
$$

Let now $n \in \mathbb{N}$, $7 \leq n \leq 59$ and $I \subset (0, 1)$ be an interval with the end points a and b, $0 < a \leq b < 1$, such that $b - a \leq \frac{1}{2n}$. The interval I may be an open, a closed, or a half-interval. Suppose also, that there exist $r, s, t, v \in \mathbb{N}$ such that we have $r = \tau(x)$, $s = \tau \left(x + \frac{h}{2} \right)$, $t = \tau \left(x + \frac{3}{4}h \right)$, $v = \tau(x + h)$, for all $x \in I$. Note that, even in the case $a \notin I$, we have $r = \tau(a)$, $s = \tau \left(a + \frac{h}{2} \right)$, $t = \tau \left(a + \frac{3}{4}h \right)$, $v = \tau(a + h)$.

Let

$$
M_1 := 1 - \eta_{n,r,v}(b), \quad M_2 := 1 - \eta_{n,r,s}(b) + \eta_{n,r,t}(a) - \eta_{n,r,v}(a). \tag{4.95}
$$

Now we consider the number

$$
\Gamma := \begin{cases} 0, & \text{if } v > n, \\ \max\{A, B, C\}, & \text{if } v \leq n \end{cases}
$$

where A, B, C are defined for $v \leq n$ as follows. Using (4.83) for $\rho = 0$, we can define

$$
A := \eta_{n,r,v}(b) \frac{1 - \frac{3}{4}h\Theta_{n,v}(a)}{\frac{3}{4}h\Theta_{n,r}(a) - 1}. \tag{4.96}
$$

Let

$$v_k := \begin{cases} h, & \text{if } k = r, \\ \frac{k-1}{n} + h - b, & \text{if } r < k \le s. \end{cases} \tag{4.97}$$

Applying (4.83) for $\rho = 0$ it follows immediately that $v_r \Theta_{n,r}(a) > 1$. Now let $r < k \le s$. If we take $\rho := \left(\frac{k-1}{n} - a\right) h^{-1} > 0$, then $\tau(a + \rho h) = k$ and we obtain from (4.83) that $\left(\frac{3}{4} + \rho\right) h \Theta_{n,k}(a) > 1$. Since $n \ge 7$ and $b - a \le \frac{1}{2n}$ we obtain $v_k \ge \left(\frac{3}{4} + \rho\right) h$. Therefore we can define

$$B := \max_{r \le k \le s} \eta_{n,k,v}(b) \frac{1 - v_k \Theta_{n,v}(a)}{v_k \Theta_{n,k}(a) - 1}. \tag{4.98}$$

Since $n \le 59$ we have $v \le t + 2$. We show that the following numbers are well defined:

$$P_t := \eta_{n,t,v}(b) \frac{1 - \frac{5}{4} h \Theta_{n,v}(a)}{\frac{5}{4} h \Theta_{n,t}(a) - 1},$$

$$P_{t+1} := \eta_{n,t+1,v}(b) \frac{1 - \left(\frac{t}{n} - b + \frac{h}{2}\right) \Theta_{n,v}(b)}{\left(\frac{t}{n} - b + \frac{h}{2}\right) \Theta_{n,t+1}(b) - 1}.$$

Indeed, applying (4.84) for $\rho = \frac{3}{4}$ we obtain immediately that $\frac{5}{4} h \Theta_{n,t}(a) > 1$. One obtains $\tau(b + \rho h) = t + 1$, for $\rho := \left(\frac{t}{n} - b\right) h^{-1} \ge \frac{3}{4}$, and $\left(\frac{t}{n} - b + \frac{h}{2}\right) \Theta_{n,t+1}(b) > 1$, from (4.84).

Set $P_{t-1} := 0$ and then put

$$C := \max_{t-1 \le k \le v-1} P_k. \tag{4.99}$$

Note that $\Gamma \ge 0$, since $C \ge 0$. Finally, define

$$E_1(a, b, n, r, s, t, v) := \frac{1}{1 + \Gamma} M_1, \quad E_2(a, b, n, r, s, t, v) := \frac{1}{1 + \Gamma} M_2. \tag{4.100}$$

Lemma 4.2.10. *Let $n \in \mathbb{N}$, $7 \le n \le 59$ and let $I \subset (0, 1)$ be an interval as above. We have*

$$\left(\frac{1}{1 + \gamma_r} \cdot \frac{M_r(0, 1]}{M(0, \infty)}\right)(x) \ge E_1(a, b, n, r, s, t, v), \text{ for all } x \in I, \tag{4.101}$$

and

$$\left(\frac{1}{1 + \gamma_r} \cdot \frac{M_r\left(0, \frac{1}{2}\right] + M_r\left(\frac{3}{4}, 1\right]}{M(0, \infty)}\right)(x) \ge E_2(a, b, n, r, s, t, v), \text{ for all } x \in I. \tag{4.102}$$

Proof. Note that the functions $\Theta_{n,k}$, $(1 \leq k \leq n)$ and $\eta_{n,l,k}$, $(1 \leq l \leq k \leq n)$ are increasing on $(0, 1)$. Let $x \in I$.

We have $\left(\frac{M_r(0,1]}{M(0,\infty)}\right)(x) = 1 - \eta_{n,r,v}(x) \geq M_1$. Also $\left(\frac{M_r\left(0,\frac{1}{2}\right]}{M(0,\infty)}\right)(x) =$

$1 - \eta_{n,r,s}(x) \geq 1 - \eta_{n,r,s}(b)$. We show that the function $x \to \left(\frac{M_r\left(\frac{3}{4},1\right]}{M(0,\infty)}\right)(x)$ is

increasing on I, in the case $t \leq \min\{v - 1, n\}$. Indeed, it suffices to prove that the

functions $x \to \frac{p_{n,k}(x)\left(\frac{k}{n}-x\right)}{(M(0,\infty))(x)}$, for $t \leq k \leq \min\{v-1, n\}$ are increasing. We have

$$\frac{d}{dx}\left(\frac{p_{n,k}(x)\left(\frac{k}{n} - x\right)}{(M(0,\infty))(x)}\right) = \frac{d}{dx}\left(\frac{n(r-1)!(n-r)!}{k!(n-k)!} \cdot x^{k-r}(1-x)^{r-k-1}\left(\frac{k}{n} - x\right)\right)$$

$$= \frac{n(r-1)!(n-r)!}{k!(n-k)!} \cdot x^{k-r-1}(1-x)^{r-k-2}\left[\left(\frac{k}{n} - x\right)(k - r + x) - x(1 - x)\right].$$

We have $\left(\frac{k}{n} - x\right)(k - r + x) - x(1 - x) \geq \frac{3}{4}h\left(\frac{3}{4}\sqrt{n} - 1 + x\right) - \frac{1}{4} \geq \frac{5}{16} - \frac{3}{4}h \geq$

$\frac{5}{16} - \frac{3}{4} \cdot \frac{1}{\sqrt{7}} > 0$. Consequently, $\left(\frac{M_r\left(0,\frac{1}{2}\right]+M_r\left(\frac{3}{4},1\right]}{M(0,\infty)}\right)(x) \geq M_2$.

It remains to show that $(\gamma_r)(x) \leq \Gamma$. If $v > n$, then $(\gamma_r)(x) = 0 = \Gamma$. Then consider that $v \leq n$. Note that, if $d > 0, \rho > 0, x \in (0, 1)$, $x + \rho h < 1$ and $k = \tau(x + \rho h)$, then

$$(\Omega_r(d, \rho))(x) = \eta_{n,k,v}(x)\frac{1 - (d + \rho)h\Theta_{n,v}(x)}{(d + \rho)h\Theta_{n,k}(x) - 1}. \tag{4.103}$$

From (4.85) and (4.103) we get immediately that $(\tilde{a}_r)(x) \leq A$.

Let $\rho \in \left(0, \frac{1}{4}\right]$ and take $k = \tau(x + \rho h)h$. We have $r \leq k \leq s$. By considering separately the cases $k = r$ and $r < k \leq s$ we get $(1 + \rho)h \geq v_k$. It follows from (4.85) and (4.103) that $(\Omega_r(1, \rho))(x) \leq B$. Consequently $(\tilde{b}_r)(x) \leq B$.

Let $\rho \in \left(\frac{3}{4}, 1\right]$ and set $k = \tau(x + \rho h)$. If $k = v$ we have $\left(\Omega_r\left(\frac{1}{2}, \rho\right)\right)(x) = -1$. Hence, if $t = v$ the inequality $(\tilde{c}_r)(x) \leq C$ is true. Let now $t < v$ and $t \leq k < v$. Since $n \leq 59$ it follows that $k \in \{t, t + 1\}$ In the case $k = t$ it follows immediately from (4.85) and (4.103) that $\left(\Omega_r\left(\frac{1}{2}, \rho\right)\right)(x) \leq P_t$. Consider now the case $k = t+1$. Choose an arbitrary number $y \in I$, $x \leq y$. We have $\tau(x+h) = \tau(y+h) = v$. Choose $\rho' > 0$ such that $x + \rho h = y + \rho'h$. Using Lemma 4.2.9 we get $\left(\Omega_r\left(\frac{1}{2}, \rho\right)\right)(x) \leq$

$\left(\Omega_r\left(\frac{1}{2}, \rho'\right)\right)(y)$. Since $\rho'h \geq \frac{t}{n} - y$, we have

$$\left(\Omega_r\left(\frac{1}{2}, \rho'\right)\right)(y) \leq \eta_{n,t+1,v}(y)\frac{1 - \left(\frac{t}{n} - y + \frac{h}{2}\right)\Theta_{n,v}(y)}{\left(\frac{t}{n} - y + \frac{t}{n}\right)\Theta_{n,t+1}(y) - 1}.$$

By passing to limit $y \to b - 0$ in the last term, we get $\left(\Omega_r\left(\frac{1}{2}, x\right)\right)(x) \leq P_{t+1}$. Consequently, $(\tilde{c}_r)(x) \leq C$. Finally we obtain $(\gamma_r)(x) \leq \Gamma$. $\square$

Proposition 4.2.2. *The inequality (4.45) holds for* $7 \leq n \leq 59$.

Proof. Let $7 \leq n \leq 59$ be fixed. The proof is based on Lemmas 4.2.8 and 4.2.10. We distinguish between two cases.

Case 1: $x \in \left(0, \frac{1}{n}\right] \cup \left[1 - \frac{1}{n}, 1\right)$. Using the symmetry we can consider only the case when $x \in \left(0, \frac{1}{n}\right]$. Then we have $M(0, \infty) = M_l(0, 1] = M_l\left(0, \frac{1}{2}\right]$. It follows that $\gamma_l = 0$. Therefore, conditions (4.88) and (4.89) are obvious and we can apply Lemma 4.2.8 .

Case 2: $x \in \left(\frac{1}{n}, 1 - \frac{1}{n}\right)$. From (4.62)–(4.68) we obtain

$$\left(\frac{1}{1 + \gamma_l}\left(M_l\left(0, \frac{1}{2}\right] + M_l\left(\frac{3}{4}, 1\right]\right)\right)(x)$$

$$= \left(\frac{1}{1 + \gamma_r}\left(M_r\left(0, \frac{1}{2}\right] + M_r\left(\frac{3}{4}, 1\right]\right)\right)(1 - x),$$

$$\left(\frac{1}{1 + \gamma_l}M_l(0, 1]\right)(x) = \left(\frac{1}{1 + \gamma_r}M_r(0, 1]\right)(1 - x),$$

and

$$(M(0, \infty))(x) = (M(0, \infty))(1 - x).$$

Therefore, in order to prove (4.45) for any $x \in \left(\frac{1}{n}, 1 - \frac{1}{n}\right)$ it is sufficient to show that we have

$$\left(\frac{1}{1 + \gamma_r} \cdot \frac{M_r(0, 1]}{M(0, \infty)}\right)(x) + \left(\frac{1}{1 + \gamma_r} \cdot \frac{M_r\left(0, \frac{1}{2}\right] + M_r\left(\frac{3}{4}, 1\right]}{M(0, \infty)}\right)(1 - x) \geq 1$$

$$(4.104)$$

for any $x \in \left(\frac{1}{n}, \frac{n-1}{n}\right)$.

Definition 4.2.1. *We call an admissible interval, any interval* $I \subset \left(\frac{1}{n}, \frac{n-1}{n}\right)$ *with* $length(I) \leq \frac{1}{2n}$ *and with the property that there exist the indices* $r_r, s_r, t_r, v_r, r_l,$ $s_l, t_l, v_l \in \mathbb{N}$ *such that we have:*

$$r_r := \tau(x), \quad s_r := \tau\left(x + \frac{h}{2}\right), \quad t_r := \tau\left(x + \frac{3}{4}h\right), \quad v_r := \tau(x + h),$$

$$(4.105)$$

$$r_l := \tau(1 - x), \quad s_l := \tau\left(1 - x + \frac{h}{2}\right),$$

$$t_l := \tau\left(1 - x + \frac{3}{4}h\right), \quad v_l := \tau(1 - x + h)$$

$$(4.106)$$

for any $x \in I$.

For an interval I as above set

$$\Lambda(I) := (r_r, s_r, t_r, v_r; r_l, s_l, t_l, v_l). \tag{4.107}$$

If $I = \{c\}$, denote $\Lambda(c)$ instead of $\Lambda(\{c\})$. Note that if I is an admissible interval, then its symmetrical interval, namely $I^- := \{1 - x | x \in I\}$, is also symmetrical and $\Lambda(I^-) = (r_l, s_l, t_l, v_l; r_r, s_r, t_r, v_r)$. If I is an admissible interval as above, having the end points $\alpha \le \beta$, and $7 \le n \le 59$, let us denote, (see (4.100))

$$E(I, n) := E_1(\alpha, \beta, n, r_r, s_r, t_r, v_r) + E_2(1 - \beta, 1 - \alpha, n, r_l, s_l, t_l, v_l). \tag{4.108}$$

By taking into account Lemma 4.2.10, in order to prove (4.104) for any x belonging to an admissible interval I, it suffices to show that

$$E(I, n) \ge 1. \tag{4.109}$$

For any $t \in \mathbb{R}$ denote by $\text{INT}(t) \in \mathbb{Z}$ and $\text{FRAC}(t) \in [0, 1)$, the unique number such that $t = \text{INT}(t) + \text{FRAC}(t)$. Denote by $0 = \delta_{n,1} < \delta_{n,2} < \cdots \delta_{n,L_n} = 1$, the elements of the set $\{0, \text{FRAC}(\sqrt{n}), \text{FRAC}\left(\frac{1}{2}\sqrt{n}\right), \text{FRAC}\left(\frac{3}{4}\sqrt{n}\right), 1 - \text{FRAC}(\sqrt{n}), 1 - \text{FRAC}\left(\frac{1}{2}\sqrt{n}\right), 1 - \text{FRAC}\left(\frac{3}{4}\sqrt{n}\right), 1\}$, written in increasing order. Set

$$I_{k,j,i}^n := \left(\frac{1}{n}\left(k + \frac{3-i}{2}\delta_{n,j} + \frac{i-1}{2}\delta_{n,j+1}\right), \frac{1}{n}\left(k + \frac{2-i}{2}\delta_{n,j} + \frac{i}{2}\delta_{n,j+1}\right)\right),$$

for $1 \le k \le n - 2$, $1 \le j \le L_n - 1$, $1 \le i \le 2$. Note that the intervals $I_{k,j,i}^n$ are admissible. Denote

$$e_n := \min\{E\left(I_{k,j,i}^n, n\right) | 1 \le k \le n - 2, \ 1 \le j \le L_n - 1, \ 1 \le i \le 2\}. \tag{4.110}$$

Consider a subroutine that computes $E_1(a, b, n, r, s, t, v)$ and $E_2(a, b, n, r, s, t, v)$ by using the formulae (4.48) and (4.93)–(4.100), where $0 < a \le b < 1$ and $r = \tau(x)$, $s = \tau\left(x + \frac{h}{2}\right)$, $t = \tau\left(x + \frac{3}{4}h\right)$, $v = \tau(x + h)$ for all $x \in [a, b]$. With this subroutine we can compute $E\left(I_{k,j,i}^n, n\right)$ by taking in (4.108) α and β to be the end points of $I_{k,j,i}^n$ and the elements of $\Lambda\left(I_{k,j,i}^n\right)$ given by (4.105) and (4.106). Note that for $u > 0$ we have $\tau(u) = \text{INT}(n \cdot u) + 1$. In order to eliminate the possible rounding errors in computing the values of $\text{INT}(\cdot)$, we take $x = \frac{\alpha+\beta}{2}$ in (4.105) and (4.106). One obtains the numerical result

$$\min_{7 \le n \le 59} e_n = e_7 = 1.04748\ldots. \tag{4.111}$$

Thus

$$E(I_{k,j,i}^n, n) > 1, \quad \text{for } 7 \le n \le 59, \ 1 \le k \le n - 2, \ 1 \le j \le L_n - 1, \ i = 1, 2. \tag{4.112}$$

It remains to verify the relation (4.104) at the end points of the intervals $I_{k,j,i}^n$, excepting the points $\frac{1}{n}$ and $\frac{n-1}{n}$. Let c be such a point. Then there exist the intervals (α, c), (c, β) belonging to the family $\{I_{k,j,i}^n, \ 1 \le k \le n - 2, \ 1 \le j \le L_n - 1, \ i = 1, 2\}$. We have to distinguish among the following cases:

Case 1. A first case appears when the vector $\Lambda((\alpha, c))$ is equal to $\Lambda((c, \beta))$, or differs from it by a single component. From this condition we obtain that $\Lambda(c) = \Lambda((\alpha, c))$ or $\Lambda(c) = \Lambda((c, \beta))$. Therefore, either $(\alpha, c]$ or $[c, \beta)$ is an admissible interval. If $(\alpha, c]$ is admissible, we have $E((\alpha, c], n) = E((\alpha, c), n)$ and if $[c, \beta)$ is admissible, we have $E([c, \beta), n) = E((c, \beta), n)$. From (4.112) it follows then that (4.104) holds for $x = c$.

Case 2. The second case appears when the vector $\Lambda((\alpha, c))$ differs from the vector $\Lambda((c, \beta))$ by at least two components. In this case we shall compute

$$E(c, n) := F(c, n; r_r, s_r, t_r, v_r; r_l, s_l, t_l, v_l), \tag{4.113}$$

where

$$F(c, n; r_r, s_r, t_r, v_r; r_l, s_l, t_l, v_l) := E_1(c, c, n, r_r, s_r, t_r, v_r)$$
$$+ E_2(1 - c, 1 - c, n, r_l, s_l, t_l, v_l),$$

and the elements of $\Lambda(c)$ are given in (4.105),(4.106) for $x = c$. We use the formula $\tau(u) = \text{INT}(n \cdot u) + 1$ for $u > 0$. However, for those indices that have different values on the intervals (α, c), (c, β), in order to eliminate the rounding errors in numerical computation of the values of $\text{INT}(\cdot)$, we give their explicit values. We have several subcases.

Subcase 2.1: n is not a square number. Then $L_n = 8$. The condition given for c in Case 2 may appear only if c is of the form $c = \frac{k}{n}$, $2 \le k \le n - 2$ when only the indices r_r and r_l are different in $\Lambda((\alpha, c))$ and $\Lambda((c, \beta))$. We compute $E\left(\frac{k}{n}, n\right)$ by using (4.113), for $c := \frac{k}{n}$, $r_r := k + 1$, $r_l := n - k + 1$ and the other indices of $\Lambda(c)$ are defined as in (4.105), (4.106) for $x = c$. Denote

$$e_n^\star := \min_{2 \le k \le n-2} E\left(\frac{k}{n}, n\right). \tag{4.114}$$

We obtain the numerical result

$$\min_{\substack{7 \le n \le 59 \\ n \ne p^2, \, p \in \mathbb{N}}} e_n^\star = e_8^\star = 1.03850\dots. \tag{4.115}$$

Subcase 2.2: $n \in \{9, 25, 49\}$. Then $L_n = 5$ and $\delta_{n,1} = 0$, $\delta_{n,2} = \frac{1}{4}$, $\delta_{n,3} = \frac{1}{2}$, $\delta_{n,4} = \frac{3}{4}$, $\delta_{n,5} = 1$. The indices r_r, r_l, v_r, v_l change their values at the points of the form $\frac{k}{n}$, the indices s_r, s_l change their values at the points of the form $\frac{1}{n}\left(k + \frac{1}{2}\right)$ and the indices t_r, t_l change their values at the points of the form $\frac{1}{n}\left(k + \frac{1}{4}\right)$ and $\frac{1}{n}\left(k + \frac{3}{4}\right)$. Thus we have to consider only the points of the form $\frac{k}{n}$, $(2 \le k \le n - 2)$ and $\frac{2k+1}{2n}$, $(1 \le k \le n - 2)$. For such points we have

$$E\left(\tfrac{k}{9}, 9\right)$$
$$= F\left(9, \tfrac{k}{9}; k+1, k+2, k+3, k+4; 10-k, 11-k, 12-k, 13-k\right),$$
$$E\left(\tfrac{k}{25}, 25\right)$$
$$= F\left(25, \tfrac{k}{25}; k+1, k+3, k+4, k+6; 26-k, 28-k, 29-k, 31-k\right),$$
$$E\left(\tfrac{k}{49}, 49\right)$$
$$= F\left(49, \tfrac{k}{49}; k+1, k+4, k+6, k+8; 50-k, 53-k, 55-k, 57-k\right).$$

Using the notation (4.114) we have: $e_9^\star = 1.38808\ldots$, $e_{25}^\star = 1.50662\ldots$, $e_{49}^\star = 1.40102\ldots$. Also we have

$$E\left(\tfrac{2k+1}{18}, 9\right)$$
$$= F\left(9, \tfrac{2k+1}{18}; k+1, k+3, k+3, k+4; 9-k, 11-k, 11-k, 12-k\right),$$
$$E\left(\tfrac{2k+1}{50}, 25\right)$$
$$= F\left(25, \tfrac{2k+1}{50}; k+1, k+4, k+5, k+6; 25-k, 28-k, 29-k, 30-k\right),$$
$$E\left(\tfrac{2k+1}{98}, 49\right)$$
$$= F\left(49, \tfrac{2k+1}{98}; k+1, k+5, k+6, k+8; 49-k, 53-k, 54-k, 56-k\right).$$

Denoting

$$e_n^{\star\star} := \min_{1 \le k \le n-2} E\left(\frac{2k+1}{2n}, n\right), \tag{4.116}$$

we obtain: $e_9^{\star\star} = 1.68430\ldots$, $e_{25}^{\star\star} = 1.45250\ldots$, $e_{49}^{\star\star} = 1.49236\ldots$.

Subcase 2.3: $n = 36$. We have $L_{36} = 3$ and $\delta_{36,1} = 0$, $\delta_{36,2} = \frac{1}{2}$, $\delta_{36,3} = 1$. We have to consider the points of the form $\frac{k}{36}$, $(2 \le k \le 34)$ and $\frac{2k+1}{72}$, $(1 \le k \le 34)$. For such points we have

$$E\left(\tfrac{k}{36}, 36\right)$$
$$= F\left(36, \tfrac{k}{36}; k+1, k+4, k+5, k+7; 37-k, 40-k, 41-k, 43-k\right),$$
$$E\left(\tfrac{2k+1}{72}, 36\right)$$
$$= F\left(36, \tfrac{2k+1}{72}; k+1, k+4, k+6, k+7; 36-k, 39-k, 41-k, 42-k\right).$$

We obtain: $e_{36}^\star = 1.54059\ldots$ and $e_{36}^{\star\star} = 1.30596\ldots$.

Subcase 2.4: $n = 16$. Then $L_{16} = 2$, $\delta_{16,1} = 0$ and $\delta_{16,2} = 1$. We need to consider only the points of the form $\frac{k}{16}$, $(2 \le k \le 14)$. For such points we have

$$E\left(\tfrac{k}{16}, 16\right)$$
$$= F\left(16, \tfrac{k}{16}; k+1, k+3, k+4, k+5; 17-k, 19-k, 20-k, 21-k\right).$$

We obtain $e_{16}^\star = 1.46743\ldots$.

This completes the proof of Proposition 4.2.2. $\qquad\qquad\qquad\square$

Proposition 4.2.3. *The inequality (4.45) holds for* $1 \le n \le 6$.

Proof. For $1 \leq n \leq 4$, Proposition 4.2.3 follows directly from Theorem 3.2.1, since $1 \leq 2h$. Let now $n \in \{5, 6\}$. Because of the symmetry we may consider without any loss of generality that $x \in \left(0, \frac{1}{2}\right]$. We distinguish between two cases.

Case 1: $x \in \left(0, \frac{h}{2}\right]$. In order to apply Theorem 3.2.1 to the functional $B_n(\cdot, x)$ we shall verify condition (3.21). Let $(u, v) \in M_{x,h}\left(\frac{0}{n}, \frac{1}{n}, \ldots, \frac{n}{n}\right)$. If $u = x$, then, as in the proof of Proposition 4.2.1, see (4.82), we arrive at the inequality $v > x + \frac{3}{4}h$. Then, it follows from Lemma 3.2.2 that the condition $ord(u, v) \geq 1$ implies $v > x + \frac{3}{4}h$. We have the following subcases.

a) $v \in (x, x + h]$. Then $x = u$, $\pi(u, v) = \tau(x)$ and from (4.83) for $\rho = 0$ we obtain

$$\sum_{j=\pi(u,v)}^{n} \left(v - \frac{j}{n}\right) p_{n,j}(x) = (v - x)T_r(0, \infty) - M_r(0, \infty) > 0.$$

b) $v \in \left(x + h, x + \frac{5}{4}h\right]$. Then $u = x$. Let $v = x + (1 + \rho)h$, with $\rho \in \left(0, \frac{1}{4}\right]$. We have $\pi(u, v) = \overline{\tau}(x + \rho h) \leq \tau(x + \rho h)$. From (4.83) it follows that

$$\sum_{j=\pi(u,v)}^{n} \left(v - \frac{j}{n}\right) p_{n,j}(x) \geq \sum_{j=\tau(x+\rho h)}^{n} \left(v - \frac{j}{n}\right) p_{n,j}(x)$$
$$= (1 + \rho)hT_r(\rho, \infty) - M_r(\rho, \infty) > 0.$$

c) $v \in \left(x + \frac{5}{4}h, 1\right)$. From Lemma 3.2.2 we obtain that $\pi(u, v) \leq \tau\left(v - \frac{h}{2}\right)$. Define $v = x + \left(\frac{1}{2} + \rho\right)h$, where $\rho > \frac{3}{4}$. Using (4.84) one has

$$\sum_{j=\pi(u,v)}^{n} \left(v - \frac{j}{n}\right) p_{n,j}(x) \geq \sum_{j=\tau\left(v-\frac{h}{2}\right)}^{n} \left(v - \frac{j}{n}\right) p_{n,j}(x)$$
$$= \left(\frac{1}{2} + \rho\right)hT_r(\rho, \infty) - M_r(\rho, \infty) > 0.$$

Case 2: $x \in \left(\frac{h}{2}, \frac{1}{2}\right]$. We prove separately for $n = 5$ and for $n = 6$.

a) Proof for $n = 5$. The inequality

$$(5x - 1) p_{5,1}(x) > 5(1 - x) p_{5,5}(x), \quad x \in \left(\frac{1}{2\sqrt{5}}, \frac{1}{2}\right], \tag{4.117}$$

holds. Indeed, this relation can be rewritten with the substitution $t = \frac{x}{1-x}$ in the following form: $4t - 1 - t^4 > 0$, $\left(t \in \left(\frac{1}{2\sqrt{5}-1}, 1\right]\right)$, which is immediate, since, for $t = \frac{1}{2\sqrt{5}-1}$ we have $4t - 1 - t^4 = 0.14514\ldots > 0$. Consider the decomposition $B_n(f, x) = G_1(f) + G_2(f)$, for $f \in B[0, 1]$, where

$$G_1(f) := x^5 \left[5 \frac{1-x}{5x-1} f\left(\frac{1}{5}\right) + f\left(\frac{5}{5}\right) \right],$$

$$G_2(f) := \sum_{k=0}^{4} p_{5,k}(x) f\left(\frac{k}{5}\right) - 5x^5 \frac{1-x}{5x-1} f\left(\frac{1}{5}\right).$$

From (4.117) it follows that the functionals G_1 and G_2 are positive. We have $G_1(e_1 - xe_0) = 0$ and hence $G_2(e_1 - xe_0) = 0$. Moreover $\frac{4}{5} < 2h$ and $x \in \left(\frac{1}{5}, \frac{1}{2}\right]$. Then we can apply Theorem 3.2.1 to the functionals G_1 and G_2. Since $G_1(e_0) + G_2(e_0) = 1$, one obtains (4.45).

b) <u>Proof for $n = 6$.</u> Consider for any $0 \le i \le 6$ the functionals

$$F_i(f) := p_{6,i}(x) \cdot f\left(\frac{i}{n}\right), \quad (f \in B[0, 1]) \tag{4.118}$$

and set

$$m_i := |F_i(e_1 - xe_0)|. \tag{4.119}$$

Using the substitution $t = \frac{x}{1-x}$, $t \in \left(\frac{1}{2\sqrt{6}-1}, 1\right]$ we get

$$m_1 = (5t - 1)m_0, \quad m_2 = |10t^2 - 5t|m_0, \ m_3 = (10t^2 - 10t^3)m_0,$$
$$m_4 = (10t^3 - 5t^4)m_0, \ m_5 = (5t^4 - t^5)m_0, \quad m_6 = t^5 m_0,$$

where $m_0 = x(1-x)^6$.

Subcase i) $x \in \left(\frac{h}{2}, \frac{1}{3}\right)$. One has

$$m_0 > m_2 + m_5 + m_6.$$

Indeed, by taking into account the above relations, this inequality can be rewritten in the form $g_1(t) > 0$, $(t \in \left(\frac{1}{2\sqrt{6}-1}, \frac{1}{2}\right))$, where $g_1(t) := \frac{1}{t} - 5 + 10t - 5t^3$. Using the derivative of g_1 we obtain for such t that $g_1(t) \ge g_1\left(\sqrt{\frac{5-\sqrt{10}}{15}}\right) = 1.14276\ldots > 0$.

Consider the decomposition $B_6(\cdot, x) = G_1 + G_2$, where

$$G_1 := \frac{m_2 + m_5 + m_6}{m_0} \cdot F_0 + F_2 + F_5 + F_6,$$

$$G_2 := \frac{m_0 - m_2 - m_5 - m_6}{m_0} \cdot F_0 + F_1 + F_3 + F_4.$$

The functionals G_1 and G_2 are positive and $G_1(e_1 - xe_0) = 0 = G_2(e_1 - xe_0)$. Since $\frac{4}{6} < 2h$, from Theorem 3.2.1 it follows that (3.22) is satisfied for $F = G_2$. In order to obtain a similar relation for G_1, let

$$G_1(f) = \sum_{i=0}^{3} v_i f(y_i), \quad f \in B[0, 1],$$

where $y_0 = 0$, $y_1 = \frac{2}{6}$, $y_2 = \frac{5}{6}$, $y_3 = \frac{6}{6}$. We have $y_0 < x < y_1$ and $x - y_0 \leq h$. The set $M_{x,h}(y_0, y_2, y_2, y_3)$ has a single element, namely $(x, 2x)$. For the pair $(x, 2x)$, condition (3.21) becomes

$$\left(2x - \frac{1}{3}\right) p_{6,2}(x) + \left(2x - \frac{5}{6}\right) p_{6,5}(x) + (2x - 1)p_{6,6}(x) \geq 0.$$

Using the substitution $t = \frac{x}{1-x}$, this inequality can be rewritten in the equivalent form $g_2(t) > 0$, $(t \in \left(\frac{1}{2\sqrt{6}-1}, \frac{1}{2}\right))$, where $g_2(t) := t^5 + 6t^4 - 5t^3 + 25t - 5$. Since g_2 is increasing on this interval we have for such t, $g_2(t) > g_2\left(\frac{1}{2\sqrt{6}-1}\right) = 1.35465\ldots > 0$. Therefore we can apply Theorem 3.2.1 to the functional G_1 and thus (4.45) follows.

Subcase ii) $x \in \left[\frac{1}{3}, \frac{7}{20}\right]$. Here the inequalities

$$m_3 > m_0, \qquad m_0 + m_1 > m_3 + m_6$$

hold. Indeed, the first one, with the substitution $t = \frac{x}{1-x}$, becomes $g_3(t) > 0$, $(t \in \left[\frac{1}{2}, \frac{7}{13}\right])$, where $g_3(t) := -10t^3 + 10t^2 - 1$. Since g_3 is concave on this interval, $g_3\left(\frac{1}{2}\right) = \frac{1}{4} > 0$, and $g_3\left(\frac{7}{13}\right) = 0.33818\ldots > 0$, the inequality is true.

The second inequality becomes, with the same substitution: $g_4(t) > 0$, $(t \in \left[\frac{1}{2}, \frac{7}{13}\right])$, where $g_4(t) := -t^4 + 10t^2 - 10t + 5$. Since g_4 is convex on the interval and $g_4'\left(\frac{1}{2}\right) = -\frac{1}{2}$, then $g_4(t) > g_4\left(\frac{1}{2}\right) + g_4'\left(\frac{1}{2}\right)\left(\frac{7}{13} - \frac{1}{2}\right) = 2.41826\ldots > 0$.

Consider the decomposition $B_n(\cdot, x) = G_1 + G_2 + G_3$, where

$$G_1 := F_0 + \frac{m_0}{m_3 + m_6} \cdot (F_3 + F_6),$$

$$G_2 := \frac{m_3 + m_6 - m_0}{m_1} \cdot F_1 + \frac{m_3 + m_6 - m_0}{m_3 + m_6}(F_3 + F_6),$$

$$G_3 := \frac{m_0 + m_1 - m_3 - m_6}{m_1} \cdot F_1 + F_2 + F_4 + F_5.$$

The functionals G_i, $i = 1, 2, 3$ are positive and satisfy the conditions $G_i(e_1 - xe_0) = 0$. The inequality (3.22) is satisfied for $F = G_3$, since $\frac{4}{6} < 2h$.

We write $G_1(f) = v_0 f(y_0) + v_1 f(y_1) + v_2 f(y_2)$, where $y_0 := 0$, $y_1 := \frac{3}{6}$, $y_2 := \frac{6}{6}$. We have $y_0 < x < y_1$ and $x - y_0 < h$. The pair $(x, 2x)$ is the unique element of $M_{x,h}(y_0, y_1, y_2)$. The condition (3.21) for G_1 and for this pair is

$$\left(2x - \frac{1}{2}\right) p_{6,3}(x) + (2x - 1)\, p_{6,6}(x) \geq 0. \qquad (4.120)$$

Employing again the substitution $t = \frac{x}{1-x}$, this inequality becomes the immediate inequality $t^4 - t^3 + 30t - 10 > 0$, $(t \in \left[\frac{1}{2}, \frac{7}{13}\right])$. Hence one may apply Theorem 3.2.1 to the functional G_1.

Furthermore we write the functional G_2 in the form $G_2(f) = \mu_0 f(z_0) + \mu_1 f(z_1) + \mu_2 f(z_2)$, where $z_0 := \frac{1}{6}$, $z_1 := \frac{3}{6}$, $z_2 := \frac{6}{6}$. We have $z_0 < x < z_1$ and $x - z_0 < h$. The unique element of the set $M_{x,h}(z_0, z_1, z_2)$ is of the form (x, v), where $v = \alpha\left(\frac{1}{6}, x, h\right)$. Since $x \in \left[\frac{1}{3}, \frac{7}{20}\right]$, it follows that $\frac{h}{4} < x - \frac{1}{6} < \frac{h}{2}$. Then we have $v = \frac{1}{6} + 4\left(x - \frac{1}{6}\right) = 4x - \frac{1}{2}$. The condition (3.21) for $F = G_2$ and the pair (x, v) is

$$(4x - 1)\, p_{6,3}(x) + \left(4x - \frac{3}{2}\right) p_{6,6}(x) \geq 0.$$

But this inequality is a consequence of (4.120), which was already proved. Then Theorem 3.2.1 may be applied to the functional G_2 too. Thus one obtains (4.45).

Subcase iii) $x \in \left(\frac{7}{20}, \frac{1}{2}\right]$. Here the inequalities

$$m_3 + m_4 > m_0, \quad \text{and} \quad m_2 > m_6$$

hold. Indeed, the first one is equivalent, by taking $t = \frac{x}{1-x}$, with the immediate inequality $10t^2 - 5t^4 - 1 > 0$, $\left(t \in \left(\frac{7}{13}, 1\right]\right)$.

The second inequality can be rewritten using the same substitution in the form $g_5(t) > 0$, $t \in \left(\frac{7}{13}, 1\right]$, where $g_5(t) := -t^4 + 10t - 5$. But g_5 is concave on this interval and we obtain $g_5(1) > 0$, $g_5\left(\frac{7}{13}\right) = 0.30055\ldots > 0$.

Because the equality $m_0 + m_1 + m_2 = m_3 + m_4 + m_5 + m_6$ is satisfied it follows that the inequality $m_0 + m_1 < m_3 + m_4 + m_5$. Consider the decomposition $B_n(\cdot, x) = G_1 + G_2 + G_3$, where

$$G_1 := F_0 + \frac{m_0}{m_3 + m_4}(F_3 + F_4),$$

$$G_2 := F_1 + \frac{m_1}{m_3 + m_4 + m_5 - m_0} \cdot \frac{m_3 + m_4 - m_0}{m_3 + m_4}(F_3 + F_4)$$
$$+ \frac{m_1}{m_3 + m_4 + m_5 - m_0} F_5,$$

$$G_3 := F_2 + \frac{m_3 + m_4 - m_0}{m_3 + m_4} \cdot \frac{m_3 + m_4 + m_5 - m_0 - m_1}{m_3 + m_4 + m_5 - m_0}(F_3 + F_4)$$
$$+ \frac{m_2 - m_6}{m_3 + m_4 + m_5 - m_0} F_5 + F_6.$$

The functionals G_i, $i = 1, 2, 3$ are positive and they satisfy the condition $G_i(e_1 - xe_0) = 0$. For all of them we can apply Theorem 3.2.1, since $\frac{4}{6} < 2h$. Consequently, the relation (4.45) holds in this subcase too. Thus the proof is complete. $\qquad\square$

4.3 Global smoothness preservation

In this section we give an application of the general estimate with the classical second order modulus to the problem of the global smoothness preservation in the case of Bernstein operators.

It is known, by a result first obtained by T. Lindvall [58], that the Bernstein operators preserve the Lipschitz classes of the first order $Lip_1(\alpha, M)$, that is

$$B_n(Lip_1(\alpha, M)) \subset Lip_1(\alpha, M), \tag{4.121}$$

where $Lip_1(\alpha, M)$ denotes the class of functions that satisfy the inequality $\omega_1(f, h) \leq M \cdot h^\alpha$ for all $h > 0$, with $M > 0$ and $0 < \alpha \leq 1$.

On the other hand, D.X.Zhou [142] obtained that the Bernstein operators do not preserve the Lipschitz classes of the second order $Lip_2(\alpha, M)$. Here $Lip_2(\alpha, M)$ denotes the set of all functions that satisfy the inequality $\omega_2(f, h) \leq M \cdot h^\alpha$ for all $h > 0$, with $M > 0$ and $0 < \alpha \leq 2$. If we do not limit ourselves to functions satisfying a second order Lipschitz condition we can consider the estimate of the form

$$\omega_2(B_n(f), h) \leq c \cdot \omega_2(f, h), \quad f \in C[0, 1], \; h > 0. \tag{4.122}$$

This type of estimate was initiated by C.Cottin and H.Gonska [23], who obtained that there are such constants c, namely we can take $c = 4.5$. We prove here, see [94], the following theorem.

Theorem 4.3.1. *We have*

$$2 \leq \sup_{n \in \mathbb{N}} \; \sup_{f \in C[0, 1] \setminus \Pi_1} \; \sup_{h \in (0, \frac{1}{2}]} \frac{\omega_2(B_n(f), h)}{\omega_2(f, h)} \leq 3. \tag{4.123}$$

Proof. Fix $n \in \mathbb{N}$, $h \in \left(0, \frac{1}{2}\right)$ and $x \in (0, 1 - 2h)$. We have, for any $f \in C[0, 1]$,

$$\Delta_h^2 B_n(f)(x) = \sum_{l=0}^{n} c_l \cdot f\left(\frac{l}{n}\right), \quad \text{where} \quad c_l := \Delta_h^2 p_{nl}(x), \; 0 \leq l \leq n. \tag{4.124}$$

Since $B_n(e_i) = e_i, \; (i = 0, 1)$, it follows

$$\sum_{l=0}^{n} c_l = 0 \qquad \text{and} \qquad \sum_{l=0}^{n} l \cdot c_l = 0. \tag{4.125}$$

For $p, q \in (0, 1)$ consider the function

$$\Psi(t) := (1 + p)^t (1 - q)^{n-t} + (1 - p)^t (1 + q)^{n-t} - 2, \quad t \in [0, n].$$

The following properties are immediate: $\Psi(0) > 0$, $\Psi(n) > 0$ and the derivative of Ψ is increasing on $[0, n]$. If we take $p := \frac{h}{x+h}$ and $q := \frac{h}{1-x-h}$ we have

$$c_l = \binom{n}{l}(x + h)^l (1 - x - h)^{n-l} \Psi(l).$$

Then $c_0 > 0$ and $c_n > 0$. From (4.125) it follows that there is $l, 0 < l < n$ such that $c_l < 0$. Hence there exist $t_0 \in (0, n)$ such that $\Psi'(t) \leq 0$, $(t \in [0, t_0])$ and $\Psi'(t) \geq 0$, $(t \in [t_0, n])$. From these it follows that there is the decomposition $\{0, 1, \ldots, n\} = I \cup J \cup K$ such that $i < j < k$ for all $i \in I$, $j \in J$, $k \in K$ and

$$c_i \geq 0, \ (i \in I), \quad c_j < 0, \ (j \in J), \quad c_k \geq 0, \ (k \in K).$$

Now denote

$$\Delta := \sum_{i \in I} c_i \cdot \sum_{k \in K} k c_k - \sum_{i \in I} i c_i \cdot \sum_{k \in K} c_k.$$

We have that $\sum_{i \in I} i c_i / \sum_{i \in I} c_i$ belongs to the convex hull of the set I. Then it is smaller than $\sum_{k \in K} k c_k / \sum_{k \in K} c_k$ which belongs to the convex hull of the set K. Consequently, we have $\Delta > 0$.

For any $j \in J$ denote

$$u_j := \frac{1}{\Delta} \sum_{k \in K} (k - j) c_k, \quad \text{and} \quad v_j := \frac{1}{\Delta} \sum_{i \in I} (j - i) c_i$$

and consider the linear positive functional $G_j : C[0, 1] \to \mathbb{R}$, given by

$$G_j(f) := u_j \sum_{i \in I} c_i \cdot f\left(\frac{i}{n}\right) + v_j \sum_{k \in K} c_k \cdot f\left(\frac{k}{n}\right), \quad f \in C[0, 1].$$

One obtains immediately that $G_j(e_0) = 1$ and $G_j(e_1) = \frac{j}{n}$.

By applying Corollary 2.2.1 (i) we obtain

$$\left| G_j(f) - f\left(\frac{j}{n}\right) \right| \leq \left[1 + \frac{1}{2} h^{-2} G_j\left(\left(e_1 - \frac{j}{n} e_0 \right)^2 \right) \right] \omega_2(f, h), \quad f \in C[0, 1].$$

$$(4.126)$$

On the other hand from (4.124) and (4.125) one can derive the representation

$$\Delta_2^h B_n(f)(x) = \sum_{j \in J} (-c_j) \left[G_j(f) - f\left(\frac{j}{n}\right) \right], \quad f \in C[0, 1]. \qquad (4.127)$$

Also we have

$$\sum_{j \in J} (-c_j) \leq \sum_{j \in J} 2 \cdot p_{nj}(x + h) \leq 2.$$

Therefore we obtain

$$|\Delta_h^2 B_n(f)(x)| \leq \left[2 + \frac{1}{2} h^{-2} \sum_{j \in J} (-c_j) G_j\left(\left(e_1 - \frac{j}{n} e_0 \right)^2 \right) \right] \omega_2(f, h), \quad f \in C[0, 1].$$

We have $G_j(e_2) - e_2\left(\frac{i}{n}\right) = G_j\left(\left(e_1 - \frac{j}{n} e_0 \right)^2 \right)$. Then, by using the well-known relation $B_n(e_2)(x) = x^2 + \frac{x(1-x)}{n}$ we deduce from (4.127)

$$\sum_{j \in J} (-c_j) G_j\left(\left(e_1 - \frac{j}{n} e_0 \right)^2 \right) = \Delta_h^2 B_n(e_2)(x) = 2h^2 \frac{n-1}{n} \leq 2h^2.$$

Consequently, we have

$$|\Delta_h^2 B_n(f)(x)| \le 3 \cdot \omega_2(f, h), \qquad f \in C[0, 1], \ h \in \left(0, \frac{1}{2}\right), \ x \in (0, 1 - 2h).$$

This last inequality can be extended by passing to the limit, for all $h \in \left(0, \frac{1}{2}\right]$ and all $x \in [0, 1 - 2h]$. Therefore we have proved the inequality

$$\omega_2(B_n(f), h) \le 3 \cdot \omega_2(f, h), \qquad f \in C[0, 1], \ h \in [0, 1].$$

Conversely, for any integer $n \ge 1$ consider the function $f_n \in C[0, 1]$ defined as follows: $f_n(0) = 0$, $f_n(1) = 0$, $f_n(k \cdot 2^{-j}) = 1 - 2^{-j}$, if $1 \le j \le n$ and k is an odd integer such that $1 \le k \le 2^j - 1$ and f_n is linear on each of the intervals of the form $[(i - 1)2^{-n}, i2^{-n}]$, for $1 \le i \le 2^n$.

We have

$$\omega_2\left(f_n, \frac{1}{2}\right) = 1, \qquad (n \ge 1). \tag{4.128}$$

Indeed, let the function $g(x, h) := |\Delta_h^2 f_n(x)|$, $h \in \left[0, \frac{1}{2}\right]$, $x \in [0, 1 - 2h]$. Then $g\left(0, \frac{1}{2}\right) = 1$. Let (x_0, h_0) be a point in which g reaches its maximum. Since g is a piecewise linear function with regard to each of his arguments, it follows that we can consider that at least two of the points x_0, $x_0 + h_0$, $x_0 + 2h_0$ belong to the set $M := \{k \cdot 2^{-n} \mid 0 \le k \le 2^n\}$. Since $f_n(t) \in \left[\frac{1}{2}, 1\right]$ when $t \in [2^{-n}, 1 - 2^{-n}]$, we have $g(x_0, h_0) \le 1$ when x_0, $x_0 + 2h_0 \in [2^{-n}, 1 - 2^{-n}]$. Then it remains to consider the case where $x_0 = 0$ or $x_0 + 2h_0 = 1$. Let for example $x_0 = 0$. If we suppose $x_0 + h_0 \notin M$ we have $x_0 + 2h_0 = k \cdot 2^{-n}$ with k odd. Consequently, $g(x_0, h_0) = |0 - 2f_n(x_0 + h_0) + 1 - 2^{-n}| < 1$ which is impossible. Then it follows that $x_0 = 0$, $x_0 + h_0 = k \cdot 2^{-j}$ and $x_0 + 2h_0 = k \cdot 2^{1-j}$, where $1 \le j \le n$ and k is odd. In this case $g(x_0, h_0) = 1$. Then (4.128) is true.

Let $n \ge 1$ and denote $q := 2^{2n}$. We have $B_q(f_{2n})(0) = B_q(f_{2n})(1) = 0$ and:

$$B_q(f_{2n})\left(\frac{1}{2}\right) = 2^{-q} \sum_{k=0}^{q} \binom{q}{k} f_{2n}\left(\frac{k}{q}\right) > 2^{-q} \sum_{\substack{0 \le k \le q \\ 2^n \nmid k}} \binom{q}{k} f_{2n}\left(\frac{k}{q}\right)$$

$$\ge 2^{-q} \sum_{\substack{0 \le k \le q \\ 2^n \nmid k}} \binom{q}{k}\left(1 - 2^{-n-1}\right) = 2^{-q}(1 - 2^{-n-1})\left[2^q - \sum_{j=0}^{2^n} \binom{q}{j}\right].$$

We have $\binom{q}{j \cdot 2^n} \le 2^{-n} \sum_{k=j \cdot 2^n}^{(j+1)2^n - 1} \binom{q}{k}$ for $0 \le j < 2^{n-1}$ and hence $\sum_{j=0}^{2^n} \binom{q}{j \cdot 2^n}$

$= 2\sum_{j=0}^{q/2-1} \binom{q}{j \cdot 2^n} + \binom{q}{q/2} \le 2^{1-n} \sum_{k=0}^{q/2-1} \binom{q}{k} + \binom{q}{q/2} \le 2^{q-n} + \binom{q}{q/2}.$

Therefore

$$|\Delta_{1/2}^2 B_q(f_{2n})(0)| \geq 2(1 - 2^{-n-1})\left(1 - 2^{-n} - \binom{q}{q/2} \cdot 2^{-q}\right).$$

Since $\lim_{q \to \infty} \binom{q}{q/2} \cdot 2^{-q} = 0$, it follows that

$$\sup_{n \in \mathbb{N}} |\Delta_{1/2}^2 B_{2^{2n}}(f_{2n})(0)| = 2$$

and therefore the left-side inequality in the theorem is also proved. $\square$

5

Two Classes of Bernstein Type Operators

5.1 Generalized Brass type operators

5.1.1 Definitions and general properties

One of the most natural extensions of the Bernstein operators was made by H. Brass [17]. These operators are of the form

$$P_n(f, x) := \sum_{k=0}^{n} f\left(\frac{k}{n}\right) q_{n,k}(x), \ f \in \mathcal{F}[0, 1], \ x \in [0, 1], \ n \in \mathbb{N}, \qquad (5.1)$$

where $q_{n,k}$ are polynomials of degree n that are positive on the interval $[0, 1]$ and are such that the following properties are true:

1) P_n is a linear positive operator,
2) P_n preserves linear functions,
3) P_n preserves the degree of any polynomial of degree at most n and
4) P_n preserves the convexity of higher order k, for any $k \geq -1$, (see Definition 1.1.1).

A concrete mode to construct these operators is given below.

First, the product of two operators with equidistant nodes is defined as follows:
If $P_n(f) := \sum_{i=0}^{n} f\left(\frac{i}{n}\right) q_{n,i}$ and $Q_m(f) := \sum_{i=0}^{m} f\left(\frac{i}{m}\right) \overline{q}_{m,i}$, where $q_{n,i} \in \Pi_n, \overline{q}_{m,i} \in \Pi_m$, $n, m \in \mathbb{N}$, $f \in C[0, 1]$, are two operators, then their product is the operator

$$R_{n+m}(f) := \sum_{i=0}^{n+m} f\left(\frac{i}{n+m}\right) r_{n+m,i}, \text{ where } r_{n+m,i} := \sum_{j=0}^{i} q_{n,j} \, \overline{q}_{m,i-j},$$

(with $q_{n,j} = 0$, if $j \notin [0, n]$ and $\overline{q}_{m,i-j} = 0$, if $i - j \notin [0, m]$). We define $P_n \oplus Q_m :=$ R_{n+m}.

Next, consider the operators $A_n : C[0, 1] \to \Pi_1$, $n \in \mathbb{N}$ of the form

$$A_n(f) := \sum_{i=0}^{n} f\left(\frac{i}{n}\right) q_{n,i}, \ f \in C[0, 1], \qquad (5.2)$$

$$q_{n,0}(x) = 1 - x, \ q_{n,i}(x) = 0, \ (1 \le i \le n - 1), \ q_{n,n}(x) = x, \ \text{for } x \in [0, 1]. \tag{5.3}$$

If the vector $\mu = (\mu_1, \dots, \mu_n)$, $n \in \mathbb{N}$, $\mu_k \in \mathbb{N} \cup \{0\}$, $(1 \le k \le n)$, satisfies the condition

$$\mu_1 + 2\mu_2 + \cdots + n\mu_n = n, \tag{5.4}$$

then μ is called an *admissible vector of order n*. The basic operators $H_n^\mu : C[0, 1] \to \Pi_m$, where $m = \mu_1 + \cdots + \mu_n$, are defined by

$$H_n^\mu := A_1 \oplus \cdots \oplus A_1 \oplus A_2 \oplus \cdots \oplus A_n \oplus A_n, \tag{5.5}$$

where, for any $1 \le k \le n$, the factor A_k has μ_k appearances.

Finally, a Brass operator of order n, $P_n : C[0, 1] \to \Pi_n$, is defined as a convex combination of the basic operators H_n^μ, i.e.,

$$P_n := \sum_\mu \alpha(\mu) \, H_n^\mu, \tag{5.6}$$

where $\alpha(\mu) \ge 0$, $\sum_\mu \alpha(\mu) = 1$, and the sum is taken over all admissible vectors of order n.

Note that, for the choice $\mu = (n, 0, 0, \dots, 0)$, we have $H_n^\mu = B_n$, where B_n denotes the Bernstein operator.

In this section we study a generalization of this type of operator, which is given in [102]. These generalized operators will maintain the basic properties 1)–4) given above.

We make the following new conventions of notation. For the integers $n \in \mathbb{N}$ and $r \in \mathbb{N} \cup \{0\}$, define $(n)_r := n(n-1)\cdots(n-r+1)$. For $k = 0$, we consider that a sum of the form $\sum\limits_{1 \le i_1 < \cdots < i_k \le m} g(y_{i_1} + \cdots + y_{i_k} + u)$ is equal to $g(u)$. If $L_1 : C[0, 1] \to C[0, 1]$, $L_2 : C[0, \infty) \to C[0, \infty)$ and $g \in C[0, \infty)$, we denote $(L_1 \circ L_2)(g)$, instead of $L_1\big((L_2(g))|_{[0,1]}\big)$. For $m \in \mathbb{N}$ set

$$\mathcal{D}_m := \{\tau = (t_1, \cdots, t_m) \in \mathbb{R} \mid t_i > 0, \ (1 \le i \le m), \ t_1 + \cdots + t_m = 1\}, \tag{5.7}$$

and if $\tau = (t_1, \dots, t_m) \in \mathcal{D}_m$, define

$$\sigma(\tau) := (t_1)^2 + \cdots + (t_m)^2. \tag{5.8}$$

For $0 \le k \le m$ and $x \in [0, 1]$, denote

$$p_{m,k}(x) := x^k (1 - x)^{m-k}. \tag{5.9}$$

Definition 5.1.1. *For any vector* $\tau = (t_1, \dots, t_m) \in \mathcal{D}_m$, $m \in \mathbb{N}$, *consider the operator* $L_\tau : C[0, 1] \to \Pi_m$, *given by*

$$L_\tau(f, x) := \sum_{k=0}^m \sum_{1 \le i_1 < \dots i_k \le m} f\left(t_{i_1} + \cdots + t_{i_k}\right) \cdot p_{m,k}(x), \tag{5.10}$$

$f \in C[0, 1]$, $x \in [0, 1]$.

Equivalently, we can write

$$L_\tau(f, x) = \sum_{I \subset \{1, \ldots, m\}} f\left(\sum_{i \in I} t_i\right) p_{m, card(I)}(x). \qquad (5.11)$$

Note that, for the choice $m = n$, $\tau = \left(\frac{1}{n}, \ldots, \frac{1}{n}\right)$, one obtains the Bernstein operator B_n.

Definition 5.1.2. *Let the vector* $\lambda = (\lambda_1, \ldots, \lambda_n)$, $n \in \mathbb{N}$, *where for any* $1 \le m \le n$, λ_m *is a positive regular Borel measure on* $\mathcal{D}_m$. *We suppose that* $\lambda_1(\mathcal{D}_1) + \cdots + \lambda_n(\mathcal{D}_n) = 1$. *Define the operator* $B_\lambda : C[0, 1] \to \Pi_n$,

$$B_\lambda(f, x) := \sum_{m=1}^{n} \int_{\mathcal{D}_m} L_\tau(f, x) \, d\lambda_m(\tau), \qquad f \in C[0, 1], \ x \in [0, 1]. \qquad (5.12)$$

For any such operator define

$$\sigma(B_\lambda) := \sum_{m=1}^{n} \int_{\mathcal{D}_m} \sigma(\tau) \, d\lambda_m(\tau). \qquad (5.13)$$

Denote by $\mathcal{B}_n$ *the class of operators* B_λ *as above.*

Theorem 5.1.1. *The class* $\mathcal{B}_n$, $n \in \mathbb{N}$ *contains the class of Brass operators of order* n.

Proof. The theorem follows from the following lemma.

Lemma 5.1.1. *For every admissible vector* $\mu = (\mu_1, \ldots, \mu_n)$, *the corresponding operator* H_n^μ *given in* (5.5), *coincides with the operator* L_τ, *defined as in* (5.10), *where* $m = \mu_1 + \cdots + \mu_n$, $\{t_1, \ldots, t_m\} \subset \left\{\frac{1}{n}, \frac{2}{n}, \ldots, \frac{n}{n}\right\}$ *and, for any* $1 \le k \le n$, *there are* μ_k *components of the vector* τ *that are equal to* $\frac{k}{n}$.

Proof. For any admissible vector $\mu = (\mu_1, \ldots, \mu_n)$, we denote $p(\mu) := \mu_1 + \cdots + \mu_n$. We prove the lemma by induction with regard to $p(\mu)$. If $p(\mu) = 1$, and $n \in \mathbb{N}$, then the lemma is obvious. Let us suppose that the lemma is valid for any admissible vector $\mu = (\mu_n, \ldots, \mu_n)$, such that $p(\mu) = m$, where n is arbitrary, and prove it for the admissible vectors with $p(\mu) = m + 1$.

Let us consider an admissible vector $\mu = (\mu_1, \ldots, \mu_n)$, with $p(\mu) = m + 1$. Consider also, the vector $\tau = (t_1, \ldots, t_{m+1})$, that is obtained from the vector μ as in the statement of the lemma. We may suppose that $t_1 \le t_2 \le \cdots \le t_{m+1}$. Let us denote $r_i := t_i \cdot n$, $(1 \le i \le m + 1)$ and $s := \max\{k \mid \mu_k \ne 0\}$. We have $r_{m+1} = s$ and $s < n$. Let $\overline{\mu} := (\overline{\mu}_1, \ldots, \overline{\mu}_{n-s})$, where $\overline{\mu}_1, \ldots, \overline{\mu}_{n-s}$ are the first $n - s$ terms of the sequence: $\mu_1, \ldots, \mu_{s-1}, \mu_s - 1, 0, 0, \ldots$. We have $\overline{\mu}_1 + 2\overline{\mu}_2 + \cdots + (n - s)\overline{\mu}_{n-s} = n - s$ and $p(\overline{\mu}) = m$. From relation (5.5) it follows that

$$H_n^{(\mu)} = H_{n-s}^{(\overline{\mu})} \oplus A_s.$$

By the assumption of induction we have $H_{n-s}^{\overline{\mu}} = L_{\overline{\tau}}$, for $\overline{\tau} = (\overline{t}_1, \ldots, \overline{t}_m) \subset$ $\left\{ \frac{1}{n-s}, \ldots, \frac{n-s}{n-s} \right\}$ and for any $1 \le k \le n - s$, the vector $\overline{\tau}$ has $\overline{\mu}_k$ components equal to $\frac{k}{n-s}$. Hence, by using (5.11) we can write

$$H_{n-s}^{\overline{\mu}}(f, x) = \sum_{j=0}^{n-s} f\left(\frac{j}{n-s}\right) r_{n-s,j}(x), \quad f \in C[0, 1], \ x \in [0, 1],$$

where

$$r_{n-s,j}(x) := \sum_{\substack{I \subset \{1,\ldots,m\} \\ \sum_{k\in I} r_k = j}} p_{m,card(I)}(x).$$

We write the operator A_s in the form $A_s(f, x) = \sum_{i=0}^{s} f\left(\frac{i}{s}\right) q_{s,i}(x)$, see (5.2), (5.3). We obtain successively:

$$H_n^{(\mu)}(f, x) = \sum_{i=0}^{n} f\left(\frac{i}{n}\right) \sum_{j=0}^{i} r_{n-s,j}(x) q_{s,i-j}(x)$$

$$= \sum_{i=0}^{n} f\left(\frac{i}{n}\right) \left(x r_{n-s,i-s}(x) + (1 - x) r_{n-s,i}(x) \right)$$

$$= \sum_{i=0}^{n} f\left(\frac{i}{n}\right) \left[\sum_{\substack{I \subset \{1,\ldots,m\} + \sum_{k\in I} r_k = i-s}} p_{m+1,card(I)+1}(x) + \sum_{\substack{I \subset \{1,\ldots,m\} \\ \sum_{k\in I} r_k = i}} p_{m+1,card(I)}(x) \right]$$

$$= \sum_{\substack{i=s \\ \sum_{k\in I} t_k = \frac{i-s}{n}}}^{n} \sum_{I \subset \{1,\ldots,m\}} f\left(\sum_{k\in I} t_k + t_{m+1}\right) p_{m+1,card(I)+1}(x)$$

$$+ \sum_{\substack{i=0 \\ \sum_{k\in I} t_k = \frac{i}{n}}}^{n} \sum_{I \subset \{1,\ldots,m\}} f\left(\sum_{k\in I} t_k\right) p_{m+1,card(I)}(x)$$

$$= \sum_{\substack{J \subset \{1,\ldots,m+1\} \\ m+1 \in J}} f\left(\sum_{k\in J} t_k\right) p_{m+1,card(J)}(x) + \sum_{I \subset \{1,\ldots,m\}} f\left(\sum_{k\in I} t_k\right) p_{m+1,card(I)}(x)$$

$$= \sum_{I \subset \{1,\ldots,m+1\}} f\left(\sum_{i\in I} t_i\right) p_{m+1,card(I)}(x) = L_{\tau}(f, x). \qquad \square$$

Remark 5.1.1. If in the place of the operator $B_\lambda \in \mathcal{B}_n$ we take a Brass operator of form (5.6), then we have

$$\sigma(P_n) = \sum_{\mu=(\mu_1,\dots,\mu_n)} \alpha(\mu) \cdot \sum_{k=1}^n \mu_k \left(\frac{k}{n}\right)^2. \tag{5.14}$$

We establish now some new notation. Let $r \in \mathbb{N} \cup \{0\}$, let the set $\{a_1, \dots, a_r\}$ of real numbers $a_i > 0$, $(1 \le i \le r)$, and let $\alpha \in \mathbb{R}$, $\alpha > 0$. We consider the operators I, T_α, $\Delta^r_{a_1,\dots,a_r}$, $\Delta^r_\alpha : C[0,\infty) \to C[0,\infty)$ defined by:

$$I(\varphi) := \varphi, \quad \varphi \in C[0,\infty), \tag{5.15}$$

$$T_\alpha(\varphi, \cdot) := \varphi(\cdot + \alpha), \quad \varphi \in C[0,\infty), \tag{5.16}$$

$$\Delta^r_{a_1,\dots,a_r} := \prod_{i=1}^r (T_{a_i} - I), \tag{5.17}$$

$$\Delta^r_\alpha := \Delta^r_{\alpha,\dots,\alpha} = \sum_{j=0}^r (-1)^{r-j} \binom{r}{j} T_{j\alpha}. \tag{5.18}$$

Let $\tau = (t_1, \dots, t_m) \in \mathcal{D}_m$ be fixed. Define the operator $E_{i,x} : C[0,\infty) \to C[0,\infty)$, by

$$E_{i,x} := (1-x)I + xT_{t_i}, \quad x \in [0,1], \ 1 \le i \le m, \tag{5.19}$$

and for any choice of the indices $1 \le i_1 < \cdots < i_r \le m$, $0 \le r \le m$, define the operator $E_x^{i_1,\dots,i_r} : C[0,\infty) \to C[0,\infty)$, by

$$E_x^{i_1,\dots,i_r} = \prod_{\substack{1 \le i \le m \\ i \notin \{i_1,\dots,i_r\}}} E_{i,x}. \tag{5.20}$$

Using the divided differences we can rewrite $\Delta^r_{a_1,\dots,a_r}$ in the following mode.

Lemma 5.1.2. *Let* $r \in \mathbb{N} \cup \{0\}$ *and the set of real numbers* $\{a_1, \dots, a_r\}$, $a_i > 0$, $(1 \le i \le r)$. *For any* $\varphi \in C[0,\infty)$ *and* $u \in [0,\infty)$ *we have*

$$\Delta^r_{a_1,\dots,a_r}(\varphi, u) = \left(\prod_{i=1}^r a_i\right) \sum_{\pi \in S_r} \left[\varphi; u, u + a_{\pi(1)}, \dots, u + a_{\pi(1)} + \cdots + a_{\pi(r)}\right], \tag{5.21}$$

where, for $r \ge 1$, S_r *denotes the set of permutations of the set* $\{1, 2, \dots, r\}$. *Also, if* $\varphi \in C^r[0,\infty)$, *then we have*

$$\Delta^r_{a_1,\dots,a_r}(\varphi, u) = \int_0^{a_1} du_1 \dots \int_0^{a_r} \varphi^{(r)}(u_1 + \cdots + u_r + u) \, du_r. \tag{5.22}$$

Proof. We prove by induction. For $r = 0$, relations (5.21) and (5.22) are obvious. We suppose the lemma true for $r-1$, $r \ge 1$. Let now φ, u, a_i, $1 \le i \le r$ be as in the hypothesis of the lemma and define the function $\psi \in C[0,\infty)$, $\psi(\cdot) = \varphi(\cdot + a_r) - \varphi(\cdot)$. By using the induction we have:

$$\Delta^r_{a_1,\ldots,a_r}(\varphi, u) = \Delta^{r-1}_{a_1,\ldots,a_{r-1}}(\psi, u)$$

$$= \sum_{\pi \in S_{r-1}} [\psi; u, u + a_{\pi(1)}, \ldots, u + a_{\pi(1)} + \cdots + a_{\pi(r-1)}] \cdot \prod_{i=1}^{r-1} a_i$$

$$= \sum_{\pi \in S_{r-1}} \Big\{ [\varphi; u + a_r, u + a_{\pi(1)} + a_r, \ldots, u + a_{\pi(1)} + \cdots + a_{\pi(r-1)} + a_r]$$

$$- [\varphi; u, u + a_{\pi(1)}, \ldots, u + a_{\pi(1)} + \ldots + a_{\pi(r-1)}] \Big\} \cdot \prod_{i=1}^{r-1} a_i$$

$$= \left(\prod_{i=1}^{r} a_i \right) \sum_{\pi \in S_{r-1}} \sum_{k=1}^{r} [\varphi; u, u + a_{\pi(1)}, \ldots, u + a_{\pi(1)} + \cdots + a_{\pi(k-1)},$$

$$u + a_{\pi(1)} + \cdots + a_{\pi(k-1)} + a_r, \cdots, u + a_{\pi(1)} + \ldots + a_{\pi(r-1)} + a_r].$$

Hence we obtained (5.21). Also, if $\varphi \in C^r[0, \infty)$, by applying the assumption of induction, we get

$$\Delta^r_{a_1,\ldots,a_r}(\varphi, u) = \int_0^{a_1} du_1 \ldots \int_0^{a_{r-1}} \psi^{(r-1)}(u_1 + \cdots + u_{r-1} + u)\, du_{r-1}.$$

This is equivalent to (5.22). □

Theorem 5.1.2. *Let $\tau = (t_1, \ldots, t_m) \in \mathcal{D}_m$, $m \in \mathbb{N}$, $f \in C[0, 1]$ and $g \in C[0, \infty)$ be an arbitrary prolongation of the function f. Let $x \in [0, 1]$. We have*

$$L_\tau(f, x) = \left(\prod_{i=1}^{m} E_{i,x} \right)(g, 0) \tag{5.23}$$

and, for $0 \le r \le m$, we have

$$(L_\tau(f, x))^{(r)} = (r!) \left(\sum_{1 \le i_1 < \cdots < i_r \le m} E_x^{i_1,\ldots,i_r} \circ \Delta^r_{t_{i_1},\ldots,t_{i_r}} \right)(g, 0). \tag{5.24}$$

Proof. Relation (5.23) follows immediately from (5.10). We prove relation (5.24) by induction with respect to r. For $r = 0$, relation (5.24) coincides with (5.23). If (5.24) is true for $r < m$, then

$$(L_\tau(f, x))^{(r+1)} = (r!) \sum_{1 \le i_1 < \cdots < i_r \le m} \frac{d}{dx} \left(\left(E_x^{i_1,\ldots,i_r} \circ \prod_{k=1}^{r} (T_{i_k} - I) \right)(g, 0) \right)$$

$$= (r!) \left(\sum_{\substack{1 \le i_1 < \cdots < i_r \le m,\, 1 \le i_{r+1} \le m \\ i_{r+1} \notin \{i_1,\ldots,i_r\}}} E_x^{i_1,\ldots,i_{r+1}} \circ \prod_{k=1}^{r+1} (T_{i_k} - I) \right)(g, 0)$$

$$= ((r+1)!) \left(\sum_{1 \le i_1 < \cdots < i_{r+1} \le m} E_x^{i_1,\ldots,i_{r+1}} \circ \prod_{k=1}^{r+1} (T_{i_k} - I) \right)(g, 0)$$

$$= ((r+1)!) \left(\sum_{1 \le i_1 < \cdots < i_{r+1} \le m} E_x^{i_1,\ldots,i_{r+1}} \circ \Delta^{r+1}_{t_{i_1},\ldots,t_{i_{r+1}}} \right)(g, 0). □$$

Corollary 5.1.1. *In the conditions of Theorem 5.1.2 we have*

$$(L_\tau(f, x))^{(r)} = (r!) \sum_{k=0}^{m-r} p_{m-r,k}(x) \sum_{\substack{1 \le i_1 < \ldots < i_r \le m \\ 1 \le j_1 < \ldots < j_k \le m \\ \{j_1, \ldots, j_k\} \cap \{i_1, \ldots, i_r\} = \emptyset}} \Delta^r_{t_{i_1}, \ldots, t_{i_r}} (g, t_{j_1} + \cdots + t_{j_k}).$$

(5.25)

Theorem 5.1.3. *For any $B_\lambda \in \mathcal{B}_n$, any $f \in C^2[0, 1]$ and any $x \in [0, 1]$ there exists a point $\xi \in [0, 1]$, such that*

$$B_\lambda(f, x) - f(x) = \frac{1}{2} x(1 - x) \sigma(B_\lambda) f''(\xi).$$

(5.26)

Proof. Take $\tau = (t_1, \ldots, t_m) \in \mathcal{D}_m$, $1 \le m \le n$. We have successively:

$$L_\tau(f, x) - f(x)$$

$$= \sum_{k=1}^m \left(E_x^{1, \ldots, k} \circ [E_{k,x} - T_{xt_k}] \circ T_{x(t_1 + \cdots + t_{k-1})} \right)(g, 0)$$

$$= \sum_{k=1}^m E_x^{1, \ldots, k} \Big((1 - x)(g(\cdot + x(t_1 + \cdots + t_{k-1})) - g(\cdot + x(t_1 + \cdots + t_k)))$$

$$+ x(g(\cdot + x(t_1 + \ldots + t_{k-1}) + t_k) - g(\cdot + x(t_1 + \cdots + t_k))), 0 \Big)$$

$$= \sum_{k=1}^m E_x^{1, \ldots, k} \Big((1 - x) \int_0^{-xt_k} ds \int_0^s g''(\cdot + x(t_1 + \cdots + t_k) + v) dv$$

$$+ x \int_0^{(1-x)t_k} ds \int_0^s g''(\cdot + x(t_1 + \cdots + t_k) + v) dv, 0 \Big)$$

$$= \sum_{k=1}^m \sum_{p=0}^{m-k} \sum_{k+1 \le i_1 < \cdots < i_p \le m} \Big[(1 - x) \int_0^{-xt_k} ds \int_0^s f'' \Big(\sum_{j=1}^p t_{i_j} + x(t_1 + \cdots + t_k) + v \Big) dv$$

$$+ x \int_0^{(1-x)t_k} ds \int_0^s f'' \Big(\sum_{j=1}^p t_{i_j} + x(t_1 + \cdots + t_k) + v \Big) dv \Big] p_{m-k,p}(x).$$

Since $B_\lambda(f, x) - f(x) = \sum_{m=1}^n \int_{\mathcal{D}_m} \Big(L_\tau(f, x) - f(x) \Big) d\lambda_m(\tau)$, it follows that for a fixed x, the expression $B_\lambda(f, x) - f(x)$ is a linear positive functional in the argument f''. By the mean value theorem, there exists a point $\xi \in [0, 1]$ such that we can write

$$B_\lambda(f, x) - f(x)$$

$$= f''(\xi) \sum_{m=1}^{n} \int_{\mathcal{D}_m} \left(\sum_{k=1}^{m} \sum_{p=0}^{m-k} \binom{m-k}{p} p_{m-k,p}(x) \right.$$

$$\left. \cdot \left(\frac{1-x}{2}(xt_k)^2 + \frac{x}{2}((1-x)t_k)^2 \right) \right) d\lambda_m(\tau)$$

$$= f''(\xi) \sum_{m=1}^{n} \int_{\mathcal{D}_m} \left(\sum_{k=1}^{m} \frac{1}{2} x(1-x)(t_k)^2 \right) d\lambda_m(\tau)$$

$$= \frac{1}{2} x(1-x)\sigma(B_\lambda)f''(\xi). \quad \square$$

Lemma 5.1.3. *For any operator $B_\lambda \in \mathcal{B}_n$, and $x \in [0, 1]$, we have*

$$B_\lambda(e_0, x) = 1, \quad B_\lambda(e_1, x) = x, \quad B_\lambda(e_2, x) = x^2 + \sigma(B_\lambda)x(1-x). \quad (5.27)$$

Proof. We obtain

$$B_\lambda(e_0, x) = \sum_{m=1}^{n} \int_{\mathcal{D}_m} \left(\sum_{k=0}^{m} \binom{m}{k} p_{m,k}(x) \right) d\lambda_m(\tau) = 1.$$

$$B_\lambda(e_1, x) = \sum_{m=1}^{n} \int_{\mathcal{D}_m} \left(\sum_{k=0}^{m} \sum_{1 \le i_1 < \cdots < i_k \le m} (t_{i_1} + \cdots + t_{i_k}) \, p_{m,k}(x) \right) d\lambda_m(\tau)$$

$$= \sum_{m=1}^{n} \int_{\mathcal{D}_m} \left(\sum_{k=0}^{m} \sum_{j=1}^{m} t_j \binom{m-1}{k-1} p_{m,k}(x) \right) d\lambda_m(\tau)$$

$$= \sum_{m=1}^{n} \int_{\mathcal{D}_m} \left(x \sum_{k=0}^{m} \binom{m-1}{k-1} p_{m-1,k-1}(x) \right) d\lambda_m(\tau) = x.$$

The value of $B_\lambda(e_2, x)$ follows from relation (5.26), for the choice $f = e_2$. $\quad \square$

The basic properties of the operators from the class $\mathcal{B}_n$ are given in the next theorem.

Theorem 5.1.4. *Let $B_\lambda \in \mathcal{B}_n$. We have:*
i) B_λ is linear and positive.
ii) If $f \in \Pi_1$, then $B_\lambda(f) = f$.
iii) If f is convex (of order 1), then $B_\lambda(f) \ge f$.
iv) $B_\lambda(\Pi_k) \subset \Pi_k$, $k \ge 0$.
v) If $f \in C[0, 1]$ is convex of order $k \ge -1$, then $B_\lambda(f)$ is convex of order k.

Proof. It is sufficient to prove these properties only for the operators L_τ, for any $\tau \in \mathcal{D}_m, 1 \le m \le n$. For these operators we have:

The point i) follows from (5.10) and the point ii) follows from Lemma 5.1.3. Also iii) is a known consequence of the properties i) and ii).

iv) If $f \in \Pi_k$, then from (5.24) and (5.22) we get $(L_\tau(f, x))^{(k+1)} = 0$, $x \in [0, 1]$. Hence $L_\tau(f) \in \Pi_k$.

v) If $f \in C[0, 1]$ is convex of order $k \geq -1$, then from (5.24) and (5.21) one has $(L_\tau(f, x))^{(k+1)} \geq 0$, $x \in [0, 1]$. Hence $L_\tau(f)$ is convex of order k. $\square$

5.1.2 Simultaneous approximation

There are known different types of estimates for the simultaneous approximation by general operators. Estimates, using the first order modulus, were obtained by Knoop and Pottinger [55] and using the second order modulus, were obtained by H. Gonska in [40]. However, we apply here specific methods that depend only on the particular form of the operators and on some general estimates obtained in Chapter 2. In the particular case of the Bernstein operators, the degree of approximation of derivatives of order r was estimated using $\omega_1(f^{(r)}, \cdot)$, first by T. Popoviciu [116], using $\omega_1(f^{(r+1)}, \cdot)$, first by D.D. Stancu [130], and using $\omega_2(f^{(r)}, \cdot)$, first by H. Gonska [40]. Estimates for simultaneous approximation by Brass operators were be obtained first in our paper [79], see also [78].

For the operators of the class $\mathcal{B}_n$, we obtain two types of estimates, each of them being better than the other in certain cases.

For any $\tau = (t_1, \ldots, t_m) \in \mathcal{D}_m$, $m \in \mathbb{N}$ and any integer $r \in \mathbb{N} \cup \{0\}$, we define

$$p_r(\tau) := \begin{cases} 1, & \text{if } r = 0, \\ (r!) \sum_{1 \leq i_1 < \cdots < i_r \leq m} t_{i_1} \cdots t_{i_r}, & \text{if } 1 \leq r \leq m, \\ 0, & \text{if } r > m. \end{cases} \qquad (5.28)$$

For $B_\lambda \in \mathcal{B}_n$, $n \in \mathbb{N}$ and $r \in \mathbb{N} \cup \{0\}$, with the notation in (5.12), define:

$$M_r^0 := \sum_{m=1}^n \int_{\mathcal{D}_m} p_r(\tau) \, d\lambda_m(\tau), \qquad (5.29)$$

$$M_{r,x}^1 := \frac{|1 - 2x|}{2} \sum_{m=1}^n \int_{\mathcal{D}_m} (p_r(\tau) - p_{r+1}(\tau)) \, d\lambda_m(\tau), \qquad (5.30)$$

$$M_{r,x}^2 := \sum_{m=1}^n \int_{\mathcal{D}_m} \left[-x(1-x)\left(p_{r+2}(\tau) - 2p_{r+1}(\tau) + p_r(\tau)\right) \right.$$
$$+ \left(\frac{1}{3} - \frac{1}{12}p_2(\tau)\right)p_r(\tau) - \frac{3r+2}{6(r+1)}p_{r+1}(\tau)$$
$$\left. + \frac{3r+1}{12(r+1)}p_{r+2}(\tau) \right] d\lambda_m(\tau). \qquad (5.31)$$

Theorem 5.1.5. *Let $B_\lambda \in \mathcal{B}_n$, $n \in \mathbb{N}$, $r \in \mathbb{N} \cup \{0\}$, $f \in C^r[0, 1]$, $h > 0$ and $x \in [0, 1]$. We have*

$$|(B_\lambda(f,x) - f(x))^{(r)}| \le (1 - M_r^0)|f^{(r)}(x)|$$
$$+ \left(M_r^0 + h^{-1}\sqrt{M_r^0 M_{r,s}^2} \right) \omega_1(f^{(r)}, h), \quad (5.32)$$

$$|(B_\lambda(f,x) - f(x))^{(r)}| \le (1 - M_r^0)|f^{(r)}(x)|$$
$$+ \left(M_r^0 + h^{-2}M_{r,s}^2 \right) \omega_1(f^{(r)}, h), \quad (5.33)$$

and for any $0 < h \le \frac{1}{2}$ *we have*

$$|(B_\lambda(f,x) - f(x))^{(r)}| \le (1 - M_r^0)|f^{(r)}(x)| + h^{-1}M_{r,x}^1 \omega_1(f^{(r)}, h)$$
$$+ \left(M_r^0 + \frac{1}{2}h^{-2}M_{r,x}^2 \right) \omega_2(f^{(r)}, h). \quad (5.34)$$

Proof. For any vector $\tau = (t_1, \ldots, t_m) \in \mathcal{D}_m$, $1 \le m \le n$, define the linear positive operator $Z_{\tau,r} : C[0,1] \to C[0,1]$, by

$$Z_{\tau,r}(\varphi, x) := (r!) \sum_{1 \le i_1 < \cdots < i_r \le n} \sum_{k=0}^{m-r} P_{m-r,k}(x) V_{i_1,\ldots,i_r}^k(\varphi),$$

where

$$V_{i_1,\ldots,i_r}^k(\varphi)$$
$$:= \sum_{\substack{1 \le j_1 < \cdots j_k \le m \\ \{j_1,\ldots,j_k\} \cap \{i_1,\ldots,i_r\} = \emptyset}} \int_0^{t_{i_1}} du_1 \ldots \int_0^{t_{i_r}} \varphi(t_{j_1} + \cdots + t_{j_k} + u_1 + \cdots + u_r)\, du_r.$$

for $\varphi \in C[0,1]$, $x \in [0,1]$. By using the conventions of our notation, we have $Z_{\tau,r} = L_\tau$, for $r = 0$ and $Z_{\tau,r} = 0$, for $r > m$. Note that, for $1 \le r \le m$ the operator $Z_{\tau,r}$ coincides with the so-called r-order Kantorovich modification of L_τ. From (5.24) and (5.22) it follows that we can write for $f \in C^r[0,1]$:

$$(L_\tau(f,x))^{(r)} = Z_{\tau,r}(f^{(r)}, x), \quad x \in [0,1]. \quad (5.35)$$

Let us compute $Z_{\tau,r}(e_i, x)$, $0 \le i \le 2$. First we have

$$Z_{\tau,r}(e_0, x) = (r!) \sum_{1 \le i_1 < \cdots < i_r \le m} \sum_{k=0}^{m-r} P_{m-r,k}(x) \binom{m-r}{k} t_{i_1} \cdots t_{i_r} = p_r(\tau).$$

For any indices $1 \le i_1 < \cdots < i_r \le m$ and $0 \le k \le m - r$ we have

$$V_{i_1,\dots,i_r}^k(e_1) = \int_0^{t_{i_1}} du_1 \dots \int_0^{t_{i_r}} \left[\binom{m-r-1}{k-1} \left(\sum_{\substack{1 \le j \le m \\ j \notin \{i_1,\dots,i_r\}}} t_j \right) \right.$$

$$\left. + \binom{m-r}{k} \sum_{j=0}^r u_j \right] du_r$$

$$= t_{i_1} \cdots t_{i_r} \left[\binom{m-r-1}{k-1} \left(\sum_{\substack{1 \le j \le m \\ j \notin \{i_1,\dots,i_r\}}} t_j \right) + \frac{1}{2} \binom{m-r}{k} \sum_{p=1}^r t_{i_p} \right].$$

Therefore

$$Z_{\tau,r}(e_1, x) = (r!) \sum_{1 \le i_1 < \cdots < i_r \le m} t_{i_1} \cdots t_{i_r} \left[x \left(\sum_{\substack{1 \le j \le m \\ j \notin \{i_1,\dots,i_r\}}} t_j \right) + \frac{1}{2} \sum_{p=1}^r t_{i_p} \right]$$

$$= \frac{2x-1}{2} p_{r+1}(\tau) + \frac{1}{2} p_r(\tau).$$

Also, for any indices $1 \le i_1 < \cdots < i_r \le m$ and $0 \le k \le m - r$ we have

$$V_{i_1,\dots,i_r}^k(e_2) = \int_0^{t_{i_1}} du_1 \dots \int_0^{t_{i_r}} \left[\binom{m-r-1}{k-1} \left(\sum_{\substack{1 \le j \le m \\ j \notin \{i_1,\dots,i_r\}}} \left((t_j)^2 + 2t_j \sum_{p=1}^r u_p \right) \right) \right.$$

$$\left. + 2\binom{m-r-2}{k-2} \left(\sum_{\substack{1 \le j_1 < j_2 \le m \\ j_1, j_2 \notin \{i_1,\dots,i_r\}}} t_{j_1} t_{j_2} \right) + \binom{m-r}{k} \left(\sum_{p=1}^r u_p \right)^2 \right] du_r$$

$$= t_{i_1} \cdots t_{i_r} \left\{ \binom{m-r-1}{k-1} \left(\sum_{\substack{1 \le j \le m \\ j \notin \{i_1,\dots,i_r\}}} (t_j)^2 + t_j \sum_{p=1}^r t_{i_p} \right) \right.$$

$$+ 2\binom{m-r-2}{k-2} \left(\sum_{\substack{1 \le j_1 < j_2 \le m \\ j_1, j_2 \notin \{i_1,\dots,i_r\}}} t_{j_1} t_{j_2} \right)$$

$$\left. + \binom{m-r}{k} \left[\frac{1}{3} \sum_{p=1}^r (t_{i_p})^2 + \frac{1}{2} \sum_{1 \le p \le q \le r} t_{i_p} t_{i_q} \right] \right\}.$$

Consequently, we obtain

$$Z_{\tau,r}(e_2, x) = (r!) \sum_{1 \le i_1 < \cdots < i_r \le m} t_{i_1} \cdots t_{i_r} \left[x \left(\sum_{\substack{1 \le j \le m \\ j \notin \{i_1,\dots,i_r\}}} \left((t_j)^2 + t_j \sum_{p=1}^r t_{i_p} \right) \right) \right.$$

$$\left. + 2x^2 \left(\sum_{\substack{1 \le j_1 < j_2 \le m \\ j_1, j_2 \notin \{i_1,\dots,i_r\}}} t_{j_1} t_{j_2} \right) + \frac{1}{3} \sum_{p=1}^r (t_{i_p})^2 + \frac{1}{2} \sum_{1 \le p < q \le r} t_{i_p} t_{i_q} \right].$$

We have:

$$\sum_{1\le p<q\le r} t_{i_p}t_{i_q} = \frac{1}{2}p_2(\tau) - \sum_{p=1}^{r} t_{i_p}\left(\sum_{\substack{1\le j\le m \\ j\notin\{i_1,\dots,i_r\}}} t_j\right) - \sum_{\substack{1\le j_1<j_2\le m \\ j_1,j_2\notin\{i_1,\dots,i_r\}}} t_{j_1}t_{j_2},$$

$$\sum_{p=1}^{r}\left(t_{i_p}\right)^2 = 1 - p_2(\tau) - \sum_{\substack{1\le j\le m \\ j\notin\{i_1,\dots,i_r\}}} (t_j)^2,$$

$$(r!)\sum_{1\le i_1<\cdots<i_r\le m} t_{i_1}\cdots t_{i_r}\left(\sum_{\substack{1\le j_1<j_2\le m \\ j_1,j_2\notin\{i_1,\dots,i_r\}}} t_{j_1}t_{j_2}\right) = \frac{1}{2}p_{r+2}(\tau),$$

$$(r!)\sum_{1\le i_1<\cdots<i_r\le m} t_{i_1}\cdots t_{i_r}\left(\sum_{\substack{1\le j\le m \\ j\notin\{i_1,\dots,i_r\}}} (t_j)^2\right) = (r!)\sum_{1\le i_1<\cdots<i_{r+1}\le m} t_{i_1}\cdots t_{i_{r+1}}\sum_{k=1}^{r+1} t_{i_k}$$

$$= \frac{1}{r+1}(p_{r+1}(\tau) - p_{r+2}(\tau)),$$

$$(r!)\sum_{1\le i_1<\cdots<i_r\le m} t_{i_1}\cdots t_{i_r}\sum_{p=1}^{r} t_{i_p}\left(\sum_{\substack{1\le j\le m \\ j\notin\{i_1,\dots,i_r\}}} t_j\right)$$

$$= r(r!)\sum_{1\le i_1<\cdots<i_{r+1}\le m} t_{i_1}\cdots t_{i_{r+1}}\sum_{k=1}^{r+1} t_{i_k}$$

$$= \frac{r}{r+1}(p_{r+1}(\tau) - p_{r+2}(\tau)).$$

Hence:

$$Z_{\tau,r}(e_2, x) = x^2 p_{r+2}(\tau) + x(p_{r+1}(\tau) - p_{r+2}(\tau)) + \frac{1}{3}p_r(\tau) - \frac{1}{12}p_2(\tau)p_r(\tau)$$

$$- \frac{3r+2}{6(r+1)}p_{r+1}(\tau) + \frac{3r+1}{12(r+1)}p_{r+2}(\tau).$$

Finally, we obtain

$$Z_{\tau,r}(e_1 - xe_0, x) = \frac{1-2x}{2}(p_r(\tau) - p_{r+1}(\tau))$$

and

$$Z_{\tau,r}((e_1 - xe_0)^2, x) = -x(1-x)\left(p_{r+2}(\tau) - 2p_{r+1}(\tau) + p_r(\tau)\right)$$

$$+ \left(\frac{1}{3} - \frac{1}{12}p_2(\tau)\right)p_r(\tau) - \frac{3r+2}{6(r+1)}p_{r+1}(\tau)$$

$$+ \frac{3r+1}{12(r+1)}p_{r+2}(\tau).$$

Now we apply estimates (1.40), (1.39) and (2.41), for $s = 2$, to the functional

$$F(g) := \sum_{m=1}^{n} \int_{\mathcal{D}_m} Z_{\tau,r}(g, x) \, d\lambda_m(\tau)$$

and to the function $g := f^{(r)}$, where $x \in [0, 1]$ is fixed. From (5.35) we obtain $F(e_0) = M_r^0$, $|F(e_1 - xe_0)| \le M_{r,x}^1$ and $F((e_1 - xe_0)^2) = M_{r,x}^2$. Thus we obtain relations (5.32), (5.33) and (5.34).   □

Remark 5.1.2. For $r = 0$ we get $M_r^0 = 1$, $M_{r,x}^1 = 0$ and $M_{r,x}^2 = x(1 - x)\sigma(B_\lambda)$. Therefore, in this case, the first term in estimates (5.32), (5.33) and the first two terms in estimate (5.34) are zero.

Corollary 5.1.2. *If $B_\lambda \in \mathcal{B}_n$, $n \in \mathbb{N}$, $f \in C^{r+1}[0, 1]$, $r \in \mathbb{N} \cup \{0\}$, and $x \in [0, 1]$, then with the notation in Theorem 5.1.5 we have*

$$|(B_\lambda(f, x) - f(x))^{(r)}| \le (1 - M_r^0)|f^{(r)}(x)| + \sqrt{M_r^0 \cdot M_{r,x}^2} \cdot \|f^{(r+1)}\|, \quad (5.36)$$

and if $f \in C^{r+2}[0, 1]$, then

$$|(B_\lambda(f, x) - f(x))^{(r)}|$$
$$\le (1 - M_r^0)|f^{(r)}(x)| + M_{r,x}^1 \cdot \|f^{(r+1)}\| + \frac{1}{2}M_{r,x}^2 \cdot \|f^{(r+2)}\|. \quad (5.37)$$

Proof. Relation (5.36) follows from (5.32) and relation (5.37) follows from (5.34). □

Corollary 5.1.3. *In the particular case where B_λ is the Bernstein operator B_n, then the estimates given in (5.32), (5.33), (5.34), (5.36) and (5.37) hold, for f, r, x and h that are like there, and*

$$M_r^0 = \frac{(n)_r}{n^r}, \quad (5.38)$$

$$M_{r,x}^1 = \frac{r(n)_r}{2n^{r+1}} \cdot |1 - 2x|, \quad (5.39)$$

$$M_{r,x}^2 = \frac{(n)_r}{n^r} \left(\frac{n - r^2 - r}{n^2} \cdot x(1 - x) + \frac{3r^2 + r}{12n^2} \right). \quad (5.40)$$

Proof. We take into account that $p_r(\tau) = \frac{(n)_r}{n^r}$.   □

Remark 5.1.3. For the Bernstein operators, we obtain from relations (5.33), (5.38), (5.40) and the inequalities $\frac{(n)_r}{n^r} \le 1$ and $1 - \frac{(n)_r}{n^r} \le \frac{r(r-1)}{2n}$, that

$$\|(B_n(f) - f)^{(r)}\| \le \frac{r(r - 1)}{2n}\|f^{(r)}\| + \left(1 + h^{-2}\frac{3n - 2r}{12n^2}\right)\omega_1(f^{(r)}, h), \quad (5.41)$$

$f \in C^r[0, 1]$, $r \ge 0$, and $h > 0$. This coincides with the Knoop-Pottinger inequality for Bernstein operators, obtained here in a different mode. An improvement of inequality (5.41) is given in [9] by I. Badea, C. Badea and H. Gonska.

An other method for obtaining estimates for simultaneous approximation is used in the following theorem.

Theorem 5.1.6. *For any operator* $B_\lambda \in \mathcal{B}_n$, $n \in \mathbb{N}$, *given in (5.12), any* $f \in C^{r+2}[0, 1]$, $r \in \mathbb{N} \cup \{0\}$, *and any* $x \in [0, 1]$, *we have*

$$\left|(B_\lambda(f, x) - f(x))^{(r)}\right| \le \sum_{m=1}^{n} \int_{\mathcal{D}_m} (1 - p_r(\tau)) \, d\lambda_m(\tau) \cdot \left|f^{(r)}(x)\right|$$

$$+ \sum_{m=1}^{n} \int_{\mathcal{D}_m} (p_r(\tau) - p_{r+1}(\tau)) \, d\lambda_m(\tau) \cdot \|f^{(r+1)}\|$$

$$+ \frac{x(1-x)}{2} \cdot \sigma(B_\lambda) \|f^{(r+2)}\|. \tag{5.42}$$

Proof. Consider that $r \ge 1$, since for $r = 0$ one can apply Theorem 5.1.3. It suffices to examine the operators L_τ, $\tau \in \mathcal{D}_{mi}$, with $m \ge r$. Let $\tau = (t_1, \ldots, t_m) \in \mathcal{D}_m$ and $x \in \left[0, \frac{1}{2}\right]$ be fixed. Denote by $g \in C^{r+2}[0, \infty)$ an arbitrary prolongation of the function f. Consider an arbitrary number $\delta > 0$ such that $x + r\delta \le 1$. We use the notation given in (5.17)–(5.20) and (5.28). In the sequel, the Landau symbol $o(\delta^r)$, $(\delta \to 0)$, is considered uniform with regard to x. Note that, from (5.22) we get for any $g \in C^r[0, \infty)$ and $t \in [0, 1]$,

$$|\Delta_\delta^r(g, t) - \delta^r g^{(r)}(t)| \le \delta^r \omega_1(g^{(r)}, r\delta)_{[0, 1+r\delta]} = o(\delta^r). \tag{5.43}$$

First, we have:

$$\left|(L_\tau(f, x) - f(x))^{(r)}\right| \le \delta^{-r} \left\{ \left|\delta^r (L_\tau(f, x))^{(r)} - p_r(\tau)(L_\tau \circ \Delta_\delta^r)(g, x)\right|\right.$$

$$+ |1 - p_r(\tau)| \cdot |(L_\tau \circ \Delta_\delta^r)(g, x)|$$

$$\left. + |(L_\tau \circ \Delta_\delta^r)(g, x) - \Delta_\delta^r(f, x)|\right\} + o(\delta^r). \tag{5.44}$$

We use (5.22), (5.23), (5.24) and (5.43). We obtain

$$(L_\tau(f, x))^{(r)} = (r!) \sum_{1 \le i_1 < \cdots < i_r \le m} E_x^{i_1, \ldots, i_r} \left(\int_0^{t_{i_1}} du_1 \ldots \int_0^{t_{i_r}} g^{(r)}(\cdot + u_1 + \ldots + u_r) \, du_r, 0\right)$$

$$= (r!) \sum_{1 \le i_1 < \cdots < i_r \le m} E_x^{i_1, \ldots, i_r} \left(t_{i_1} \cdots t_{i_r} \cdot g^{(r)}(\cdot)\right.$$

$$\left. + \int_0^{t_{i_1}} du_1 \ldots \int_0^{t_{i_r}} du_r \int_0^{u_1 + \cdots + u_r} g^{(r+1)}(\cdot + u) \, du, 0\right).$$

On the other hand, we have

$$p_r(\tau)(L_\tau \circ \Delta_\delta^r)(g, x) = p_r(\tau)\delta^r \left(\prod_{i=1}^m E_{i,x} \right)(g^{(r)}, 0) + o(\delta^r)$$

$$= \delta^r (r!) \sum_{1 \le i_1 < \cdots < i_r \le m} t_{i_1} \cdots t_{i_r} E_x^{i_1, \ldots, i_r} \left(g^{(r)}(\cdot) \right.$$

$$+ \sum_{k=1}^r p_{r,k}(x) \sum_{\{j_1, \ldots, j_k\} \subset \{i_1, \ldots, i_r\}} \int_0^{t_{j_1} + \cdots + t_{j_k}} g^{(r+1)}(\cdot + u)\, du, \, 0 \Bigg) + o(\delta^r).$$

From the relations above we get

$$\left| \delta^r (L_\tau(f, x))^{(r)} - p_r(\tau)(L_\tau \circ \Delta_\delta^r)(g, x) \right|$$

$$\le \delta^r (r!) \sum_{1 \le i_1 < \cdots < i_r \le m} \|f^{(r+1)}\| E_x^{t_{i_1}, \ldots, t_{i_r}} \left(\int_0^{t_{i_1}} du_1 \ldots \int_0^{t_{i_r}} du_r \int_0^{u_1 + \cdots + u_r} du \right.$$

$$+ t_{i_1} \cdots t_{i_r} \sum_{k=1}^r p_{r,k}(x) \sum_{\{j_1, \ldots, j_k\} \subset \{i_1, \ldots, i_r\}} \int_0^{t_{j_1} + \cdots + t_{j_k}} du, \, 0 \Bigg) + o(\delta)$$

$$\le \delta^r \|f^{(r+1)}\| (r!) \sum_{1 \le i_1 < \cdots < i_r \le m} t_{i_1} \cdots t_{i_r} \left(\sum_{k=1}^r t_{i_k} \right)$$

$$\cdot \left[\frac{1}{2} + x \sum_{k=1}^r \binom{r-1}{k-1} p_{r-1,k-1}(x) \right] + o(\delta^r)$$

$$\le \delta^r (p_r(\tau) - p_{r+1}(\tau)) \|f^{(r+1)}\| + o(\delta^r). \tag{5.45}$$

Also, using relation (5.43) we get:

$$|1 - p_r(\tau)| \cdot |(L_\tau \circ \Delta_\delta^r)(g, x)| \le |1 - p_r(\tau)| \cdot \delta^r \|f^{(r)}\| + o(\delta^r). \tag{5.46}$$

Now, from relation (5.43) and Theorem 5.1.3 it follows that

$$|(L_\tau \circ \Delta_\delta^r)(g, x) - \Delta_\delta^r(f, x)| = \delta^r |L_\tau(g^{(r)}, x) - g^{(r)}(x)| + o(\delta^r)$$

$$\le \delta^r \cdot \frac{x(1-x)}{2} \cdot \sigma(L_\lambda) \|f^{(r+2)}\| + o(\delta^r). \tag{5.47}$$

Finally, by passing to limit $\delta \to 0$, from relations (5.44), (5.45), (5.46), (5.47) we get for $0 \le x \le \frac{1}{2}$, the estimate

$$\left| (L_\tau(f, x) - f(x))^{(r)} \right|$$

$$\le \left| f^{(r)}(x) \right| (1 - p_r(\tau)) + \|f^{(r+1)}\|$$

$$\cdot (p_r(\tau) - p_{r+1}(\tau)) + \frac{x(1-x)}{2} \cdot \sigma(L_\tau) \|f^{(r+2)}\|.$$

By symmetry, this relation is also true for $x \in \left[\frac{1}{2}, 1\right]$. Consequently we obtain (5.42).
∎

Corollary 5.1.4. *For the Bernstein operators B_n, we have the following estimate:*

$$|(B_n(f, x) - f(x))^{(r)}| \leq \left(1 - \frac{(n)_r}{n^r}\right)|f^{(r)}(x)| + \frac{r(n)_r}{n^{r+1}} \cdot \|f^{(r+1)}\|$$

$$+ \frac{x(1-x)}{2n}\|f^{(r+2)}\|, \tag{5.48}$$

$f \in C^{r+2}[0, 1], r \in \mathbb{N} \cup \{0\}, x \in [0, 1]$.

Finally, note an extremal property of the Bernstein operator B_n in the class $\mathcal{B}_n$.

Theorem 5.1.7. *For any operator $B_\lambda \in \mathcal{B}_n$, $n \in \mathbb{N}$, and any convex function $f \in C[0, 1]$, we have*

$$|B_\lambda(f, x) - f(x)| \geq |B_n(f, x) - f(x)|, \quad x \in [0, 1]. \tag{5.49}$$

Proof. For the proof, we use the so-called "Sturm's principle", that says that if we have a continuous symmetrical function $F : \mathcal{D}_m \to \mathbb{R}$, $m \geq 2$, with the property

$$F(t_1, \ldots, t_m) \leq F\left(t_1, \ldots, t_{i-1}, \frac{t_i + t_j}{2}, t_{i+1}, \ldots, t_{j-1}, \frac{t_i + t_j}{2}, t_{j+1}, \ldots, t_m\right),$$

for any $(t_1, \ldots, t_m) \in \mathcal{D}_m$ and any $1 \leq i < j \leq m$, then we have

$$\max_{(t_1, \ldots, t_m) \in \mathcal{D}_m} F(t_1, \ldots, t_m) = F\left(\frac{1}{m}, \ldots, \frac{1}{m}\right).$$

For $1 \leq m \leq n$, we take the function $F(t_1, \ldots, t_m) := L_\tau(f, x)$, $\tau = (t_1, \ldots, t_m) \in \mathcal{D}_m$. For a vector $\tau = (t_1, \ldots, t_m) \in \mathcal{D}_m, m \geq 2$ and for the indices $0 \leq i < j \leq m$, denote $\tau_{i,j} := \left(t_1, \ldots, t_{i-1}, \frac{t_i+t_j}{2}, t_{i+1}, \ldots, t_{j-1}, \frac{t_i+t_j}{2}, t_{j+1}, \ldots, t_m\right)$.
We have:

$$L_\tau(f, x) - L_{\tau_{i,j}}(f, x) = \sum_{k=1}^{m-1} \sum_{1 \leq i_1 < \cdots < i_{k-1} \leq m} \sum_{\substack{1 \leq i < j \leq m \\ i, j \notin \{i_1, \ldots, i_{k-1}\}}} \left[f(t_{i_1} + \cdots + t_{i_{k-1}} + t_i)\right.$$

$$+ f(t_{i_1} + \cdots + t_{i_{k-1}} + t_j) - 2f\left(t_{i_1} + \cdots + t_{i_{k-1}} + \frac{t_i + t_j}{2}\right)\bigg] p_{m,k}(x) \geq 0.$$

Hence, it follows that $L_\tau(f, x) \geq B_m(f, x)$, $\tau \in \mathcal{D}_m$. Also, from Aramă's formula (4.7), we have $B_m(f, x) \geq B_n(f, x)$. Consequently, $B_\lambda(f, x) \geq B_n(f, x)$. By taking into account Theorem 5.1.4 (iii), the proof is finished. ∎

5.2 Generalized Durrmeyer type operators

5.2.1 Durrmeyer type operators with general weights

Denote by Γ, the gamma function and by B, the beta function. For any real numbers $0 < s < t$, define

$$\binom{t}{s} := \frac{\Gamma(t+1)}{\Gamma(s+1)\Gamma(t-s+1)} = B(t+1, t-s+1)^{-1}\frac{1}{t+1} \tag{5.50}$$

and

$$p_{t,s}(x) := \binom{t}{s}x^s(1-x)^{t-s}, \ x \in [0, 1]. \tag{5.51}$$

We consider the following sequence of positive linear functions on the space $C[0, 1]$:

Definition 5.2.1. *Let $a, b \in \mathbb{R}$ and let the function $h \in C[0, 1]$, $h(t) > 0$, for $t \in [0, 1]$. Put $\mu(t) = t^a(1-t)^b h(t)$, $t \in (0, 1)$. Put $p := \max\{0, [-a]\}$ and $q := \max\{0, [-b]\}$. For any $n \in \mathbb{N}$, $n \geq p + q$, for $f \in C[0, 1]$ and $x \in [0, 1]$, define*

$$L_n^\mu(f, x) := \sum_{k=0}^n p_{n,k}(x) \cdot H_{n,k}^\mu(f), \tag{5.52}$$

where

$$H_{n,k}^\mu(f) := \lim_{\rho \searrow 0, \ \eta \searrow 0} \frac{\int_\rho^{1-\eta} f(t)h(t)p_{a+b+n,a+k}(t)\,dt}{\int_\rho^{1-\eta} h(t)p_{a+b+n,a+k}(t)\,dt}. \tag{5.53}$$

Remark 5.2.1. The operators L_n^μ were introduced in [106]. In the case $a = b = 0$ and $h(t) = 1$, (the Lagrange weight), the operators L_n^μ become the well-known Durrmeyer operators [31], introduced independently also by A. Lupaş [62]. M. Dierrennic [25] obtained that, in this case, the operators L_n^μ admit another representation as modified Fourier–Lagrange sums. For $a > -1$, $b > -1$ and $h(t) = 1$, (the Jacobi weight), the operators L_n^μ were introduced both in the Bernstein form as well as modified Fourier–Jacobi sums in our paper [77] and independently by Berens and Xu [13]. In [13] it is pointed out that in the second form these operators coincide with the operators introduced by Kogbetliantz as the de la Vallée Poussin means of the ultraspheric series. Moreover, for $a = b = -\frac{1}{2}$, $h(t) = 1$ the operators L_n^μ coincide with the classical singular integrals of de la Vallée Poussin (via the transform $x \to \Theta = \arccos(1 - 2x)$). The Durrmeyer operators with Jacobi weight were extensively studied by many authors and they were extended in the many-dimensional case; see, for instance, Mache [67], Derriennic [26], Ditzian [29], Waldron [139]. Linear combinations of Durrmeyer operators with Jacobi weights in the case $a = b$ were considered by Lupaş [64] in order to improve their power of approximation.

In a similar manner in which the Bernstein operators were modified by Durrmeyer, many other operators were modified afterwards and Durrmeyer versions of them were obtained.

Proposition 5.2.1. *The operators L_n^μ are well defined and can be expressed in the form*

$$L_n^\mu(f, x) := f(0) \sum_{k=0}^{p-1} p_{n,k}(x) + f(1) \sum_{k=n-q+1}^{n} p_{n,k}(x)$$

$$+ \sum_{k=p}^{n-q} p_{n,k}(x) \frac{\int_0^1 p_{n,k}(t) t^a (1-t)^b h(t) f(t) \, dt}{\int_0^1 p_{n,k}(t) t^a (1-t)^b h(t) \, dt}, \tag{5.54}$$

where $p := \max\{0, [-a]\}$, $q := \max\{0, [-b]\}$, $f \in C[0,1]$, $x \in [0,1]$, $n \in \mathbb{N}$, $n \geq p + q$.

Proof. We have to show that the terms $H_{n,k}^\mu(f)$, $0 \leq k \leq n$ have the form given in (5.54). Let $0 \leq k \leq p-1$. Then $n-k > q$ and hence the integrals $\int_\rho^{1-0} p_{n,k}(t) t^a (1-t)^b h(t) f(t) \, dt$ and $\int_\rho^{1-0} p_{n,k}(t) t^a (1-t)^b h(t) \, dt$ are well defined for any $0 < \rho < 1$. For any $\varepsilon > 0$, there exists $0 < \rho_\varepsilon < 1$, such that $|f(y) - f(0)| < \varepsilon$, for any $y \in (0, \rho_\varepsilon)$. Then, for $0 < \rho < \rho_\varepsilon$, we have:

$$\left| \frac{\int_\rho^{1-0} p_{n,k}(t) t^a (1-t)^b h(t) f(t) \, dt}{\int_\rho^{1-0} p_{n,k}(t) t^a (1-t)^b h(t) \, dt} - f(0) \right|$$

$$\leq \frac{\varepsilon \int_\rho^{\rho_\varepsilon} p_{n,k}(t) t^a (1-t)^b h(t) \, dt + 2\|f\| \int_{\rho_\varepsilon}^{1-0} p_{n,k}(t) t^a (1-t)^b h(t) \, dt}{\int_\rho^{1-0} p_{n,k}(t) t^a (1-t)^b h(t) \, dt}$$

$$= \frac{\varepsilon + 2\|f\| T(\rho)}{1 + T(\rho)},$$

where

$$T(\rho) := \frac{\int_{\rho_\varepsilon}^{1-0} p_{n,k}(t) t^a (1-t)^b h(t) \, dt}{\int_\rho^{\rho_\varepsilon} p_{n,k}(t) t^a (1-t)^b h(t) \, dt}.$$

Since h is continuous and $h(0) > 0$, we have $\lim_{\rho \searrow 0} T(\rho) = 0$. Hence, there is $0 < \delta_\varepsilon < \rho_\varepsilon$, such that

$$\left| \frac{\int_\rho^{1-0} p_{n,k}(t) t^a (1-t)^b h(t) f(t) \, dt}{\int_\rho^{1-0} p_{n,k}(t) t^a (1-t)^b h(t) \, dt} - f(0) \right| \leq 2\varepsilon,$$

for any $0 < \rho < \delta_\varepsilon$. We get $H_{n,k}^\mu(f) = f(0)$.

For the case where $k \geq n - q + 1$, the proof is similar. For $p \leq k \leq n - q$ the proof is obvious. $\qquad \square$

First we consider the problem of approximation of continuous functions.

Theorem 5.2.1. *The sequence* $(L_n^\mu(f))_n$ *is uniformly convergent to* f *on the interval* $[0, 1]$, *for any* $f \in C[0, 1]$.

Proof. It suffices to show that $\lim_{n\to\infty} L_n^\mu(e_j) = e_j$, $j = 0, 1, 2$, uniformly.

Let $0 < \varepsilon < 1$. There is $m \in \mathbb{N}$, $m > -a - b$, such that $\|B_m(h) - h\| < \varepsilon \min_{t\in[0,1]} |h(t)|$, where $B_m(h)$ is the Bernstein polynomial of degree m corresponding to h. Define

$$\sigma(t) := B_m(t)t^a(1 - t)^b, \quad t \in [0, 1].$$

We have for any $n \geq p + q$ and $j \geq 0$:

$$L_n^\sigma(e_j) \leq \frac{1 + \varepsilon}{1 - \varepsilon} \sum_{k=p}^{n-q} \frac{\int_0^1 t^{a+k+j}(1 - t)^{b+n-k}h(t)\,dt}{\int_0^1 t^{a+k}(1 - t)^{b+n-k}h(t)\,dt} p_{n,k} = \frac{1 + \varepsilon}{1 - \varepsilon} L_n^\mu(e_j)$$

$$\leq L_n^\mu(e_j) + \frac{2\varepsilon}{1 - \varepsilon}.$$

Similarly, we have $L_n^\sigma(e_j) \geq L_n^\mu(e_j) - \frac{2\varepsilon}{1+\varepsilon}$. It remains to show that

$$\lim_{n\to\infty} L_n^\sigma(e_j) = e_j, \quad j = 0, 1, 2, \quad \text{uniformly.}$$

Put $D_i := h\left(\frac{i}{m}\right)\binom{m}{i}$, $0 \leq i \leq m$. For $n \geq p + q + 2$, $j \geq 0$, $j \geq 0$ and $x \in [0, 1]$ we have

$$L_n^\sigma(e_j, x) = \sum_{k=p}^{n-q} \frac{\sum_{i=0}^{m} D_i B(k + i + a + 1 + j, n - k + m - i + b + 1)}{\sum_{i=0}^{m} D_i B(k + i + a + 1, n - k + m - i + b + 1)} \cdot p_{n,k}(x).$$

(5.55)

First we have $L_n^\sigma(e_0) = e_0$. Then from (5.55) it follows that

$$L_n^\sigma(e_1, x) = \frac{1}{n + m + a + b + 2} \sum_{k=0}^{n} k p_{n,k}(x) + \sum_{k=0}^{n} c_{n,k}^1 p_{n,k}(x)$$

$$= \frac{n}{n + m + a + b + 2} \cdot x + \sum_{k=0}^{n} c_{n,k}^1 p_{n,k}(x),$$

where

$$c_{n,k}^1 := \begin{cases} -\frac{k}{n+m+a+b+2}, & 0 \leq k \leq p - 1, \\[2mm] \dfrac{\sum_{i=0}^{m} D_i B(k+i+a+1, n-k+m-i+b+1)(i+a+1)}{\sum_{i=0}^{m} D_i B(k+i+a+1, n-k+m-i+b+1)(n+m+a+b+2)}, & p \leq k \leq n - q, \\[2mm] \frac{n-k+m+a+b+2}{n+m+a+b+2}, & n - q + 1 \leq k \leq n. \end{cases}$$

It remains to show that $\lim_{n\to\infty} \sum_{k=0}^{n} c_{n,k}^1 p_{n,k}(x) = 0$, uniformly on $[0, 1]$. This follows from the inequalities

$$\left| \sum_{k=0}^{p-1} c_{n,k}^1 p_{n,k}(x) \right| \le \sum_{k=0}^{p-1} |c_{n,k}^1| \le \frac{p^2}{n+m+a+b+2},$$

$$\left| \sum_{k=n-q+1}^{n} c_{n,k}^1 p_{n,k}(x) \right| \le \sum_{k=n-q+1}^{n} |c_{n,k}^1| \le \frac{q(m+a+b+q+1)}{n+m+a+b+2},$$

$$\left| \sum_{k=p}^{n-q} c_{n,k}^1 p_{n,k}(x) \right| \le \frac{\max\{|a+1|, |m+a+1|\}}{n+m+a+b+2}.$$

Also, denoting $d_n := (n+m+a+b+2)(n+m+a+b+3)$, in analogous mode, we get

$$L_n^\sigma(e_2, x) = \frac{1}{d_n} \sum_{k=0}^{n} k(k-1) p_{n,k}(x) + \sum_{k=0}^{n} c_{n,k}^2 p_{n,k}(x)$$

$$= \frac{n(n-1)}{d_n} \cdot x^2 + \sum_{k=0}^{n} c_{n,k}^2 p_{n,k}(x),$$

where

$$c_{n,k}^2 := \begin{cases} -\frac{k(k-1)}{d_n}, & 0 \le k \le p-1, \\[2mm] \dfrac{\sum\limits_{i=0}^{m} D_i B(k+i+a+1, n-k+m-i+b+1)[k(2i+2a+4)+(i+a+1)(i+a+2)]}{d_n \sum\limits_{i=0}^{m} D_i B(k+i+a+1, n-k+m-i+b+1)}, & p \le k \le n-q, \\[4mm] 1 - \frac{k(k-1)}{d_n}, & n-q+1 \le k \\ & \le n. \end{cases}$$

Since $\lim\limits_{n\to\infty} \frac{n(n-1)}{d_n} x^2 = x^2$, uniformly on $[0, 1]$, it remains to show the uniform limit

$$\lim_{n\to\infty} \sum_{k=0}^{n} c_{n,k}^2 p_{n,k}(x) = 0.$$

This follows from the inequalities below.

$$\left| \sum_{k=0}^{p-1} c_{n,k}^2 p_{n,k}(x) \right| \le \frac{p^3}{d_n},$$

$$\left| \sum_{k=n-q+1}^{n} c_{n,k}^2 p_{n,k}(x) \right| \le q \cdot \frac{(n+m+a+b+3)^2 - (k-1)^2}{d_n}$$

$$\le q \cdot \frac{(m+a+b+q+3)(2n+m+a+b+2)}{d_n},$$

$$\left| \sum_{k=p}^{n-q} c_{n,k}^2 p_{n,k}(x) \right| \le \frac{n(2m+|2a+4|) + (m+|a+1|)(m+|a+2|)}{d_n}. \qquad \square$$

5.2.2 Durrmeyer type operators with generalized Jacobi weights

We consider now the particular case of the operators (5.52), (5.53), when $h(t) = 1$, $t \in [0, 1]$ and $a, b \in \mathbb{R}$. Denote by $M_n^{a,b}$ this particular operator. Using Proposition 5.2.1 and notation (5.51) we can rewrite this operator in the form

$$M_n^{a,b}(f, x) = f(0) \sum_{k=0}^{p-1} p_{n,k}(x) + f(1) \sum_{k=n-q+1}^{n} p_{n,k}(x)$$

$$+ (a + b + n + 1) \sum_{k=p}^{n-q} p_{n,k}(x) \int_0^1 f(t) \, p_{a+b+n,a+k}(t) \, dt, (5.56)$$

for $f \in C[0, 1]$ and $x \in [0, 1]$ and where $p := \max\{0, [-a]\}$ and $q := \max\{0, [-b]\}$. Also, we use the notation

$$M_n^{a,b}(f, x) = \sum_{k=0}^{n} H_{n,k}^{a,b}(f) \cdot p_{n,k}(x), \tag{5.57}$$

where the functionals $H_{n,k}^{a,b}$, $0 \le k \le n$ are defined in a clear mode.

In order to estimate the degree of approximation by the operators $M_n^{a,b}$ we need the following lemma.

Lemma 5.2.1. *For any $a, b \in \mathbb{R}$, $n \in \mathbb{N}$, $n \ge p + q$ and $x \in [0, 1]$, we have*

$$M_n^{a,b}(e_0, x) = 1 \tag{5.58}$$

and for $m \in \mathbb{N}$ we have

$$M_n^{a,b}(e_m, x) =$$

$$\sum_{k=p}^{n-q} \frac{(a + k + 1)(a + k + 2) \dots (a + k + m)}{(a + b + n + 2)(a + b + n + 3) \dots (a + b + n + m + 1)} \cdot p_{n,k}(x)$$

$$+ \sum_{k=n-q+1}^{n} p_{n,k}(x). \tag{5.59}$$

Proof. The relation follows immediately from (5.56). □

Theorem 5.2.2. *For $a, b \in \mathbb{R}$, $n \ge p + q + 2$, $f \in C[0, 1]$, $h > 0$ and $x \in [0, 1]$, we have*

$$|M_n^{a,b}(f, x) - f(x)| \le \frac{c}{n - p - q} \cdot h^{-1} \omega_1(f, h) + \left(1 + \frac{2c}{n - p - q} \cdot h^{-2}\right) \omega_2(f, h), \tag{5.60}$$

where $c := \max\{|a + 1|, |a + 2|, |b + 1|, |b + 2|\}$.

Proof. We use the identities

$$\sum_{k=0}^{n}\left(\frac{k}{n}-x\right)p_{n,k}(x)=0,\ \sum_{k=0}^{n}\left(\frac{k(k-1)}{n(n-1)}-2x\cdot\frac{k}{n}+x^2\right)p_{n,k}(x)=0.$$

In order to apply Corollary 2.2.1 (ii) we make the estimates

$$|M_n^{a,b}(e_1-xe_0,x)|$$

$$=\left|-\sum_{k=0}^{p-1}\frac{k}{n}\cdot p_{n,k}(x)+\sum_{k=p}^{n-q}\frac{(a+1)(n-k)-(b+1)k}{n(a+b+n+2)}\cdot p_{n,k}(x)\right.$$

$$\left.+\sum_{k=n-q+1}^{n}\frac{n-k}{n}\cdot p_{n,k}(x)\right|$$

$$\le\frac{|p-1|}{n}\sum_{k=0}^{p-1}p_{n,k}(x)+\frac{c}{a+b+n+2}\sum_{k=0}^{n-q}p_{n,k}(x)+\frac{|q-1|}{n}\sum_{k=n-q+1}^{n}p_{n,k}(x)$$

$$\le\frac{c}{n-p-q}.$$

Also we have:

$$M_n^{a,b}((e_1-xe_0)^2,x)=x^2\sum_{k=0}^{p-1}p_{n,k}(x)+(1-x)^2\sum_{k=n-q+1}^{n}p_{n,k}(x)$$

$$+\sum_{k=p}^{n-q}\left[\frac{(a+k+1)(a+k+2)}{(a+b+n+2)(a+b+n+3)}-2x\cdot\frac{a+k+1}{a+b+n+2}+x^2\right]p_{n,k}(x)$$

$$=\sum_{k=0}^{p-1}\left[2x\cdot\frac{k}{n}-\frac{k(k-1)}{n(n-1)}\right]p_{n,k}(x)+\sum_{k=n-q+1}^{n}\left[1-2x\cdot\frac{n-k}{n}-\frac{k(k-1)}{n(n-1)}\right]p_{n,k}(x)$$

$$+\sum_{k=p}^{n-q}\left[\frac{(a+k+1)(a+k+2)}{(a+b+n+2)(a+b+n+3)}-\frac{k(k-1)}{n(n-1)}\right.$$

$$\left.+2x\left(\frac{k}{n}-\frac{a+k+1}{a+b+n+2}\right)\right]p_{n,k}(x)$$

$$=:T_1+T_2+T_3.$$

Then we have:

$$T_1\le 2x\sum_{k=0}^{p-1}\frac{k}{n}\cdot p_{n,k}(x)\le 2x\cdot\frac{|p-1|}{n}\sum_{k=0}^{p-1}p_{n,k}(x)\le\frac{2c}{n}\sum_{k=0}^{p-1}p_{n,k}(x),$$

and in a symmetrical mode

$$T_2=\sum_{k=n-q+1}^{n}\left[2(1-x)\frac{n-k}{n}-\frac{(n-k)(n-k-1)}{n(n-1)}\right]p_{n,k}(x)\le\frac{2c}{n}\sum_{k=n-q+1}^{n}p_{n,k}(x).$$

Finally

$$T_3 = \sum_{k=p}^{n-q} \left[\frac{a+k+1}{a+b+n+2} \cdot \frac{(a+2)(n-k)-k(b+1)}{n(a+b+n+3)} \right.$$

$$+ \frac{k[(a+2)(n-k)-(k-1)(b+1)]}{n(n-1)(a+b+n+2)} + 2x \cdot \left. \frac{k(b+1)-(n-k)(a+1)}{n(a+b+n+2)} \right] p_{n,k}(x)$$

$$\leq \sum_{k=p}^{n-q} \left[\frac{c(a+k+1)}{(a+b+n+2)(a+b+n+3)} \right.$$

$$+ \frac{ck}{n(a+b+n+2)} + \left. \frac{2cx}{a+b+n+2} \right] p_{n,k}(x) \leq \frac{4c}{n-p-q} \sum_{k=p}^{n-q} p_{n,k}(x).$$

Consequently $M_n^{a,b}((e_1 - xe_0)^2, x) \leq \frac{4c}{n-p-q}$. The theorem is proved. □

In the case $a > -1$, $b > -1$, we have a better estimate:

Theorem 5.2.3. *If $a > -1$, $b > -1$, then for any $f \in C[0, 1]$, $n \in \mathbb{N}$, $0 < h \leq \frac{1}{2}$ and $x \in [0, 1]$ we have*

$$|M_n^{a,b}(f, x) - f(x)| \leq \sigma_n^1(x)h^{-1}\omega_1(f, h) + \left(1 + \frac{1}{2} \cdot \sigma_n^2(x)h^{-2}\right) \omega_2(f, h),$$

$$(5.61)$$

where

$$\sigma_{n,a,b}^1(x) := \left| \frac{(a+1) - (a+b+2)x}{a+b+n+2} \right|,$$

$$\sigma_{n,a,b}^2(x)$$

$$:= \frac{2nx(1-x) + x^2(a+b+2)(a+b+3) - 2x(a+1)(a+b+3) + (a+1)(a+2)}{(a+b+n+2)(a+b+n+3)}.$$

Proof. From relations (5.59) and from the identities

$$\sum_{k=0}^{n} p_{n,k}(x) = 1, \quad \sum_{k=0}^{n} \frac{k}{n} \cdot p_{n,k}(x) = x, \quad \sum_{k=0}^{n} \frac{k(k-1)}{n(n-1)} p_{n,k}(x) = x^2,$$

we get that

$$M_n^{a,b}(e_0, x) = 1, \quad |M_n^{a,b}(e_1 - xe_0, x)| = \sigma_{n,a,b}^1(x), \quad M_n^{a,b}((e_1 - xe_0)^2, x) = \sigma_{n,a,b}^2(x).$$

Then we apply Corollary 2.2.1, for $s = 2$. □

The case $a = -1$ and $b = -1$ is worth noticing. These operators were considered by Goodman and Sharma [47], Parvanov and Popov [76] and Gavrea [34]. In [99] the limit properties given in the next two theorems is proved.

Theorem 5.2.4. *For any fixed $f \in C[0, 1]$, $x \in [0, 1]$ and $n \in \mathbb{N}$, we have*

$$\lim_{\substack{a \searrow -1 \\ b \searrow -1}} M_n^{a,b}(f, x) = M_n^{-1,-1}(f, x), \quad \text{uniformly with respect to } x \in [0, 1]. \quad (5.62)$$

Proof. We have the relation

$$\lim_{a\searrow-1} (a+1) \int_0^1 g(t)t^a \, dt = g(0), \quad \text{for } g \in C[0,1]. \tag{5.63}$$

Indeed, let $\varepsilon > 0$ arbitrarily. There exists $0 < \rho_\varepsilon < 1$, such that $|g(t) - g(0)| < \varepsilon$, for any $t \in [0, \rho_\varepsilon]$. Then, for any $a \in (-1, 0)$ we have

$$\left| (a+1) \int_0^1 g(t)t^a \, dt - g(0) \right| = \left| (a+1) \int_0^1 (g(t) - g(0))t^a \, dt \right|$$

$$\leq \varepsilon(a+1) \int_0^{\rho_\varepsilon} t^a \, dt + 2(a+1)\|g\| \leq \varepsilon + 2(a+1)\|g\|.$$

From relation (5.63) it follows immediately that

$$\lim_{\substack{a\searrow-1\\b\searrow-1}} \int_0^1 g(t)t^a(1-t)^{n+b} \, dt = g(0), \quad \text{for } g \in C[0,1], \ n \in \mathbb{N}. \tag{5.64}$$

Fix f and n. We use the notation (5.57). Note that, for any $a > -1, b > -1$ and any $x \in [0,1]$ we have

$$\left| M_n^{a,b}(f,x) - M_n^{-1,-1}(f,x) \right| \leq \max_{0 \leq k \leq n} \left| H_{n,k}^{a,b}(f) - H_{n,k}^{-1,-1}(f) \right|.$$

It remains to show that

$$\lim_{\substack{a\searrow-1\\b\searrow-1}} H_{n,k}^{a,b}(f) = H_{n,k}^{-1,-1}(f), \quad \text{for } 0 \leq k \leq n. \tag{5.65}$$

For $1 \leq k \leq n-1$ this limit is immediate. For $k = 0$, using relation (5.64) we obtain

$$\lim_{\substack{a\searrow-1\\b\searrow-1}} H_{n,0}^{a,b}(f) = \lim_{\substack{a\searrow-1\\b\searrow-1}} \frac{(a+1)\int_0^1 f(t)t^a(1-t)^{n+b}\,dt}{(a+1)\int_0^1 t^a(1-t)^{n+b}\,dt} = f(0) = H_{n,0}^{-1,-1}(f).$$

In a similar mode we can prove (5.65) for $k = n$. $\qquad\square$

Theorem 5.2.5. *The limit* $\lim_{\substack{a\searrow-1\\b\searrow-1}} M_n^{a,b} = M_n^{-1,-1}$, $n \geq 1$ *does not exist with regard to the operator norm.*

Proof. (Reductio ad absurdum) Suppose that there is $\eta > 0$ such that for any $-1 < a < -1+\eta$ we have $\|M_n^{-1,-1} - M_n^{a,a}\| < \frac{1}{2}$. That is $\|M_n^{-1,-1}(f) - M_n^{a,a}(f)\| < \frac{1}{2}$, for any $f \in C[0,1]$, with $\|f\| \leq 1$. Fix $a \in (-1, \frac{1}{2})$ and $n \geq 1$. For any $0 < \delta < 1$, define the function $f_\delta \in C[0,1]$, by $f_\delta(0) = 1$, $f_\delta(x) = 0$, $(x \in [\delta, 1])$ and f_δ is linear on the interval $[0, \delta]$. Then $\|f_\delta\| = 1$, $M_n^{-1,-1}(f_\delta, 0) = 1$ and $\lim_{\delta\to0} M_n^{a,a}(f_\delta, 0) = 0$. Contradiction. $\qquad\square$

A characteristic property of the operators $M_n^{-1,-1}$, which is not satisfied by the operators $M_n^{a,b}$, with $(a,b) \neq (-1,-1)$, is given by the following proposition.

Proposition 5.2.2. *The operators $M_n^{-1,-1}$ preserve linear functions.*

Proof. From Theorem 5.2.4 and the relations given in the statement of Theorem 5.2.3 we get for any $n \in \mathbb{N}$ and $x \in [0, 1]$:

$$M_n^{-1,-1}(e_0, x) = \lim_{\substack{a\searrow-1 \\ b\searrow-1}} M_n^{a,b}(e_0, x) = 1, \text{ and}$$

$$M_n^{-1,-1}(e_1 - xe, x) = \lim_{\substack{a\searrow-1 \\ b\searrow-1}} M_n^{a,b}(e_1 - xe_0, x) = 0. \quad \square$$

In the particular case of the operators $M_n^{-1,-1}$ we obtain the following estimates.

Theorem 5.2.6. *For any $f \in C[0, 1]$, $n \in \mathbb{N}$, $h > 0$ and $x \in [0, 1]$ we have*

$$|M_n^{-1,-1}(f, x) - f(x)| \le \left(1 + h^{-2} \cdot \frac{x(1 - x)}{n + 1}\right) \omega_2(f, h). \tag{5.66}$$

Proof. We compute $\lim_{\substack{a\searrow-1 \\ b\searrow-1}} \sigma_{n,a,b}^j(x)$, for $j = 1, 2$, see Theorem 5.2.3, and then we apply Corollary 2.2.1, for $s = 2$. $\quad \square$

Corollary 5.2.1. *For $f \in C[0, 1]$, $n \in \mathbb{N}$, $x \in (0, 1)$, we have*

$$|M_n^{-1,-1}(f, x) - f(x)| \le 2\omega_2\left(f, \sqrt{\frac{x(1 - x)}{n + 1}}\right) \tag{5.67}$$

and

$$\|M_n^{-1,-1}(f) - f\| \le \frac{5}{4} \cdot \omega_2\left(f, \frac{1}{\sqrt{n + 1}}\right). \tag{5.68}$$

Now we consider the problem of the simultaneous approximation.

Theorem 5.2.7. *For any $a, b \in \mathbb{R}$, any $f \in C^r[0, 1]$, and for any $0 < \alpha < \beta < 1$, we have*

$$\lim_{n\to\infty} \|(M_n^{a,b}(f))^{(r)} - f^{(r)}\|_{[\alpha,\beta]} = 0. \tag{5.69}$$

Proof. Let $f \in C^r[0, 1]$, $n > p + q + r$ and $x \in [\alpha, \beta]$. Using formula (5.57), we have successively

$$(M_n^{a,b}(f, x))^{(r)} = \sum_{k=0}^{n} \binom{n}{k} \sum_{j=0}^{r} \binom{r}{j}(x^k)^{(j)}((1 - x)^{n-k})^{(r-j)} H_{n,k}^{a,b}(f)$$

$$= \frac{n!}{(n - r)!} \sum_{k=0}^{n} \sum_{j=\max\{0,r-n+k\}}^{\min\{r,k\}} (-1)^{r-j} \binom{r}{j} p_{n-r,k-j}(x) H_{n,k}^{a,b}(f)$$

$$= \frac{n!}{(n - r)!} \sum_{k=0}^{n-r} p_{n-r,k}(x) \sum_{j=0}^{r}(-1)^{r+j} \binom{r}{j} H_{n,k+j}^{a,b}(f).$$

For $p \leq k \leq n - q - r$, we get

$$\sum_{j=0}^{r}(-1)^{r+j}\binom{r}{j}H_{n,k+j}^{a,b}(f)$$

$$= (a+b+n+1)\int_{0}^{1}f(t)\left[\sum_{j=0}^{r}(-1)^{r+j}\binom{r}{j}P_{a+b+n,a+k+j}(t)\right]dt$$

$$= (-1)^{r}\frac{\Gamma(a+b+n+2)}{\Gamma(a+b+n+r+1)}\int_{0}^{1}f(t)(p_{a+b+n+r,a+k+r})^{(r)}(t)\,dt.$$

Using the integration r- times by parts, it follows that

$$(-1)^{r}\int_{0}^{1}f(t)(p_{a+b+n+r,a+k+r})^{(r)}(t) = \int_{0}^{1}f^{(r)}(t)p_{a+b+n+r,a+k+r}(t)\,dt.$$

Consequently, one obtains

$$(M_{n}^{a,b}(f,x))^{(r)} = \frac{n!\Gamma(a+b+n+2)}{(n-r)!\Gamma(a+b+n+r+2)}\cdot M_{n-r}^{a+r,b+r}(f^{(r)},x) + R_{n},$$

$$(5.70)$$

where

$$R_{n} := \frac{n!}{(n-r)!}\left(\sum_{k=0}^{p-1}S_{k} + \sum_{k=n-q-r+1}^{n-r}S_{k}\right),$$

$$S_{k} := p_{n-r,k}(x)\left[\sum_{j=0}^{r}(-1)^{r+j}\binom{r}{j}H_{n,k+j}^{a,b}(f)\right.$$

$$\left. - \frac{\Gamma(a+b+n+2)}{\Gamma(a+b+n+r+2)}\cdot H_{n-r,k}^{a+r,b+r}(f^{(r)})\right].$$

From Theorem 5.2.1, it follows that

$$\lim_{n\to\infty}\frac{n!\Gamma(a+b+n+2)}{(n-r)!\Gamma(a+b+n+r+2)}\cdot M_{n-r}^{a+r,b+r}(f^{(r)},x) = f^{(r)}(x),$$

uniformly related to $x \in [0, 1]$. Also, we have

$$|R_{n}| \leq \left(2^{r}\frac{n!}{(n-r)!}\|f\|_{[0,1]} + \frac{n!\Gamma(a+b+n+2)}{(n-r)!\Gamma(a+b+n+r+2)}\cdot\|f^{(r)}\|_{[0,1]}\right)$$

$$\times\left[\sum_{k=0}^{p-1}p_{n-r,k}(x) + \sum_{k=n-q-r+1}^{n-r}p_{n-r,k}(x)\right].$$

$$(5.71)$$

Since

$$\lim_{n\to\infty}n^{r}\left(\sum_{k=0}^{p-1}p_{n-r,k}(x) + \sum_{k=n-q-r+1}^{n-r}p_{n-r,k}(x)\right) = 0, \quad \text{uniformly related to } x \in [\alpha, \beta],$$

the theorem is proved. $\qquad\square$

Remark 5.2.2. In Theorem 5.2.7, in the general case, the interval $[\alpha, \beta]$ cannot be replaced by the whole interval $[0, 1]$. Indeed, if we take $a = -3$, $b = 0$ and $f_0(t) := t$, $t \in [0, 1]$, then we have

$$M_n^{-3,0}(f_0, x) = \sum_{k=3}^{n} \frac{k-2}{n-1} p_{n,k}(x)$$

$$= \frac{n}{n-1} x - \frac{2}{n-1} + \frac{2}{n-1}(1-x)^n + \frac{n}{n-1} x(1-x)^{n-1}$$

and

$$(M_n^{-3,0}(f_0, x))' = \frac{n}{n-1} - \frac{n}{n-1}(1-x)^{n-1} - nx(1-x)^{n-2}.$$

Hence

$$\left(M_n^{-3,0}\left(f_0, \frac{1}{n} \right) \right)'$$

$$= \frac{n}{n-1} - \frac{n}{n-1}\left(1 - \frac{1}{n} \right)^{n-1} - \left(1 - \frac{1}{n} \right)^{n-2} \rightarrow 1 - \frac{2}{e}, \ (n \rightarrow \infty).$$

In the particular case $a > -1$, $b > -1$ we have

Theorem 5.2.8. If $a > -1$, $b > -1$, then for any $f \in C^r[0, 1]$, and for any $n \in \mathbb{N}$, $r \in \mathbb{N}$, $r < n$, $x \in (0, 1)$ and $0 \le h \le \frac{1}{2}$ we have

$$|(M_n^{a,b}(f, x))^{(r)} - f^{(r)}(x)| \le |q_{n,a,b,r} - 1| \cdot |f^{(r)}(x)|$$
$$+ q_{n,a,b,r} \cdot \sigma_{n-r,a+r,b+r}^1(x)h^{-1}\omega_1(f, h) \tag{5.72}$$
$$+ q_{n,a,b,r}\left(1 + \frac{1}{2} \cdot \sigma_{n-r,a+r,b+r}^2(x)h^{-2} \right)\omega_2(f, h),$$

where $\sigma_{n-r,a+r,b+r}^1(x)$ and $\sigma_{n-r,a+r,b+r}^2(x)$ are defined as in Theorem 5.2.3 and

$$q_{n,a,b,r} := \frac{n!\Gamma(a+b+n+2)}{(n-r)!\Gamma(a+b+n+r+2)}. \tag{5.73}$$

Proof. From relations (5.70), since $p = 0$ and $q = 0$, it follows that

$$(M_n^{a,b}(f, x))^{(r)} = q_{n,a,b,r} \cdot M_{n-r}^{a+r,b+r}(f^{(r)}, x). \tag{5.74}$$

Then we apply Theorem 5.2.3. □

At the end, we briefly discuss the eigenvalues of the operators $M_n^{a,b}$, $a \ge -1$, $b \ge -1$. For $a, b > -1$ a main property of the operators $M_n^{a,b}$ is their representation as modified Fourier finite sums. Let us denote by $< \cdot, \cdot >_{a,b}$ the inner product

$$< f, g >_{a,b} := \int_0^1 f(t)g(t)t^a(1-t)^b \, dt, \ f, g \in C[0, 1]. \tag{5.75}$$

Denote by $P_n^{a,b}$ the Jacobi polynomials, for the weight function $t^a(1-t)^b$, $t \in [0, 1]$. Then we have, see [77],

$$M_n^{a,b}(f,x) = \sum_{m=0}^{n} \lambda_{n,m}^{a,b} < f, P_m^{a,b} >_{a,b} \cdot P_m^{a,b}(x), \quad f \in C[0,1], \ x \in [0,1], \ n \in \mathbb{N},$$

$$(5.76)$$

where

$$\lambda_{n,m}^{a,b} := \frac{\Gamma(n+1)\Gamma(a+b+n+2)}{\Gamma(n-m+1)\Gamma(a+b+n+m+2)}, \quad 0 \le m \le n. \quad (5.77)$$

From relation (5.76) it follows that the operator $M_n^{a,b}$ is self-adjoint with respect to the inner product $< \cdot, \cdot >_{a,b}$ and preserves the degree of the polynomials with degree not greater than n. Conversely, if an operator $L : C[0,1] \to C[0,1]$ is self-adjoint with respect to the inner product $< \cdot, \cdot >_{a,b}$ and preserves the degree of the polynomials with degree not greater than n, then it is of the form

$$L(f,x) = \sum_{m=0}^{n} \gamma_m < f, P_m^{a,b} >_{a,b} \cdot P_m^{a,b}(x), \quad f \in C[0,1], \ x \in [0,1]. \quad (5.78)$$

We mention the following property of extremality, see [80]:

Theorem 5.2.9. *Define the set* $\mathcal{U}_{a,b}$ *of linear operators* $L : C[0,1] \to C[0,1]$ *of the form (5.78) that are convex of order* k, *for any* $0 \le k \le n$, *(see Definition 1.1.2) and such that* $L(e_0) = e_0$. *We have* $M_n^{a,b} \in \mathcal{U}_{a,b}$, *and for any operator* $L \in \mathcal{U}_{a,b}$, *of the form (5.78) we have*

$$\gamma_m \le \lambda_{n,m}^{a,b}. \quad (5.79)$$

When we pass from the class of the operators $M_n^{a,b}$, $a > -1$, $b > -1$, to the limit operator $M_n^{-1,-1}$, we gain, on the one hand, the ability to reproduce linear functions, but on the other hand we lose the ability to represent them as modified Fourier finite sums, similar to (5.76) with regard to any inner product. This fact is a consequence of the following result, see [93].

Theorem 5.2.10. *Let* X *be a Hausdorff compact space and let* $< \cdot, \cdot >$, *be an inner product on the space* $C(X)$ *of real continuous functions on* X. *Let* $U \subset C(X)$ *be a finite-dimensional subspace and let* $L : C(X) \to C(X)$ *be a linear operator of the form* $L(f) = \sum_{j=1}^{n} \lambda_j < f, p_j > r_j$, $f \in C(X)$, $p_j, r_j \in U$, $\lambda_j \in \mathbb{R}$, $(1 \le j \le n)$. *Define* $V := \{u \in U | L(u) = u\}$. *Suppose the following three conditions: a)* L *is a positive operator, b) for any* $u_1, u_2 \in U$, *such that* $u_1 \ne 0$, $u_2 \ne 0$, *we have* $u_1 \cdot u_2 \ne 0$, *and c) there is a strictly positive function* $v \in V$. *Then* $\dim V = 1$.

However, it is possible to extend the relation (5.76) in a weaker form. Define

$$c_{m,k} := (-1)^k \binom{m}{k} \frac{((m-1)!)^2}{(k-1)!(m-k-1)!}, \quad 1 \le k \le m-1, \ m \ge 2. \quad (5.80)$$

Consider the polynomials $\tilde{Q}_m$, defined by

$$\tilde{Q}_0(x) := 1, \quad \tilde{Q}_1(x) := 1 - 2x, \quad \tilde{Q}_m(x) := \sum_{k=1}^{m-1} c_{n,k} x^k (1-x)^{m-k}, \ m \geq 2.$$

$$(5.81)$$

Also, consider the functionals $F_m : C[0, 1] \to \mathbb{R}$, given by:

$$F_0(f) := f(0) + f(1), \quad F_1(f) := f(0) - f(1), \quad \text{and for } m \geq 2 : \quad (5.82)$$

$$F_m(f) := (m-1)![f(0) + (-1)^m f(1)] + \sum_{k=1}^{m-1} c_{m,k} \int_0^1 t^{k-1}(1-t)^{m-k-1} f(t)\, dt,$$

where $f \in C[0, 1]$. The relation

$$F_m(\tilde{Q}_j) = 0, \text{ for any } m, j \in \mathbb{N} \cup \{0\}, \ m \neq j \quad (5.83)$$

holds. With this notation we have, see [99]:

Theorem 5.2.11. *For any $f \in C[0, 1]$ any $n \in \mathbb{N}$, $n \geq 1$ and any $x \in [0, 1]$ we have*

$$M_n^{-1,-1}(f, x) = \sum_{m=0}^n \tilde{\lambda}_{n,m} \frac{F_m(f)}{F_m(\tilde{Q}_m)} \tilde{Q}_m(x), \quad (5.84)$$

where

$$\tilde{\lambda}_{n,m} := \frac{n!(n-1)!}{(n-m)!(n+m-1)!}, \quad 0 \leq m \leq n. \quad (5.85)$$

Remark 5.2.3. From Theorem 5.2.11 it follows that $\tilde{Q}_m$, $0 \leq m \leq n$ are the eigenvectors, and $\tilde{\lambda}_{n,m}$, $0 \leq m \leq n$ are the corresponding eigenvalues of the operator $M_n^{-1,-1}$. We have $\tilde{\lambda}_{n,0} = 1$, and $\tilde{\lambda}_{n,1} = 1$, which is equivalent with the fact that the operator $M_n^{-1,-1}$ preserves linear functions.

6

Approximation Operators for Vector-Valued Functions

6.1 Approximation of functions with real argument

6.1.1 Introduction. Generalized positive and convex operators

Let $(E, < \cdot, \cdot >)$ be a Euclidean space with the norm denoted by $\| \cdot \|$. Let $I = [a, b]$, $a < b$ be a real interval. Let $C(I, E)$ be the space of continuous functions, endowed with the sup-norm denoted by $\| \cdot \|_I$ and denote by $C^k(I, E)$, $k \geq 1$, the subspace of functions having a continuous derivative of order k on I. If $\varphi : I \to \mathbb{R}$ and $w \in E$ we denote by φw, the function $(\varphi w)(x) = \varphi(x)w$.

If $L : C(I, E) \to C(I, E)$ is a linear operator, denote $\|L\| := \sup_{\|f\|_I \leq 1} \|L(f)\|_I$. For any $f \in C(I, E)$ and any distinct point $y_1, \ldots, y_p$ of I, the divided difference of f on these points, denoted by $[f; y_1, \ldots, y_p]$, is defined as in the scalar case, see relation (1.5)

Definition 6.1.1. ([86]), ([89]) *A function $f : I \to E$ is called* **convex of order** k, $k \geq -1$ *if for any strictly ordered (increasing or decreasing) points $x_1, \ldots, x_{k+3}$ of I we have*

$$< [f; x_1, \ldots, x_{k+2}], [f; x_2, \ldots, x_{k+3}] >\geq 0, \tag{6.1}$$

or in an equivalent mode,

$$\|[f; x_1, \ldots, x_{k+2}] + t[f; x_2, \ldots, x_{k+3}]\| \geq \|[f; x_1, \ldots, x_{k+2}]\|, \quad \textit{for all } t > 0. \tag{6.2}$$

Denote by $K^k(I, E)$ the space of convex functions of order k.

Remark 6.1.1. ([86]) In the case $E = \mathbb{R}$, $f \in K^k(I, \mathbb{R})$ iff f or $-f$ is usual convex of order k.

Definition 6.1.2. *A linear operator $L : C(I, E) \to C^{k+1}(I, E)$, $k \geq -1$ is called* **convex of order** k, *if for any $f \in C^{k+1}(I, E)$ such that $f^{(k+1)} \in K^{-1}(I, E)$, we have $(L(f))^{(k+1)} \in K^{-1}(I, E)$. In the case $k = -1$ the operator L is named* **positive.**

We define now a more restrictive type of convexity of higher order.

Definition 6.1.3. A function $f \in \mathcal{F}(I, E)$ is said to be **completely convex of order** $k \geq -1$, if for any choice of two sequences of distinct points $x_1, \ldots, x_{k+2}$ and $y_1, \ldots, y_{k+2}$ of I we have

$$< [f; x_1, \ldots, x_{k+2}], [f; y_1, \ldots, y_{k+2}] > \geq 0. \tag{6.3}$$

Denote by $K_c^k(I, E)$ the space of functions that are completely convex of order k.

Correspondingly, we introduce the following class of (generally, nonlinear) operators:

Definition 6.1.4. *An operator $L : V \to \mathcal{F}(I, E)$, $V \subset \mathcal{F}(I, E)$ is said to be k-completely convex, $k \geq -1$ if*

$$L(f) \in K_c^k(I, E), \text{ for any } f \in V \cap K_c^k(I, E), \tag{6.4}$$

$$L(f) - L(g) \in K_c^k(I, E) \text{ for any } f, g \in V, \ f - g \in K_c^k(I, E). \tag{6.5}$$

6.1.2 A Korovkin type theorem

We prove a uniform boundedness type principle and a Korovkin type theorem for sequences of linear positive operators, see Definition 6.1.2.

Theorem 6.1.1. *[108] Let $(L_n)_n$ be a sequence of positive linear operators, $L_n : C(I, E) \to C(I, E)$ with the property:*

$$\lim_{n \to \infty} \|L_n(e_j w) - e_j w\|_I = 0, \text{ for all } w \in E, \text{ and } j = 0, 1. \tag{6.6}$$

Then, there is a positive integer n_0 such that $\|L_n\| \leq 8$, for any $n \geq n_0$.

Proof. Denote $\rho := \min\left\{\frac{1}{240}, \frac{1}{120} \cdot \frac{b-a}{1+|a|}\right\}$. Since the space E is Euclidean it follows that the limit in (6.6) is unifom related to w with $\|w\| = 1$ and j. Hence there is a natural number n_0 such that

$$\|L_n(e_j w) - e_j w\|_I < \rho, \text{ for all } w \in E, \|w\| = 1, \ j = 0, 1, \ n \geq n_0.$$

Fix a natural number $n \geq n_0$. Let $f \in C(I, E)$, $\|f\|_I \leq 1$. Suppose that $\|L_n(f)\|_I > 8$. Let $z \in I$ such that $\|L_n(f, z)\| = \|L_n(f)\|_I$. We distinguish between two cases.

  Case 1: For all $x \in I$ we have $\|L_n(f, x) - L_n(f, z)\| < \frac{1}{32}\|L_n(f)\|_I$.
It follows that $\|L_n(f, x) - L_n(f, y)\| \leq \frac{1}{16}\|L_n(f)\|_I$, for all $x, y \in I$ and also $\|L_n(f, x)\| \geq \frac{31}{32}\|L_n(f)\|_I$, for all $x \in I$. Define

$$w := \frac{L_n(f, a)}{\|L_n(f, a)\|}, \quad \mu := \frac{4}{5}\|L_n(f, a)\|, \quad g := \frac{\mu}{2(b-a)}((2b - 3a)e_0 + e_1)w - f.$$

We show that $g \in K^{-1}(I, E)$. Let $x, y \in I$ and $t > 0$. Note that $\mu > 6$. We have successively:

$$\|g(x) + tg(y)\| - \|g(x)\| = \|(1+t)g(x) + t(g(y) - g(x))\| - \|g(x)\|$$
$$\geq t(\|g(x)\| - \|g(x) - g(y)\|)$$
$$\geq t\left[\mu\left(1 + \frac{x-a}{2(b-a)}\right)\|w\| - 2\|f(x)\| - \|f(y)\| - \mu\frac{|y-x|}{2(b-a)}\|w\|\right]$$
$$\geq t\left(\frac{1}{2}\mu - 3\right) \geq 0.$$

Since L_n is positive it follows that $L_n(g) \in K^{-1}(I, E)$. But one obtains a contradiction:

$$\|L_n(g,a) + L_n(g,b)\| - \|L_n(g,a)\|$$
$$\leq \left\|\frac{5}{2}\mu w - L_n(f,a) - L_n(f,b)\right\| - \|\mu w - L_n(f,a)\|$$
$$+\mu\left[\left(3 + \frac{3|a|}{2(b-a)}\right)\|L_n(e_0 w) - e_0 w\|_I + \frac{3}{2(b-a)}\|L_n(e_1 w) - e_1 w\|_I\right]$$
$$\leq \|L_n(f,a) - L_n(f,b)\| - \frac{1}{4}\mu + \frac{\mu}{40} \leq \frac{1}{16}\|L_n(f)\|_I - \frac{9}{50}\|L_n(f,a)\| < 0.$$

<u>Case 2:</u> There exists $x \in I$ such that $\|L_n(f,x) - L_n(f,z)\| \geq \frac{1}{32}\|L_n(f)\|_I$. Define

$$\mu := \frac{1}{5}\|L_n(f,x) + 4L_n(f,z)\|, \quad w := \frac{1}{5\mu}(L_n(f,x) + 4L_n(f,z)), \quad g := \mu e_0 w - f.$$

We have $\mu > \frac{4}{5}\|L_n(f,z)\| - \frac{1}{5}\|L_n(f,x)\| \geq \frac{3}{5}\|L_n(f)\|_I \geq \frac{24}{5}$. For any $x_1, x_2 \in I$ and $t > 0$ we have

$$\|g(x_1) + tg(x_2)\| - \|g(x_1)\| \geq t(\|g(x_1)\| - \|g(x_1) - g(x_2)\|)$$
$$\geq t(\mu\|w\| - 2\|f(x_1)\| - \|f(x_2)\|) \geq t(\mu - 3) \geq 0.$$

Hence $g \in K^{-1}(I, E)$. Then $L_n(g) \in K^{-1}(I, E)$. But one obtains a contradiction:

$$\|L_n(g,x) + 4L_n(g,z)\| - \|L_n(g,x)\| \leq 6\mu\|L_n(e_0 w) - e_0 w\|_I - \|\mu w - L_n(f,x)\|$$
$$< 6\mu\rho - \frac{4}{5}\|L_n(f,z) - L_n(f,x)\| \leq \|L_n(f)\|_I\left(6\rho - \frac{1}{40}\right) \leq 0.$$

The contradictions obtained in both cases prove that we must have $\|L_n(f)\|_I \leq 8$. Since f, with $\|f\|_I \leq 1$ was given arbitrarily, the theorem is proved. $\qquad\square$

Theorem 6.1.2. [108] *Let* $(L_n)_n$ *be a sequence of positive linear operators* $L_n : C(I, E) \to C(I, E)$ *with the property*

$$\lim_{n\to\infty} \|L_n(e_j w) - e_j w\|_I = 0, \quad \text{for all } w \in E, \text{ and } j = 0, 1, 2. \quad (6.7)$$

Then, for any $f \in C(I, E)$ *we have*

$$\lim_{n\to\infty} \|L_n(f) - f\|_I = 0. \quad (6.8)$$

Proof. First, we consider the particular case where $f \in C^2(I, E)$. Let n_0 be the integer assured by Theorem 6.1.1. Take $x \in I$ and $0 < \varepsilon < 1$. Put $v := f'(x)$. We have

$$f(y) = f(x) + (y - x)v + \int_x^y (y - u)f''(u)du, \ y \in I.$$

Choose a number μ such that

$$\mu :> \max\left\{2\|f''\|_I, \ \frac{16}{(b-a)^2}(11\|f\|_I + (b-a)\|v\| + 1)\right\}.$$

For any $w \in E$ with $\|w\| = 1$ consider the functions

$$h_w := e_0 f(x) + (e_1 - xe_0)v + \mu(e_1 - xe_0)^2 w, \ \text{and} \ g_w := h_w - f.$$

We have $g_w \in K^{-1}(I, E)$. Indeed, let $y_1, y_2 \in I$. First consider that $y_1 \neq x$, $y_2 \neq x$. For $i = 1, 2$, put $T_i := (y_i - x)^{-2} \int_x^{y_i} (y_i - u)f''(u) \, du$. We have $\|T_i\| \leq \frac{1}{2}\|f''\|_I$. One obtains

$$< g(y_1), g(y_2) > = (y_1 - x)^2(y_2 - x)^2[\mu^2 - \mu(< w, T_1 > + < w, T_2 >) + < T_1, T_2 >]$$
$$\geq (y_1 - x)^2(y_2 - x)^2\left[\mu^2 - \mu\|f''\|_I - \frac{1}{4}\|f''\|_I^2\right] \geq 0.$$

For $y_1 = x$ or $y_2 = x$ the inequality above is immediate. Consequently, $L_n(g_w)$ is positive.

Since E is Euclidean, the limit (6.7) is uniform with respect to w, $\|w\| = 1$ and $j = 0, 1, 2$. Consequently there exists $n_x \in \mathbb{N}$, $n_x \geq n_0$, such that

$$\|L_n(h_w) - h_w\| < \frac{\varepsilon}{4}, \ \text{for all} \ w \in E, \ \|w\| = 1, \ n \geq n_x.$$

Suppose that there is $n \geq n_x$ such that $\|L_n(f, x) - f(x)\| \geq \frac{\varepsilon}{2}$. Consider, for a choice that $x \leq \frac{1}{2}(a + b)$. Choose

$$w := \frac{L_n(f, x) - f(x)}{\|L_n(f, x) - f(x)\|}, \ \text{and put} \ \lambda := \frac{\|L_n(f, x) - f(x)\|}{\mu(b - x)^2}.$$

Note that

$$h_w(x) = f(x), \ \text{and} \ \lambda \leq \frac{\|f\|_I + \|L_n(f)\|_I}{\mu(b - x)^2} \leq \frac{9\|f\|_I}{\mu(b - x)^2} < 1.$$

Also we have

$$\|g_w(b)\| \geq \mu(b - x)^2 - ((b - x)\|v\| + \|f(x)\| + \|f(b)\|).$$

It follows that

$$\|L_n(g_w, b) + L_n(g_w, x)\| - \|L_n(g_w, b)\| \le \|\lambda L_n(g_w, b) + L_n(g_w, x)\|$$
$$- \lambda \|L_n(g_w, b)\|$$
$$\le 2\lambda \|L_n(g_w, b) - g_w(b)\| + \|L_n(h_w, x) - h_w(x)\| + \|\lambda g_w(b) - L_n(f, x) + f(x)\|$$
$$- \lambda \|g_w(b)\|$$
$$\le 2\lambda (\|L_n(h_w, b) - h_w(b)\| + \|L_n(f, b)\| + \|f(b)\|) + \frac{\varepsilon}{4}$$
$$+ 2\lambda ((b - x)\|v\| + \|f(x)\| + \|f(b)\|) - \lambda (b - x)^2 \mu$$
$$\le 2\lambda \left(\frac{\varepsilon}{4} + 11\|f\|_I + (b - a)\|v\| \right) + \frac{\varepsilon}{4} - \lambda (b - x)^2 \mu < \frac{\varepsilon}{4} - \frac{1}{2}\lambda (b - x)^2 \mu \le 0.$$

The contradiction that we obtained proves that $\|L_n(f, x) - f(x)\| < \frac{\varepsilon}{2}$. From the continuity of the function $L_n(f) - f$, it follows that there is a neighbourhood V_x of x such that

$$\|L_n(f, y) - f(y)\| < \varepsilon, \text{ for all } y \in I \cap V_x, \ n \ge n_x.$$

Since I is compact, there are the points $x_1, \dots, x_m \in I$ such that $I \subset V_{x_1} \cup \cdots \cup V_{x_m}$. Set $n_\varepsilon := \max\{n_{x_1}, \dots, n_{x_m}\}$. Then we have $\|L_n(f) - f\|_I < \varepsilon$, for $n \ge n_\varepsilon$.

Consider now the general case, when $f \in C(I, E)$. Let $0 < \varepsilon < 1$. Choose $\tilde{f} \in C^2(I, E)$, such that $\|f - \tilde{f}\|_I < \frac{\varepsilon}{18}$. There is $n_\varepsilon \in \mathbb{N}$, $n_\varepsilon > n_0$, such that $\|L_n(\tilde{f}) - \tilde{f}\| < \frac{\varepsilon}{2}$, for $n \ge n_\varepsilon$. Then, for such integers n we have

$$\|L_n(f) - f\|_I \le \|L_n(f - \tilde{f})\|_I + \|L_n(\tilde{f}) - \tilde{f}\|_I + \|\tilde{f} - f\|_I$$
$$\le 9\|f - \tilde{f}\|_I + \|L_n(\tilde{f}) - \tilde{f}\|_I < \varepsilon. \quad \square$$

6.1.3 Simultaneous approximation

In this subsection we give generalizations of the theorem of Sendov and Popov [124] on simultaneous approximation. First we present a generalization, given in [108], for sequences of linear convex operators, see Definition 6.1.2.

Lemma 6.1.1. *If $a, b, v \in E$ are such that $\|a\| = \|b\| = \|v\| = 1$ and $< a, b > \le 0$, then $\max\{< a, v >, < b, v >\} \ge -\frac{1}{\sqrt{2}}$.*

Proof. Let $m := \dim E$. Suppose that $m \ge 2$ and $a \ne -b$, since otherwise, the lemma is obvious. Let $\{u_1, \dots, u_m\}$ be an orthonormal base such that $u_1 = \frac{a+b}{\|a+b\|}$, $u_2 = \frac{a-b}{\|a-b\|}$. We have the representation $v = \sum_{i=1}^{m} \lambda_i u_i$, where $\sum_{i=1}^{m} \lambda_i^2 = 1$. We get

$$\max\{< a, v >, < b, v >\}$$
$$= \max \left\{ \frac{\lambda_1(1 + < a, b >)}{\|a + b\|} + \frac{\lambda_2(1 - < a, b >)}{\|a - b\|}, \frac{\lambda_1(< a, b > + 1)}{\|a + b\|} \right.$$
$$\left. + \frac{\lambda_2(< a, b > - 1)}{\|a - b\|} \right\}$$
$$= \frac{\lambda_1(1 + < a, b >)}{\|a + b\|} + \frac{|\lambda_2|(1 - < a, b >)}{\|a - b\|}.$$

In this expression, let us consider a, b fixed and v variable. The minimum value is obtained for $\lambda_1 = -1$ and $\lambda_i = 0$, for $2 \le i \le m$ and it is equal to

$$-\frac{1+ <a, b>}{\|a+b\|} = -\frac{1}{\sqrt{2}} \cdot \sqrt{1+ <a, b>} \ge -\frac{1}{\sqrt{2}}. \qquad \square$$

Lemma 6.1.2. *If the sequence of functions $(f_n)_n$, $f_n \in C^1(I, E)$ is uniformly convergent on I to the function $f \in C^1(I, E)$, and if $(f_n)' \in K^0(I, E)$, $n \in \mathbb{N}$, then for any subinterval $[c, d] \subset (a, b)$, the sequence $((f_n)')_n$ is uniformly convergent on $[c, d]$ to f'.*

Proof. Suppose the contrary. Then there are a number $\lambda > 0$, a sequence $(x_k)_k$, $x_k \in [c, d]$ and a sequence of natural numbers $(n_k)_k$ such that

$$\|f'(x_k) - (f_{n_k})'(x_k)\| > \lambda, \ k \in \mathbb{N}.$$

There is $\delta_1 > 0$ such that $\|f'(x) - f'(y)\| < \frac{\lambda}{4}$, for all $x, y \in I$, $|x - y| < \delta_1$. Put $\delta := \min\{\delta_1, c-a, b-d\}$ and $\rho := \frac{1}{8}\lambda\delta$. Let k be fixed such that $\|f - f_{n_k}\|_{[a,b]} < \rho$. Put $g := f_{n_k}$ and

$$T_1 := \int_{x_k-\delta}^{x_k} (g'(t) - g'(x_k)) \, dt, \ T_2 := \int_{x_k}^{x_k+\delta} (g'(t) - g'(x_k)) \, dt.$$

Since $g' \in K^0(I, E)$ it follows that

$$< g'(t_1) - g'(x_k), \ g(t_2) - g(x_k) > \le 0, \ \text{for all } t_1 \in [a, x_k), \ t_2 \in (x_k, b].$$

If we approximate the integrals T_1 and T_2 by Riemann sums we get from above that

$$< T_1, \ T_2 > \le 0.$$

First consider that $T_i \ne 0$, $i = 1, 2$. Define $\alpha := \frac{T_1}{\|T_1\|}$, $\beta := \frac{T_2}{\|T_2\|}$, $v := \frac{u}{\|u\|}$, where $u := \delta(g'(x_k) - f'(x_k))$. By Lemma 6.1.1 we have that $\max\{< \alpha, v >, < \beta, v >\} \ge -\frac{1}{\sqrt{2}}$. Suppose, for a choice that $< \beta, v > \ge -\frac{1}{\sqrt{2}}$. We have successively

$$\|g(x_k + \delta) - f(x_k + \delta)\| = \|g(x_k) + \int_{x_k}^{x_k+\delta} g'(t) \, dt - f(x_k) - \int_{x_k}^{x_k+\delta} f'(t) \, dt\|$$

$$\ge \|\int_{x_k}^{x_k+\delta} (g'(t) - f'(t)) \, dt\| - \rho$$

$$\ge \|\int_{x_k}^{x_k+\delta} (g'(t) - f'(x_k)) \, dt\| - \|\int_{x_k}^{x_k+\delta} (f'(x_k) - f'(t)) \, dt\| - \rho$$

$$\ge \|\int_{x_k}^{x_k+\delta} (g'(t) - f'(x_k)) \, dt\| - 3\rho = \|T_2 + u\| - 3\rho$$

$$= \sqrt{\|T_2\|^2 + \|u\|^2 + 2 < T_2, u >} - 3\rho \ge \sqrt{\|T_2\|^2 + \|u\|^2 - \sqrt{2}\|T_2\| \, \|u\|} - 3\rho$$

$$\ge \frac{1}{\sqrt{2}}\|u\| - 3\rho > \frac{1}{\sqrt{2}}\lambda\delta - 3\rho > \rho.$$

Therefore we have obtained a contradiction. In the case $T_2 = 0$, then as above we obtain $\|g(x_k + \delta) - f(x_k + \delta)\| \geq \|u\| - 3\rho > \rho$. Contradiction. The lemma is proved. $\qquad\square$

Theorem 6.1.3. *Let* $(L_n)_n$ *be a sequence of linear operators* $L_n : C(I, E) \to C^{r+1}(I, E)$, $r \geq 1$, *with the properties:*

1) L_n *are convex of order* k *for* $-1 \leq k \leq r$.

2) $\lim\limits_{n \to \infty} \|L_n(e_j w) - e_j w\|_{[a,b]} = 0$, *for all* $w \in E$ *and* $j = 0, 1, 2$.

Then, for any $f \in C^r(I, E)$ *and any subinterval* $[c, d] \subset (a, b)$, *we have*

$$\lim_{n \to \infty} \|(L_n(f))^{(r)} - f^{(r)}\|_{[c,d]} = 0. \tag{6.9}$$

Proof. First we consider the case where $[a, b] \subset [0, \infty)$. Fix a subinterval $[c, d] \subset [a, b]$. For $0 \leq k \leq r$, consider the points $c_k := a + \frac{k}{r}(c - a)$ and $d_k := b - \frac{k}{r}(b - d)$. We prove by induction with respect to $0 \leq k \leq r$ the relations

$$\lim_{n \to \infty} \|(L_n(e_j w))^{(k)} - (e_j w)^{(k)}\|_{[c_k, d_k]} = 0, \text{ for all } w \in E, \ 0 \leq j \leq r + 2, \tag{6.10}$$

For $k = 0$ relations (6.10) are assured by the condition 2) of the theorem, by applying Theorem 6.1.2. Suppose now that relations (6.10) are true for $k < r$ and prove it for $k + 1$. Fix w and j. Since $[a, b] \subset [0, \infty)$, one can obtain immediately that $(e_j w)^{(k+2)} \in K^{-1}(I, E)$. Because L_n is convex of order $k+1$, then $(L_n(e_j w))^{(k+2)} \in K^{-1}(I, E)$. It follows that $(L_n(e_j w))^{(k+1)} \in K^0(I, E)$. Indeed, let us denote $g := (L_n(e_j w))^{(k+1)}$ and take the points $x_1 < x_2 < x_3$ of I. For any $s \in [x_1, x_2]$ and $t \in [x_2, x_3]$ we have $< g'(s), g'(t) >\geq 0$. Then, approximating the integrals $\int_{x_1}^{x_2} g'(s)\,ds$ and $\int_{x_2}^{x_3} g'(t)\,dt$ by Riemann sums, we get $< \int_{x_1}^{x_2} g'(s)\,ds, \int_{x_2}^{x_3} g'(t)\,dt >\geq 0$, that is $< g(x_2) - g(x_1), g(x_3) - g(x_2) >\geq 0$. Then using Lemma 6.1.2, we get relation (6.10) for $k + 1$.

Now, consider the operators

$$U(g, x) := \int_a^x \frac{(x - t)^{r-1}}{(r - 1)!} g(t)\,dt, \ g \in C(I, E), \ x \in I; \quad R_n := (L_n \circ U)^{(r)}, \ n \in \mathbb{N}.$$

If $f \in C^r[a, b]$, the Taylor's formula yields

$$f(x) = \sum_{p=0}^{r-1} \frac{(x - a)^p}{p!} \cdot f^{(p)}(a) + U(f^{(r)}, x), \quad x \in [a, b].$$

Consequently, for any $n \in \mathbb{N}$ we have

$$(L_n(f))^{(r)} = \sum_{p=0}^{r-1} \sum_{j=0}^{p} \binom{p}{j} \frac{(-a)^{p-j}}{p!} (L_n(e_j f^{(p)}(a)))^{(r)} + R_n(f^{(r)}). \tag{6.11}$$

From relations (6.11) and (6.10),(for $k = r$ and $0 \le j \le r - 1$), it follows that in order to prove the theorem it suffices to show that $\lim\limits_{n\to\infty} \|R_n(f^{(r)}) - f^{(r)}\|_{[c,d]} = 0$.
For this we apply Theorem 6.1.2. Since $(U(g))^{(r)} = g$ for any $g \in C(I, E)$ and L_n is convex of order $r - 1$, it follows that the operator R_n is positive. It remains to show that

$$\lim_{n\to\infty} \|R_n(e_j w) - e_j w\|_{[c,d]} = 0, \text{ for all } w \in E \text{ and } j = 0, 1, 2. \tag{6.12}$$

For such $w \in E$ and j we have, after a short calculation, that

$$R_n(e_j w) = \left[L_n \left(\left(\sum_{i=0}^{j} \frac{i!}{(r+i)!} \binom{j}{i} a^{j-i} (e_1 - ae_0)^{r+i} \right) w \right) \right]^{(r)}. \tag{6.13}$$

Now from (6.13) we can remark that relations (6.10) imply relations (6.12).

Finally, let us show that the case of a general interval $I = [a, b]$ can be reduced to the case where $I \subset [0, \infty)$. Indeed, we can consider the function $\Psi(t) = t - a$, $t \in [a, b]$, and the sequences of linear positive operators $\tilde{L}_n : C([0, b - a], E) \to C^{r+1}([0, b - a], E)$,

$$\tilde{L}_n(g, x) := L_n(g \circ \Psi, \Psi^{-1}(x)), \quad g \in C[0, b - a], \ x \in [0, b - a].$$

We can see immediately that the operators $\tilde{L}_n$ are convex of order k, for $-1 \le k \le r$. Also relations

$$\lim_{n\to\infty} \|\tilde{L}_n(e_j w) - e_j w\|_{[0,b-a]} = 0, \text{ for all } w \in E, \text{ and } 0 \le j \le r + 2,$$

are equivalent to relations

$$\lim_{n\to\infty} \|L_n((e_1 - a)^j w) - (e_1 - a)^j w\|_{[a,b]} = 0, \text{ for all } w \in E, \text{ and } 0 \le j \le r + 2.$$

But these last inequalities are implied by condition 2) in the theorem, if we take into account the linearity of the operators L_n. It follows that for any subinterval $[c, d] \subset (a, b)$ and any $g \in C^r[0, b - a]$, we have

$$\lim_{n\to\infty} \|(\tilde{L}_n(g))^{(r)} - g^{(r)}\|_{[c-a,d-a]} = 0, \text{ for all } g \in C^r[0, b - a],$$

and this is equivalent to relation (6.9).   □

In the second part of this subsection we consider the simultaneous approximation for the class of nonlinear completely convex operators, given in Definition 6.1.4. First note that from Lemma 6.1.2, since $K_c^0(I, E) \subset K^0(I, E)$, we obtain:

Corollary 6.1.1. *If the sequence of functions $(f_n)_n$, $f_n \in C^1(I, E)$ is uniformly convergent to the function $f \in C^1(I, E)$ on I and if $f_n' \in K_c^0(I, F)$, $(n \in \mathbb{N})$, then for any subinterval $[c, d] \subset (a, b)$ the sequence $(f_n')_n$ is uniformly convergent on $[c, d]$ to the function f'.*

We have also:

Lemma 6.1.3. *Let* $f \in \mathcal{F}(I, E)$ *Let* $x_i, \in \mathbb{R}$, $(1 \le i \le n)$ *and* $t \in \mathbb{R}$ *be such that* x_i *and* $x_i + t$, $(1 \le i \le n)$ *are* $2n$ *distinct points of the interval* I. *Define the function* $f_t(x) := (f(x + t) - f(x))/t$, $x \in [\max\{a, a - t\}, \min\{b, b - t\}]$. *We have*

$$[f_t; x_1, \ldots, x_n] = \sum_{k=1}^{n} [f; x_1 + t, \ldots, x_{k-1} + t, x_k + t, x_k, x_{k+1}, \ldots, x_n]. \quad (6.14)$$

Proof. For $p \in \mathbb{N}$, $u \in \mathbb{R}$, $u \ne 0$, $y_i \in \mathbb{R}$, $(1 \le i \le p)$ such that $y_1 \ne y_i$, $y_1 \ne y_i - u$, $2 \le i \le p$, define,

$$\Theta_u^p(y_1, \ldots, y_p) := \sum_{j=1}^{p} \prod_{i=2}^{j} (y_1 - y_i)^{-1} \cdot \prod_{i=j}^{p} (y_1 + u - y_i)^{-1}.$$

(For $j = 1$ take $\prod_{i=2}^{j} = 1$). Using the relation

$$\Theta_u^{p+1}(y_1, \ldots, y_{p+1}) = (y_1 + u - y_{p+1})^{-1} \left[\Theta_u^p(y_1, \ldots, y_p) + \prod_{i=2}^{p+1} (y_1 - y_i)^{-1} \right],$$

we can prove by induction with regard to p that

$$\Theta_u^p(y_1, \ldots, y_p) = \left(u \cdot \prod_{i=2}^{p} (y_1 - y_i) \right)^{-1}.$$

Then we have

$$[f_t; x_1, \ldots, x_n] = \sum_{k=1}^{n} \frac{f(x_k + t) - f(x_k)}{t} \prod_{1 \le i \le n, i \ne k} (x_k - x_i)^{-1}$$

$$= \sum_{k=1}^{n} f(x_k + t) \cdot \Theta_t^{n-k+1}(x_k, \ldots, x_n) \cdot \prod_{i=1}^{k-1} (x_k - x_i)^{-1}$$

$$+ \sum_{k=1}^{n} f(x_k) \cdot \Theta_{-t}^k(x_k + t, \ldots, x_1 + t) \cdot \prod_{i=k+1}^{n} (x_k - x_i)^{-1}$$

$$= \sum_{k=1}^{n} [f; x_1 + t, \ldots, x_{k-1} + t, x_k + t, x_k, x_{k+1}, \ldots, x_n]. \quad \square$$

Lemma 6.1.4. *If* $f \in C^k(I, E) \cap K_c^m(I, E)$, $k \ge 1$, $m - k + 1 \ge 0$, *then* $f^{(k)} \in K_c^{m-k}(I, E)$.

Proof. Using Lemma 6.1.3 it is easy to obtain that $f' \in \cap K_c^{n-2}(I, E)$, if $f \in C^1(I, E) \cap K_c^{n-1}(I, E)$, $n \ge 1$. Then the theorem follows by induction. $\square$

The main result is the following one.

Theorem 6.1.4. *Let* $(L_n)_n$, $L_n : C^{k+1}(I, E) \to C^{k+1}(I, E)$, $k \geq 1$ *be a sequence of operators that are j-completely convex for $1 \leq j \leq k$. If we have*

$$\lim_{n \to \infty} \|L_n(f) - f\|_{[a,b]} = 0, \quad \text{for all } f \in C^{k+1}(I, E), \tag{6.15}$$

then for any $f \in C^{k+1}(I, E)$, for any subinterval $[c, d] \subset (a, b)$ and for any $1 \leq j \leq k$,

$$\lim_{n \to \infty} \|(L_n(f))^{(j)} - f^{(j)}\|_{[c,d]} = 0. \tag{6.16}$$

Proof. Fix the interval $[c, d]$ and let the subintervals $[c_j, d_j]$, $(1 \leq j \leq k)$ be such that $[c_0, d_0] = [a, b]$, $(c_j, d_j) \supset [c_{j+1}, d_{j+1}]$ and $[c_k, d_k] = [c, d]$. First note that for any function $g \in C^m([a, b], F)$, $m \geq 0$ and for any points $y_0 < \dots < y_m$ of $[a, b]$ we have $(m!)\|[g; y_0, \cdots, y_m]\| \leq \|g^{(m)}\|_{[a,b]}$. This one is a consequence of Peano's formula

$$[g; y_0, \dots, y_m] = \int_{y_0}^{y_m} \phi(t) \cdot g^{(m)}(t) \, dt,$$

where $\phi : [y_0, y_m] \to \mathbf{R}$ is a continuous positive function independent of g. Moreover, if $g \in C^m([a, b])$, then there is a point $\xi \in [\min\{y_0, \dots, y_m\}, \max\{y_0, \dots, y_m\}]$ such that

$$[g; y_0, \dots, y_m] = (m!)^{-1} g^{(m)}(\xi).$$

In the particular case where $g = e_p$, $p \in \mathbb{N}$, we get

$$[e_p; y_0, \dots, y_m] = \begin{cases} 0, & \text{if } p < m \\ \binom{p}{m}\xi^{p-m}, & \text{if } p \geq m. \end{cases} \tag{6.17}$$

Fix now $f \in C^{k+1}(I, E)$. Denote $\rho := \max\{|a|, |b|\}$. We can choose by induction the numbers $\lambda_j > 0$, $2 \leq j \leq k + 1$ such that

$$(\lambda_j)^2 \geq 2\lambda_j \left(\sum_{i=j+1}^{k+1} \binom{i}{j} \rho^{i-j} \cdot \lambda_i + (j!)^{-1}\|f^{(j)}\|_I \right)$$
$$+ \left(\sum_{i=j+1}^{k+1} \binom{i}{j} \rho^{i-j} \cdot \lambda_i + (j!)^{-1}\|f^{(j)}\|_I \right)^2,$$

$2 \leq j \leq k + 1$, (for $j = k + 1$ take $\sum_{i=j+1}^{k+1} = 0$). Let $v \in E$ with $\|v\| = 1$, and consider the function

$$h(x) := f(x) + \left(\sum_{i=2}^{k+1} \lambda_i \cdot x^i \right) v, \quad x \in I.$$

We have $h, h - f \in K_c^j(I, E)$, $1 \leq j \leq k$. Indeed, let $2 \leq j \leq k + 1$ and two sets of distinct points of I: $x_0, \dots, x_j$ and $y_0, \dots, y_j$. One obtains:

$$< [h; x_0, \ldots, x_j], [h; y_0, \ldots, y_j] >$$

$$= \left\langle \left(\sum_{i=j}^{k+1} \binom{i}{j} \lambda_i \xi_i^{i-j} \right) v + [f; x_0, \ldots, x_j], \left(\sum_{i=j}^{k+1} \binom{i}{j} \lambda_i \eta_i^{i-j} \right) v + [f; y_0, \ldots, y_j] \right\rangle$$

$$\geq 0,$$

where $\xi_i, \eta_i \in [a.b]$, for $j \leq i \leq k+1$. Hence $h \in K_c^{j-1}(I, E)$. In a similar mode we can see that $h - f \in K_c^{j-1}(I, E)$.

From Lemma 6.1.4 it follows that $(L_n(h))^{(j)}, (L_n(h) - L_n(f))^{(j)} \in K_c^0(I, E)$, $1 \leq j \leq k$, $n \geq 1$, and from Corollary 6.1.1 one can deduce by induction, for $1 \leq j \leq k$:

$$\lim_{n \to \infty} \|(L_n(h))^{(j)} - (h)^{(j)}\|_{[c_j, d_j]} = 0 = \lim_{n \to \infty} \|(L_n(h)) - L_n(f))^{(j)} - (h-f)^{(j)}\|_{[c_j, d_j]}.$$

From these limits follows (6.16). □

6.2 Approximation of functions with vector argument

6.2.1 Introduction. Linear functionals and operators induced by positive measures

In this section we extend in the vector case some of the estimates with the second order moduli, given in Chapter 2. The content of this section is based on [107] and [111].

Let $(X, \|\cdot\|_X)$ be a normed space and $(Y, \|\cdot\|_Y)$ be a Banach space. Let $D \subset X$ be relatively compact and let μ be a Borel positive measure on D. Suppose $\mu(D) > 0$.

For $x \in X$ and the number $r > 0$, denote by $B(x, r)$ the ball with the centre x and the radius r. If $x, y \in X$, then (x, y) denotes the open segment with the end points x and y and $[x, y]$, the closed segment with these end points. For $A \subset X$, denote by $Sp \, A$, the linear subspace generated by A and by $conv \, A$, the convex hull of the set A.

Denote by $\mathcal{F}(D, Y)$ the space of functions from D to Y, by $C(D, Y)$ the subspace of continuous functions and by $UC(D, Y)$, the subspace of uniformly continuous functions. If $f \in C(D, Y)$, then $\|f\|_\infty$ denotes its sup-norm.

With these data we have:

Definition 6.2.1. *Given a number $\delta > 0$, a partition $\{D_1, \ldots, D_m\}$ of D, i.e., $D = D_1 \cup \cdots \cup D_m$, $D_i \cap D_j = \emptyset$, for $i \neq j$, is called δ-compatible with the measure μ if for any $1 \leq i \leq m$, $D_i \neq \emptyset$, D_i is μ-measurable and diameter(D_i) < δ.*

Remark 6.2.1. Note that for any $\delta > 0$ there is a partition of D that is δ-compatible with μ. Indeed, since D is relatively compact, there is a minimal integer n and the elements $x_1, \cdots, x_n \in D$, such that $D \subset B(x_1, \delta) \cup \cdots \cup B(x_n, \delta)$. Then we can put $D_1 = B(x_1, \delta) \cap D$ and $D_{i+1} = B(x_{i+1}, \delta) \cap D \setminus (D_1 \cup \cdots \cup D_i)$, $1 \leq i \leq n - 1$.

Theorem 6.2.1. *For any $f \in UC(D, Y)$, there is a unique element $b \in Y$ having the following property: for any $\varepsilon > 0$, there is $\delta_{\varepsilon, f} > 0$, such that for any partition $\{D_1, \ldots, D_m\}$ of D that is $\delta_{\varepsilon, f}$-compatible with μ, and for any choice of the elements $x_i \in D_i$, $1 \le i \le m$, we have*

$$\left\| b - \sum_{i=1}^{m} f(x_i) \mu(D_i) \right\|_Y < \varepsilon. \tag{6.18}$$

Proof. Let $(\varepsilon_k)_k$ be a sequence of numbers $\varepsilon_k > 0$, with the limit 0. Since f is uniformly continuous on D, there is a sequence of numbers $(\delta_k)_k$, $\delta_k > 0$, such that for any points $x, y \in D$ with $\|x - y\|_X \le \delta_k$, we have $\|f(x) - f(y)\|_Y < \varepsilon_k$. Let $\{D_1^k, \ldots, D_{m_k}^k\}$ be a partition of D that is δ_k-compatible with μ. Choose $x_i^k \in D_i^k$, for $1 \le i \le m_k$. Put

$$b_k = \sum_{i=1}^{m_k} f(x_i^k) \mu(D_i^k), \quad k \in \mathbb{N}.$$

For two indices $k, l \in \mathbb{N}$, denote by $\{E_1, \ldots, E_p\}$, the partition of D, formed by the nonempty sets of the family $\{D_i^k \cap D_j^l \mid 1 \le i \le m_k, \ 1 \le j \le m_l\}$. Choose the points $y_j \in E_j$, $1 \le j \le p$. We have

$$\left\| b_k - \sum_{j=1}^{p} f(y_j) \mu(E_j) \right\|_Y \le \sum_{i=1}^{m_k} \sum_{1 \le j \le p, \ E_j \subset D_i^k} \|f(y_j) - f(x_i^k)\|_Y \mu(E_j) < \varepsilon_k \mu(D).$$

An analogous relation exists for b_l. It follows that $\|b_k - b_l\|_Y < (\varepsilon_k + \varepsilon_l) \mu(D)$. Since $\lim_{k \to \infty} \varepsilon_k = 0$, the sequence $(b_k)_k$ is fundamental. Therefore we obtain an element $b \in Y$ such that $b = \lim_{k \to \infty} b_k$.

Let now $\varepsilon > 0$. We can choose an index k such that $\|b_k - b\|_Y < \varepsilon/2$. Then we choose $\delta_{\varepsilon, f} > 0$ to be such that $\|f(x) - f(y)\|_Y < \varepsilon/(2\mu(D))$, for any points $x, y \in D$, with $\|x - y\|_X < \delta_{\varepsilon, f}$. Let an arbitrary partition of D, $\{D_1, \ldots, D_m\}$ be $\delta_{\varepsilon, f}$-compatible with μ. Choose the points $x_i \in D_i$. $1 \le i \le m$. By a similar argument as above we obtain

$$\left\| b_k - \sum_{i=1}^{m} f(x_i) \mu(D_i) \right\|_Y < \frac{\varepsilon}{2\mu(D)} \cdot \mu(D) = \frac{\varepsilon}{2}.$$

Consequently, we obtain (6.18). The uniqueness of the point b in this relation is immediate. $\qquad\square$

We use the notion of "functional" in the following extended sense. If A is a linear subspace of $\mathcal{F}(D, Y)$, then a linear mapping $F : A \to Y$ will be called a **functional** on A with values in Y. Theorem 6.2.1 makes possible the following definition, which introduces a Bochner type integral.

Definition 6.2.2. *For any Banach space $(Y, \|\cdot\|_Y)$, the functional $F_Y : UC(D, Y) \to Y$ is defined by $F_Y(f) = b$, $f \in UC(D, Y)$, where b is the unique element in Y that satisfies the condition (6.18). We call the functional F_Y, the **functional induced by the measure μ on $UC(D, Y)$.**

When it is necessary we denote this functional by F_Y^μ. Another notation of the functional F_Y is:

$$F_Y(f) = \int_D f \, d\mu. \tag{6.19}$$

The simplest functionals of this type are the discrete functionals, of the form

$$F_Y(f) := \sum_{i=1}^{n} \rho_i f(a_i), \quad f \in UC(D, Y), \tag{6.20}$$

where $a_i \in D$, and $\rho_i \geq 0$, $(1 \leq i \leq n)$. Here we have $\mu = \sum_{i=1}^{n} \rho_i \delta_{a_i}$, where δ_{a_i} denotes the Dirac functional of the point a_i. The discrete functionals can be used for approximation of functionals induced by general Borel positive measures. So, a great part of the properties of the functionals given in Definition 6.2.2 can be established by proving them in the case of the discrete functionals and then by passing to the limit in a sequence of discrete functionals. In this direction we have the following auxiliary result.

Lemma 6.2.1. *Let X, Y, D, μ be fixed as above. For any $f \in UC(D, Y)$ and any number $\varepsilon > 0$, there exists a number $\delta_{f,\varepsilon} > 0$, such that, for any partition $\{D_1, \ldots, D_n\}$ that is $\delta_{f,\varepsilon}$- compatible with the measure μ, if we denote $\rho_i := \mu(D_i)$, and we choose the points $a_i \in D_i$, $1 \leq i \leq n$, then we have*

$$\left\| F_Y(f) - \sum_{i=1}^{n} \rho_i f(a_i) \right\|_Y < \varepsilon.$$

The proof of this lemma is immediate. In the next theorem we give some basic properties of the functionals F_Y. The proof of them can be reduced to discrete functionals, using Lemma 6.2.1.

Theorem 6.2.2. *We have:*

i) The functional F_Y is linear, i.e., $F_Y(\alpha f + \beta g) = \alpha F_Y(f) + \beta F_Y(g)$, for all $f, g \in UC(D, Y)$ and all $\alpha, \beta \in \mathbb{R}$.

ii) In the case $Y = \mathbb{R}$, the functional $F_\mathbb{R}$ is positive, i.e., $F_\mathbb{R}(f) \geq 0$, for all $f \in UC(D, \mathbb{R})$, $f \geq 0$. Consequently it is also monotone.

iii) More generally, if $K \subset Y$ is a closed cone in Y and if $f(D) \subset K$, then $F_Y(f) \in K$.

iv) Let $\{D_1, \ldots, D_n\}$ be a partition of D consisting of measurable sets. Denote by μ_{D_k}, the measure induced by μ on D_k, $1 \leq k \leq n$. We have

$$\int_D f \, d\mu = \sum_{k=1}^{n} \int_{D_k} (f|_{D_k}) \, d\mu_{D_k}, \quad f \in UC(D, Y).$$

v) For any $f \in UC(D, Y)$ we have $\|F_Y(f)\|_Y \leq F_\mathbb{R}(\|f\|_\infty) = \|f\|_\infty F_\mathbb{R}(e_0)$.

The point iii) expresses a property of generalized positivity.

Let id $\in \mathcal{F}(X, X)$ be the identity function and let $e_0 \in \mathcal{F}(X, \mathbb{R})$, $e_0(t) = 1$, ($t \in D$). For any $y \in Y$, we denote also by y, the constant function on D equal to y at each point.

Definition 6.2.3. *The element $z_\mu = F_X(\mathrm{id})/\mu(D)$ is the* **barycentre** *of the measure* μ.

Denote by $X^\star$ the space of linear continuous functionals $\varphi : X \to \mathbb{R}$.

Lemma 6.2.2. *The point $z \in D$ is the barycentre of the measure $\mu \neq 0$, if and only if for any $\varphi \in X^\star$, we have*

$$\int_D \varphi(t)\, d\mu(t) = \varphi(z), \quad \text{or equivalently} \quad F_{\mathbb{R}}^\mu(\varphi) = \varphi(z).$$

Starting from the functionals we consider the following class of "positive" linear operators.

Definition 6.2.4. *Let $\{\mu_x\}_{x \in D}$ be a family of Borel positive measures on D. The* **linear operator induced by the family** $\{\mu_x\}_{x \in D}$, $L_Y : UC(D, Y) \to \mathcal{F}(D, Y)$, *is given by*

$$L_Y(f, x) = F_Y^{\mu_x}(f), \quad f \in UC(D, Y), \ x \in D. \tag{6.21}$$

★

As in the real case we can restrict ourselves to estimating the functionals. Then one obtains directly, pointwise estimates for operators.

In what follows until the end of this section, there will be fixed, a Banach space $(X, \| \cdot \|_X)$, a convex compact subset $D \subset X$ and a Borel measure μ on D. Suppose $\mu(D) > 0$. The Banach space $(Y, \| \cdot \|_Y)$ will be taken arbitrarily. Important cases will be the choices $Y = X$ and $Y = \mathbb{R}$.

If we use the first order modulus of continuity, given by

$$\omega_1(f, h) := \sup\{\| f(x) - f(y) \|_Y \mid x, y \in D, \ \|x - y\|_X \leq h\}, \tag{6.22}$$
$$f \in C(D, Y), \ h > 0,$$

we obtain a result that generalizes the Mond estimate, [69], see (1.39).

Theorem 6.2.3. *For any $f \in C(D, Y)$, $x \in D$, $h > 0$, the inequality*

$$\| F_Y(f) - f(x) \|_Y \leq \| f(x) \|_Y \, |F_{\mathbb{R}}(e_0) - 1| \tag{6.23}$$
$$+ \left(F_{\mathbb{R}}(e_0) + h^{-2} F_{\mathbb{R}}(\| \cdot - x \|_X^2) \right) \omega_1(f, h)$$

holds.

Proof. We have successively

$$\|F_Y(f) - f(x)\|_Y \le \|F_Y(f) - F_Y(f(x)e_0)\|_Y + \|F_Y(f(x)e_0) - f(x)\|_Y$$
$$= \|F_Y(f - f(x))\|_Y + \|f(x)\|_Y\,|F_\mathbb{R}(e_0) - 1|$$
$$\le F_\mathbb{R}(\|f - f(x)\|_Y) + \|f(x)\|_Y\,|F_\mathbb{R}(e_0) - 1|$$
$$\le F_\mathbb{R}(e_0 + h^{-2}\| \cdot - x\|_X^2)\omega_1(f, h) + \|f(x)\|_Y\,|F_\mathbb{R}(e_0) - 1|$$
$$= \Big(F_\mathbb{R}(e_0) + h^{-2} F_\mathbb{R}(\| \cdot - x\|_X^2)\Big)\omega_1(f, h) + \|f(x)\|_Y\,|F_\mathbb{R}(e_0) - 1|.\ \square$$

The estimates with second order moduli will be treated below.

6.2.2 Auxilliary results

We give some decompositions of the convex combinations.

Lemma 6.2.3. *Let the points $a_1 \in D, \dots, a_n \in D$ and the real numbers $\rho_1 > 0, \dots, \rho_n > 0$ with $\rho_1 + \cdots + \rho_n = 1$, $n \ge 1$. Put $x := \rho_1 a_1 + \cdots + \rho_n a_n$. Then there exist the real numbers $\eta_{i,j} > 0$, $1 \le i \le s$, $1 \le j \le p + 1$, $(s \ge 1, 1 \le i \le s)$, such that we can rewrite the set $\{a_1, \dots, a_n\}$ as*

$$\{a_1, \dots, a_n\} = \{b_{i,j} \mid 1 \le i \le s,\ 1 \le j \le p + 1\},$$

and the points $b_{i,j}$, satisfy the conditions below:
1) For any $1 \le i \le s$,

$$\left(\sum_{j=1}^{p+1} \eta_{i,j}\right) x = \sum_{j=1}^{p+1} \eta_{i,j} b_{i,j}.$$

2) For any $1 \le k \le n$, ρ_k is equal to the sum of the numbers $\eta_{i,j}$ with the property that $b_{i,j} = a_k$.

Proof. Define $X_1 = Sp\{a_1 - x, \dots, a_n - x\}$. Put $p = dim\, X_1$. We prove by induction with respect to n. For $n = 1$ it follows that $x = a_1$ and $\rho_1 = 1$. Then we can take $s = 1$, $\eta_{1,j} = \frac{1}{p+1}$, $b_{1,j} = a_1$, $1 \le j \le p + 1$.

Consider Lemma 6.2.3 true for $n - 1$, $n \ge 2$ and prove it for n. Since x belongs to the convex hull of the set $\{a_1, \dots, a_n\}$, from the Caratheodory theorem, there exist at least $p + 1$ extremal points of this set, (not necessarily distinct), denoted by $b_1, \dots, b_{p+1}$ and there exist the real numbers $\eta_1 \ge 0, \dots, \eta_{p+1} \ge 0$ such that $\eta_1 + \cdots + \eta_{p+1} = 1$ and $x = \eta_1 b_1 + \cdots + \eta_{p+1} b_{p+1}$. For any integer $1 \le k \le n$, denote by ν_k the sum of the numbers η_j for the indices j with the property $b_j = a_k$. Set

$$t = \max_{1 \le k \le n} \frac{\nu_k}{\rho_k}.$$

Hence $t > 0$. Define

$$b_{1,j} := b_j, \text{ and } \eta_{1,j} := \frac{1}{t}\eta_j,\ 1 \le j \le p + 1.$$

Set, also

$$\tilde{\rho}_k := \frac{\rho_k - \frac{1}{t} v_k}{1 - \sum_{k=1}^{n} \frac{1}{t} v_k}, \quad 1 \le k \le n \quad \text{(with the convention } 0/0 = 0).$$

We have $\tilde{\rho}_k \ge 0$ for all indices $1 \le k \le n$, and there exists at least an index k_0 such that $\tilde{\rho}_{k_0} = 0$. Suppose, for a choice that $\tilde{\rho}_1 > 0, \dots, \tilde{\rho}_{n-r} > 0$ and $\tilde{\rho}_{n-r+1} = \cdots = \tilde{\rho}_n = 0$, where $1 \le r \le n$. If $r = n$, then the proof is finished, by taking $s = 1$. If $r < n$, note that we have

$$x = \sum_{k=1}^{n-r} \tilde{\rho}_k a_k.$$

By using the hypothesis of induction we find the points $b_{i,j}$ and the numbers $\tilde{\eta}_{i,j} \ge 0$, $2 \le i \le s, 1 \le j \le p + 1$, such that $\{a_1, \dots, a_{n-r}\} = \{b_{i,j}, \mid 2 \le i \le s, \ 1 \le j \le p + 1\}$ and we have

1') $\left(\sum_{j=1}^{p+1} \tilde{\eta}_{i,j} \right) x = \sum_{j=1}^{p+1} \tilde{\eta}_{i,j} b_{i,j}$, for all $2 \le i \le s$.

2') For any $1 \le k \le n - r$, $\tilde{\rho}_k$ is equal to the sum of the numbers $\tilde{\eta}_{i,j}$ with the property that $b_{i,j} = a_k$.

Now we can define the numbers

$$\eta_{i,j} := \left(1 - \sum_{k=1}^{n} \frac{1}{t} v_k \right) \tilde{\eta}_{i,j}, \quad \text{for } 2 \le i \le s, \ 1 \le j \le p + 1.$$

With these choices the lemma is proved. $\qquad \square$

A stronger form of this lemma is the following one.

Lemma 6.2.4. *Let the points $a_1 \in D, \dots, a_n \in D$ and the real numbers $\rho_1 > 0, \dots, \rho_n > 0$ with $\rho_1 + \cdots + \rho_n = 1$, $n \ge 1$. Put $x := \rho_1 a_1 + \cdots + \rho_n a_n$. Then there are the real numbers $\eta_{i,j} > 0$, $1 \le i \le s$, $1 \le j \le r_i + 1$, $(s \ge 1, r_i \ge 0, 1 \le i \le s)$, such that we can rewrite the set $\{a_1, \dots, a_n\}$ as*

$$\{a_1, \dots, a_n\} = \{b_{i,j}, \mid 1 \le i \le s, \ 1 \le j \le r_i + 1\},$$

and the points $b_{i,j}$, satisfy the following conditions:
1) For any $1 \le i \le s$,

$$\left(\sum_{j=1}^{r_i+1} \eta_{i,j} \right) x = \sum_{j=1}^{r_i+1} \eta_{i,j} b_{i,j}.$$

2) For any $1 \le k \le n$, ρ_k is equal to the sum of the numbers $\eta_{i,j}$ with the property that $b_{i,j} = a_k$.
3) For any $1 \le i \le s$, if $r_i \ge 1$, then the vectors

$$b_{i,1} - x, \dots, b_{i,r_i} - x$$

are linearly independent.

Proof. By applying Lemma 6.2.3 we obtain the families $\{b_{i,j}, \mid 1 \leq i \leq s, \ 1 \leq j \leq r_i + 1\}$ and $\{\eta_{i,j}, \mid 1 \leq i \leq s, \ 1 \leq j \leq r_i + 1\}$ $(s \geq 1, \ r_i \geq 0, 1 \leq i \leq s)$, where $r_i = p$, such that $\{a_1, \ldots, a_n\} = \{b_{i,j}, \mid 1 \leq i \leq s, \ 1 \leq j \leq r_i + 1\}$, and conditions 1) and 2) of the present lemma are satisfied. It remains to obtain condition 3).

We apply an inductive procedure as follows. Fix an index $1 \leq i \leq s$. For simplicity put $r := r_i$, $b_j := b_{i,j}$, $v_j := b_{i,j} - x$, $\eta_j := \eta_{i,j}$, $1 \leq j \leq r + 1$. We have $\eta_1 v_1 + \cdots + \eta_{r+1} v_{r+1} = 0$. Consider that $r \geq 1$ and the vectors $v_1, \ldots, v_r$ are linearly dependent. There are the numbers $\beta_1, \ldots, \beta_r$ such that $\beta_1 v_1 + \cdots + \beta_r v_r = 0$. We can suppose that there is at least an index $1 \leq j \leq r$, such that $\beta_j < 0$. Define

$$M := \max\{\eta_j / \beta_j \mid 1 \leq j \leq r, \ \beta_j < 0\}$$

and let $j_0 \in \{1, \ldots, r\}$ such that $\eta_{j_0} / \beta_{j_0} = M, \ \beta_{j_0} < 0$. Define

$$\lambda_j := \eta_j - M\beta_j, \ (1 \leq j \leq r), \ \lambda_{r+1} = \eta_{r+1}.$$

It follows that $\lambda_1 v_1 + \ldots + \lambda_{r+1} v_{r+1} = 0, \ \lambda_j \geq 0, \ (1 \leq j \leq r + 1), \ \lambda_{j_0} = 0$ and $\lambda_{r+1} > 0$.

Now define

$$m := \min\{\eta_j / \lambda_j \mid 1 \leq j \leq r + 1, \ \lambda_j > 0\}.$$

Let $j_1 \in \{1, \ldots, r + 1\}$, such that $m = \eta_{j_1} / \lambda_{j_1}$. Define

$$\nu_j := \eta_j - m\lambda_j, \ 1 \leq j \leq r + 1.$$

Then we have $\nu_1 v_1 + \cdots + \nu_{r+1} v_{r+1} = 0, \ \nu_j \geq 0, \ (1 \leq j \leq r + 1), \ \nu_{j_1} = 0$ and $\nu_{j_0} > 0$.

Then we can replace the pair

$$\{b_1, \ldots, b_{r+1}\}, \quad \{\eta_1, \ldots, \eta_{r+1}\}$$

by the two pairs

$$\{b_j \mid 1 \leq j \leq r + 1, \ \lambda_j > 0\}, \quad \{m\lambda_j \mid 1 \leq j \leq r + 1, \ \lambda_j > 0\}$$

and

$$\{b_j \mid 1 \leq j \leq r + 1, \ \nu_j > 0\}, \quad \{\nu_j \mid 1 \leq j \leq r + 1, \ \nu_j > 0\}.$$

Note that $j_0 \notin \{j \mid \lambda_j > 0\}$ and $j_1 \notin \{j \mid \nu_j > 0\}$. Using induction, after a finite number of applications of this procedure we obtain systems for which the vectors $v_1, \ldots, v_r$ are linearly independent (for each i). $\qquad \square$

Now we give an extremal property of the barycentre of the Borel positive measure μ on D, in the particular case when X is a Hilbert space.

Lemma 6.2.5. *Suppose that X is Hilbert space. Let $a_i \in X$, $1 \leq i \leq m$ and let the real numbers $\rho_i \geq 0$, $\eta_i \geq 0$, $1 \leq i \leq m$ be such that $\rho_1 + \cdots + \rho_m = 1$ and $\eta_1 + \cdots + \eta_m = 1$. Put $x := \rho_1 a_1 + \cdots + \rho_m a_m$ and $y := \eta_1 a_1 + \cdots + \eta_m a_m$. We have*

$$\sum_{k=1}^{m} \rho_k \|y - a_k\|_X^2 \geq \sum_{k=1}^{m} \rho_k \|x - a_k\|_X^2 = \sum_{i=1}^{m} \sum_{j=1}^{m} \rho_i \rho_j \|a_i - a_j\|_X^2. \tag{6.24}$$

Proof. We have

$$\sum_{k=1}^{m} \rho_k \|y - a_k\|_X^2 = \sum_{k=1}^{m} \rho_k \left\| \sum_{1 \le i \le m,\, i \ne k} \eta_i a_i - (1 - \eta_k) a_k \right\|_X^2$$

$$= \sum_{k=1}^{m} \rho_k \Bigg[\sum_{1 \le i \le m,\, i \ne k} \eta_i^2 \|a_i\|_X^2 + (1 - \eta_k)^2 \|a_k\|_X^2 - 2 \sum_{1 \le i \le m,\, i \ne k} (1 - \eta_k)\eta_i < a_k, a_i >$$

$$+ 2 \sum_{1 \le i < j \le m,\, i,j \ne k} \eta_i \eta_j < a_i, a_j > \Bigg]$$

$$= \sum_{j=1}^{m} (\eta_j^2 - 2\eta_j \rho_j + \rho_j) \|a_j\|_X^2 + 2 \sum_{1 \le i < j \le m} (\eta_i \eta_j - \eta_i \rho_j - \eta_j \rho_i) < a_i, a_j >$$

$$= \left(\sum_{j=1}^{m} \rho_j (1 - \rho_j) \|a_j\|_X^2 - 2 \sum_{1 \le i < j \le m} \rho_i \rho_j < a_i, a_j > \right)$$

$$+ \left(\sum_{j=1}^{m} (\eta_j - \rho_j)^2 \|a_j\|_X^2 + 2 \sum_{1 \le i < j \le m} (\eta_i - \rho_i)(\eta_j - \rho_j) < a_i, a_j > \right)$$

$$= \sum_{1 \le i < j \le m} \rho_i \rho_j (\|a_i\|_X^2 + \|a_j\|_X^2 - 2 < a_i, a_j >) + \left\| \sum_{j=1}^{m} (\eta_j - \rho_j) a_j \right\|_X^2$$

$$= \sum_{1 \le i < j \le m} \rho_i \rho_j \|a_i - a_j\|_X^2 + \left\| \sum_{j=1}^{m} (\eta_j - \rho_j) a_j \right\|_X^2$$

$$\ge \sum_{1 \le i < j \le m} \rho_i \rho_j \|a_i - a_j\|_X^2.$$

Moreover, the last inequality becomes equality in the case when $\eta_i = \rho_i$, $1 \le i \le m$. $\qquad \square$

This lemma can be extended to an arbitrary Borel positive measure, as follows.

Lemma 6.2.6. *Let X be a Hilbert space, and $D \subset X$ be compact and convex. Let μ be a positive Borel measure on D and let z be the barycentre of μ. For any $y \in D$ we have*

$$\int_D \|t - y\|_X^2 \, d\mu(t) \ge \int_D \|t - z\|_X^2 \, du(t) = \iint_{D \times D} \|t_1 - t_2\|_X^2 \, d\mu(t_1) \times d\mu(t_2).$$

$$(6.25)$$

Proof. We apply Lemmas 6.2.1 and 6.2.5 to a sequence of discrete measure $(\mu_k)_k$, which can be constructed by using δ_k-compatible partitions with the measure μ, where $\delta_k \to 0$, $(k \to \infty)$. Then the sequence of the barycentres of the measures μ_k tends to the barycentre of μ. $\qquad \square$

6.2.3 Estimates with moduli ω_2 and ω_2^*

In this subsection X will be a **Euclidean** space, Y will be an arbitrary Banach space, $D \subset X$ will be convex and compact and μ will be a positive Borel measure on D.

$$p := \dim X \tag{6.26}$$

Consider the extended second order moduli , ω_2 and ω_2^* which can be constructed in a natural way on the space $C(D, Y)$. As in the real case, we denote for $f \in C(D, Y), a, b \in D, a \neq b$ and $x = (1 - t)a + tb, t \in [0, 1]$,

$$\Delta(f; a, x, b) := (1 - t)f(a) + tf(b) - f((1 - t)a + tb). \tag{6.27}$$

Definition 6.2.5. Let $f \in C(D, Y)$ and the number $h > 0$. Define

$$\omega_2(f, h) := \sup \left\{ \left\| f(u) - 2f\left(\frac{u + v}{2}\right) + f(v) \right\|_Y, \right.$$

$$\left. u, v \in D, \ \|u - v\|_X \le 2h \right\}, \tag{6.28}$$

$$\omega_2^*(f, h) := \sup \left\{ \|\Delta(f; u, y, v)\|_Y, \ u, v \in I, \ u \neq v, \ y \in [u, v], \ \|y - u\|_X, \right.$$

$$\left. \le h \|v - y\|_X \le h \right\}. \tag{6.29}$$

First we treat the estimates with ω_2. The same method can be used for ω_2^*.

Lemma 6.2.7. For any $f \in C(D, Y), a \in D, b \in D, t \in [0, 1]$ and $h > 0$, we have

$$\|\Delta(f; a, (1 - t)a + tb, b)\|_Y \le \left(1 + \frac{1}{2}h^{-2}\|x - a\|_X \|x - b\|_X \right) \omega_2(f, h),$$

Proof. We can consider that $a \neq b$. We have

$$\|\Delta(g; x_1, (1 - t)x_1 + tx_2, x_2)\|_Y \le \left(1 + \frac{1}{2}\rho^{-2}t(1 - t)(x_1 - x_2)^2\right) \omega_2(g, \rho),$$

where $g \in C([0, 1], Y), x_1, x_2 \in [0, 1], t \in [0, 1]$ and $\rho > 0$. Indeed, in the proof of Theorem 2.2.1 this inequality is established in the particular case $Y = \mathbb{R}$, see relation (2.45) for $\lambda = 0$. All the steps in the proof of this inequality can be generalized by replacing the absolute value with the norm, see also the proofs of Lemmas 2.2.1, 2.2.2, 2.2.3 and Theorem 2.2.1.

Define the function $\varphi : [0, 1] \to D, \varphi(t) := (1 - t)a + tb, t \in [0, 1]$. Put $g := f \circ \varphi$ and $\rho := \frac{h}{\|b-a\|_X}$. We have

$$\|(1 - t)f(a) + tf(b) - f((1 - t)a + tb)\|_Y = \|(1 - t)g(0) + tg(1) - g(t)\|_Y$$

$$\le \left(1 + \frac{1}{2}\rho^{-2}t(1 - t)\right) \omega_2(g, \rho) = \left(1 + \frac{1}{2}h^{-2}\|x - a\|_X \|x - b\|_X \right) \omega_2(g, \rho)$$

$$= \left(1 + \frac{1}{2}h^{-2}\|x - a\|_X \|x - b\|_X \right) \omega_2(f, h). \qquad \square$$

Lemma 6.2.8. *Let* $m \geq 2$, $a_1 \in D, \ldots, a_m \in D$ *and* $\rho_1 > 0, \ldots, \rho_m > 0$, $\rho_1 + \cdots + \rho_m = 1$. *Put* $x := \rho_1 a_1 + \cdots + \rho_m a_m$. *For any* $f \in C(D, Y)$ *and* $h > 0$, *we have*

$$\left\| \sum_{i=1}^{m} \rho_i f(a_i) - f(x) \right\|_Y \leq \left(m - 1 + \frac{1}{2} h^{-2} \sum_{i=1}^{m} \rho_i \|x - a_i\|_X^2 \right) \omega_2(f, h). \quad (6.30)$$

Proof. We prove relation (6.30) by induction with respect to m. For $m = 2$, using Lemma 6.2.7 and Lemma 6.2.5 we get

$$\left\| \sum_{i=1}^{2} \rho_i f(a_i) - f(x) \right\|_Y \leq \left(1 + \frac{1}{2} h^{-2} \rho_1 \rho_2 \|a_1 - a_2\|_X^2 \right) \omega_2(f, h)$$

$$= \left(1 + \frac{1}{2} h^{-2} \sum_{i=1}^{2} \rho_i \|x - a_i\|_X^2 \right) \omega_2(f, h).$$

Now suppose relation (6.30) true for $m - 1$, $m \geq 3$ and prove it for m. For any $1 \leq k \leq m$, define

$$d_k := \frac{1}{1 - \rho_k} \sum_{1 \leq i \leq m, \, i \neq k} \rho_i a_i.$$

Note that $\rho_k a_k + (1 - \rho_k) d_k = x$. From Lemma 6.2.7 it follows that

$$\|\rho_k f(a_k) + (1 - \rho_k) f(d_k) - f(x)\|_Y \leq \left(1 + \frac{1}{2} h^{-2} \|x - d_k\|_X \|x - a_k\|_X \right) \omega_2(f, h).$$

But

$$\|x - d_k\|_X = \left\| \sum_{j=1}^{m} \rho_j a_j - \frac{1}{1 - \rho_k} \sum_{1 \leq j \leq m, \, j \neq k} \rho_j a_j \right\|_X$$

$$= \frac{1}{1 - \rho_k} \left\| (1 - \rho_k) \sum_{j=1}^{m} \rho_j a_j - \sum_{1 \leq j \leq m, \, j \neq k} \rho_j a_j \right\|_X$$

$$= \frac{1}{1 - \rho_k} \left\| (1 - \rho_k) \rho_k a_k - \sum_{1 \leq j \leq m, \, j \neq k} \rho_j \rho_k a_j \right\|_X$$

$$= \frac{\rho_k}{1 - \rho_k} \|a_k - x\|_X.$$

Therefore

$$\|\rho_k f(a_k) + (1 - \rho_k) f(d_k) - f(x)\|_Y \leq \left(1 + \frac{1}{2} \frac{\rho_k}{1 - \rho_k} h^{-2} \|x - a_k\|_X^2 \right) \omega_2(f, h).$$
$$(6.31)$$

From the hypothesis of induction and from Lemma 6.2.5 we get

$$\left\| \sum_{1 \leq i \leq m, \, i \neq k} \frac{\rho_i}{1 - \rho_k} f(a_i) - f(d_k) \right\|_X$$

$$\leq \left[m - 2 + \frac{1}{2} h^{-2} \sum_{1 \leq i \leq m, \, i \neq k} \frac{\rho_i}{1 - \rho_k} \|d_k - a_i\|_X^2 \right] \omega_2(f, h)$$

$$= \left[m - 2 + \frac{1}{2} h^{-2} \sum_{1 \leq i < j \leq m, \, i, j \neq k} \frac{\rho_i \rho_j}{(1 - \rho_k)^2} \|a_i - a_j\|_X^2 \right] \omega_2(f, h). \quad (6.32)$$

Using the decomposition

$$\sum_{i=1}^m \rho_i f(a_i) - f(x) = \left[\rho_k f(a_k) + (1 - \rho_k) f(d_k) - f(x) \right]$$

$$+ (1 - \rho_k) \left[\sum_{1 \leq i \leq m, \, i \neq k} \frac{\rho_i}{1 - \rho_k} f(a_i) - f(d_k) \right],$$

we obtain from (6.31) and (6.32) that

$$\left\| \sum_{i=1}^m \rho_i f(a_i) - f(x) \right\|_Y \leq T_k, \quad 1 \leq k \leq m,$$

where

$$T_k := \left[1 + (m - 2)(1 - \rho_k) \right.$$

$$\left. + \frac{1}{2} \frac{1}{1 - \rho_k} h^{-2} \left(\rho_k \|x - a_k\|_X^2 + \sum_{1 \leq i < j \leq m, \, i, j \neq k} \rho_i \rho_j \|a_i - a_j\|_X^2 \right) \right] \omega_2(f, h).$$

Since $\sum_{k=1}^m (1 - \rho_k) = m - 1$ we have

$$\left\| \sum_{i=1}^m \rho_i f(a_i) - f(x) \right\|_Y \leq \frac{1}{m - 1} \sum_{k=1}^m (1 - \rho_k) T_k$$

$$= \frac{1}{m - 1} \left[m - 1 + (m - 2) \sum_{k=1}^m (1 - \rho_k)^2 \right.$$

$$\left. + \frac{1}{2} h^{-2} \left(\sum_{k=1}^m \rho_k \|x - a_k\|_X^2 + \sum_{k=1}^m \sum_{1 \leq i < j \leq m, \, i, j \neq k} \rho_i \rho_j \|a_i - a_j\|_X^2 \right) \right] \omega_2(f, h).$$

From Lemma 6.2.5 we obtain

$$\sum_{k=1}^m \sum_{1 \leq i < j \leq m, \, i, j \neq k} \rho_i \rho_j \|a_i - a_j\|_X^2 = (m - 2) \sum_{1 \leq i < j \leq m} \rho_i \rho_j \|a_i - a_j\|_X^2$$

$$= (m - 2) \sum_{k=1}^m \rho_k \|x - a_k\|_X^2.$$

It follows that

$$\left\| \sum_{i=1}^{m} \rho_i f(a_i) - f(x) \right\|_Y$$

$$\leq \left(1 + \frac{m-2}{m-1} \sum_{k=1}^{m} (1 - \rho_k)^2 + \frac{1}{2} h^{-2} \sum_{k=1}^{m} \rho_k \|x - a_k\|_X^2 \right) \omega_2(f, h).$$

Finally, note that

$$\sum_{k=1}^{m} (1 - \rho_k)^2 = \sum_{k=1}^{m} \left(\sum_{1 \leq i \leq m, \, i \neq k} \rho_i \right)^2$$

$$= (m - 1) \sum_{i=1}^{m} \rho_i^2 + 2(m - 2) \sum_{1 \leq i < j \leq m} \rho_i \rho_j$$

$$= \sum_{i=1}^{m} \rho_i^2 + (m - 2) \left(\sum_{i=1}^{m} \rho_i \right)^2 \leq m - 1.$$

Relation (6.30) is proved. $\qquad \square$

Theorem 6.2.4. *If $x \in D$ is the barycentre of the measure μ, then the inequality*

$$\|F_Y(f) - f(x)\|_Y \leq \|f(x)\|_Y \, |F_{\mathbb{R}}(e_0) - 1|$$

$$+ \left[p F_{\mathbb{R}}(e_0) + \frac{1}{2} h^{-2} F_{\mathbb{R}}(\| \cdot - x \|_X^2) \right] \omega_2(f, h) \quad (6.33)$$

holds for all $f \in C(D, Y)$ and $h > 0$.

Proof. Since we have

$$\|F_Y(f) - f(x)\|_Y \leq \|f(x)\|_Y \, |F_{\mathbb{R}}(e_0) - 1| + \|F_Y(f) - F_{\mathbb{R}}(e_0) f(x)\|_Y,$$

it remains to show that

$$\|F_Y(f) - F_{\mathbb{R}}(e_0) f(x)\|_Y \leq \left[p F_{\mathbb{R}}(e_0) + \frac{1}{2} h^{-2} F_{\mathbb{R}}(\| \cdot - x \|_X^2) \right] \omega_2(f, h). \quad (6.34)$$

First we prove relation (6.34) in the particular case where F_Y is a discrete operator of the form (6.20). In this case we have $L_{\mathbb{R}}(e_0) = \sum_{i=1}^{n} \rho_i$. For $1 \leq i \leq n$, put $\rho_i' := \rho_i(x) / F_{\mathbb{R}}(e_0)$. We apply Lemma 6.2.3 with these data and obtain the points $b_{i,j} \in D$ and the numbers $\eta_{i,j} \geq 0, \, 1 \leq i \leq s, \, 1 \leq j \leq p + 1$ that satisfy the conditions in this lemma, (with ρ_i' instead of ρ_i.) Consequently, from Lemma 6.2.8 we get

$$\|F_Y(f) - F_\mathbb{R}(e_0)f(x)\|_Y = F_\mathbb{R}(e_0)\Big\| \sum_{i=1}^{s} \Big(\sum_{j=1}^{p+1} \eta_{i,j} f(b_{i,j}) - \Big(\sum_{j=1}^{p+1} \eta_{i,j} \Big) f(x) \Big) \Big\|_Y$$

$$\leq F_\mathbb{R}(e_0) \sum_{i=1}^{s} \Big(\sum_{j=1}^{p+1} \eta_{i,j} \Big) \Big\| \sum_{j=1}^{p+1} \frac{\eta_{i,j}}{\Big(\sum\limits_{j=1}^{p+1} \eta_{i,j} \Big)} f(b_{i,j}) - f(x) \Big\|_Y$$

$$\leq F_\mathbb{R}(e_0) \sum_{i=1}^{s} \Big(\sum_{j=1}^{p+1} \eta_{i,j} \Big) \Big[p + \frac{1}{2} h^{-2} \sum_{j=1}^{p+1} \frac{\eta_{i,j}}{\Big(\sum\limits_{j=1}^{p+1} \eta_{i,j} \Big)} \|x - b_{i,j}\|_X^2 \Big] \omega_2(f, h)$$

$$= F_\mathbb{R}(e_0) \Big[p + \frac{1}{2} h^{-2} \sum_{k=1}^{n} \rho_k \|a_k - x\|_X^2 \Big] \omega_2(f, h)$$

$$= \Big[p F_\mathbb{R}(e_0) + \frac{1}{2} h^{-2} F_\mathbb{R}(\| \cdot - x\|_X^2) \Big] \omega_2(f, h). \tag{6.35}$$

Now the proof of relation (6.34) for general Borel positive measure μ follows from above, approximating the functionals F_Y and $F_\mathbb{R}$ by discrete functionals and using Lemma 6.2.1. □

When we revert to the condition that x is the barycentre of μ we derive a weaker estimate.

Theorem 6.2.5. *We have*

$$\|F_Y(f) - f(x)\|_Y \leq \|f(x)\|_Y |F_\mathbb{R}(e_0) - 1| + (F_\mathbb{R}(e_0) + h^{-1}\|F_X(\cdot - x)\|_X)\omega_1(f, h)$$

$$+ \Big[p F_\mathbb{R}(e_0) + \frac{1}{2} h^{-2} F_\mathbb{R}(\| \cdot - x\|_X^2) \Big] \omega_2(f, h), \tag{6.36}$$

for all $f \in C(D, Y)$, $x \in D$ and $h > 0$.

Proof. Denote by z the barycentre of the measure μ. Using Theorem 6.2.4, and Lemma 6.2.6 we have

$$\|F_Y(f) - f(x)\|_Y \leq \|f(x)\|_Y |F_\mathbb{R}(e_0) - 1| + \|F_Y(f) - F_\mathbb{R}(e_0)f(x)\|_Y$$

$$\leq \|f(x)\|_Y |F_\mathbb{R}(e_0) - 1| + F_\mathbb{R}(e_0)\|f(x) - f(z)\|_Y + \|F_Y(f) - F_\mathbb{R}(e_0)f(z)\|_Y$$

$$\leq \|f(x)\|_Y |F_\mathbb{R}(e_0) - 1| + F_\mathbb{R}(e_0)(1 + h^{-1}\|x - z\|_X)\omega_1(f, h)$$

$$+ \Big[p F_\mathbb{R}(e_0) + \frac{1}{2} h^{-2} F_\mathbb{R}(\| \cdot - z\|_X^2) \Big] \omega_2(f, h)$$

$$\leq \|f(x)\|_Y |F_\mathbb{R}(e_0) - 1| + (F_\mathbb{R}(e_0) + h^{-1}\|F_X(\cdot - x)\|_X)\omega_1(f, h)$$

$$+ \Big[p F_\mathbb{R}(e_0) + \frac{1}{2} h^{-2} F_\mathbb{R}(\| \cdot - x\|_X^2) \Big] \omega_2(f, h). \quad □$$

For the modulus ω_2^*, in a similar way, starting with the generalization of the inequalities given in Lemmas 2.2.1, 2.2.4 and relation (2.54) given in Theorem 2.2.3, when $\mathbb{R}$ is replaced with the Banach space Y, we arrive at the following estimates.

Theorem 6.2.6. *With the above notation we have:*
i) If x is the barycentre of the measure μ, then

$$\|F_Y(f) - f(x)\|_Y \leq \|f(x)\|_Y |F_\mathbb{R}(e_0) - 1|$$
$$+ \left[pF_\mathbb{R}(e_0) + h^{-2} F_\mathbb{R}(\| \cdot - x\|_X^2) \right] \omega_2^\star(f, h) \quad (6.37)$$

holds for all $f \in C(D, Y)$ and $h > 0$.
ii) In the general case we have

$$\|F_Y(f) - f(x)\|_Y \leq \|f(x)\|_Y |F_\mathbb{R}(e_0) - 1|$$
$$+ (F_\mathbb{R}(e_0) + h^{-1}\|F_X(\cdot - x)\|_X)\omega_1(f, h)$$
$$+ \left[pF_\mathbb{R}(e_0) + h^{-2} F_\mathbb{R}(\| \cdot - x\|_X^2) \right] \omega_2^\star(f, h) \quad (6.38)$$

holds, for all $f \in C(D, Y)$, $x \in D$ and $h > 0$.

Remark 6.2.2. In the case where $X = \mathbb{R}$, $D = [a, b]$, $Y = \mathbb{R}$, and $F(e_0) = 1$, $F(e_1) = x$, estimate (6.33) coincides with (2.48) and estimate (6.37) coincides with (2.51), for $s = 2$. Consequently, the constants $\frac{1}{2}$ and 1, which appear respectively, in front of the term $h^{-2} F_\mathbb{R}(\| \cdot - x\|_X^2)$ are optimal.

From these theorems and Lemma 6.2.2, we obtain immediate corollaries for the linear operators $L_Y : C(D, Y) \to \mathcal{F}(D, Y)$, induced by a family of positive Borel measures $\{\mu_x\}_{x \in D}$. We illustrate only for the modulus ω_2.

Corollary 6.2.1. *Suppose that the conditions*
i) $L_\mathbb{R}(e_0) = e_0$ and
ii) $L_\mathbb{R}(\varphi) = \varphi$, for any $\varphi \in X^\star$,
are satisfied. Then,

$$\|L_Y(f, x) - f(x)\|_Y \leq \left(pL_\mathbb{R}(e_0, x) + \frac{1}{2} h^{-2} L_\mathbb{R}(\| \cdot - x\|_X^2, x) \right) \omega_2(f, h) \quad (6.39)$$

for $f \in C(D, Y)$, $h > 0$, $x \in D$.

Corollary 6.2.2. *In the general case we have*

$$\|L_Y(f, x) - f(x)\|_Y \leq \|f(x)\|_Y \cdot |L_\mathbb{R}(e_0, x) - 1|$$
$$+ \left(L_\mathbb{R}(e_0, x) + h^{-1}\|L_X((\cdot - x), x)\|_X \right) \omega_1(f, h)$$
$$+ \left(pL_\mathbb{R}(e_0, x) + \frac{1}{2} h^{-2} L_\mathbb{R}(\| \cdot - x\|_X^2, x) \right) \omega_2(f, h) \quad (6.40)$$

for $f \in C(D, Y)$, $h > 0$, $x \in D$.

Now let there be a sequence of linear positive operators $((L_n)_Y)_n$, induced on the space $C(D, Y)$ by a family of Borel positive measure $\{\mu_x^n\}_{x \in D, n \in \mathbb{N}}$. Define

$$\Psi_n(x) := (L_n)_\mathbb{R}(\| \cdot - x\|_X^2, x), \ x \in D. \quad (6.41)$$

We derive the following theorems of convergence of Popoviciu type.

Theorem 6.2.7. *Suppose that the conditions*

    i) $(L_n)_{\mathbb{R}}(e_0) = e_0$,

    ii) $(L_n)_{\mathbb{R}}(\varphi) = \varphi$, *for all* $\varphi \in X^*$,

    iii) $\lim_{n\to\infty} \|\Psi_n\|_\infty = 0$

hold. Then we have

$$\lim_{n\to\infty} \|(L_n)_Y(f) - f\|_\infty = 0, \ for \ all \ f \in C(D, Y). \tag{6.42}$$

Proof. We apply Corollary 6.2.1 by taking $L = L_n$ and $h = \sqrt{\|\Psi_n\|_\infty}$. $\qquad\square$

Theorem 6.2.8. *Suppose that*

    i) $\lim_{n\to\infty} \|(L_n)_{\mathbb{R}}(e_0) - e_0\|_\infty = 0$,

    ii) $\lim_{n\to\infty} \|\Psi_n\|_\infty = 0$.

Then relation (6.42) holds.

Proof. From condition i) it follows that there exists a constant $M > 1$ such that $\|L_n(e_0)\|_\infty \leq M$, for all $n \in \mathbb{N}$. From the Schwartz inequality we obtain for any $x \in D$:

$$\|(L_n)_X(\cdot - x, x)\|_X \leq (L_n)_{\mathbb{R}}(\|\cdot - x\|_X, x) \leq \sqrt{\Psi_n(x)}\sqrt{(L_n)_{\mathbb{R}}(e_0, x)} \leq \sqrt{M}\sqrt{\|\Psi_n\|_\infty}.$$

We apply inequality (6.40) for $L := L_n$ and $h := \sqrt{\|\Psi_n\|_\infty}$ and we get

$$\|(L_n)_Y(f) - f\|_\infty \leq \|f\|_\infty \|L_n(e_0) - e_0\|_\infty + 2M\omega_1\left(f, \sqrt{\|\Psi_n\|_\infty}\right)$$

$$+(pM+1)\omega_2\left(f, \sqrt{\|\Psi_n\|_\infty}\right).$$

If we allow $n \to \infty$ we obtain relation (6.42). $\qquad\square$

From Theorem 6.2.8 we can deduce as corollaries, Bohman–Korovkin type theorems, which use a finite number of test functions. For this, let $\{u_1, \ldots, u_p\}$ be a fixed orthonormal base of the space X. Consider additionally the functions $\sigma : X \to \mathbb{R}$, and $\pi_i : X \to \mathbb{R}$, $1 \leq i \leq p$, defined by

$$\sigma(x) := \sum_{i=1}^{p} (x_i)^2, \ \text{ and } \ \pi_i(x) := x_i, \ \ \left(x = \sum_{i=1}^{p} x_i u_i \in X\right). \tag{6.43}$$

We have

Corollary 6.2.3. *If the conditions*

    i) $\lim_{n\to\infty} \|(L_n)_{\mathbb{R}}(e_0) - e_0\|_\infty = 0$,

    ii) $\lim_{n\to\infty} \|(L_n)_{\mathbb{R}}(\pi_i) - \pi_i\|_\infty = 0, \ (1 \leq i \leq p)$

    iii) $\lim_{n\to\infty} \|(L_n)_{\mathbb{R}}(\sigma) - \sigma\|_\infty = 0$

are satisfied, then relation (6.42) holds.

Proof. Conditions i) ii) and iii) imply condition ii) in Theorem 6.2.8. Indeed, let $R > 0$ be such that $D \subset B(0, R)$. Then for any $x = \sum_{i=1}^{p} x_i u_i \in E$ we have

$$|\Psi_n(x)| = \left| (L_n)_{\mathbb{R}}(\sigma, x) - 2 \sum_{i=1}^{p} x_i (L_n)_{\mathbb{R}}(\pi_i, x) + \sum_{i=1}^{p} (x_i)^2 (L_n)_{\mathbb{R}}(e_0, x) \right|$$

$$\leq |(L_n)_{\mathbb{R}}(\sigma, x) - \sigma(x)| + 2 \sum_{i=1}^{p} |x_i| \, |(L_n)_{\mathbb{R}}(\pi_i, x) - \pi_i(x)|$$

$$+ \sum_{i=1}^{p} (x_i)^2 |(L_n)_{\mathbb{R}}(e_0, x) - e_0(x)|$$

$$\leq \|(L_n)_{\mathbb{R}}(\sigma) - \sigma\|_\infty + 2R \sum_{i=1}^{p} \|L_n(\pi_i) - \pi_i\|_\infty + R^2 \sum_{i=1}^{p} \|(L_n)_{\mathbb{R}}(e_0) - e_0\|_\infty. \quad \square$$

Another form of Corollary 6.2.3 with a stronger hypothesis is the following.

Corollary 6.2.4. *If the conditions*

$$\lim_{n \to \infty} \|(L_n)_{\mathbb{R}}((\pi_i)^j) - (\pi_i)^j\|_\infty = 0, \ 1 \leq i \leq p, \ 0 \leq j \leq 2, \tag{6.44}$$

hold, then relation (6.42) holds.

Examples 6.2.1. Let $X = \mathbb{R}^p$, $D_1 := [0, 1]^p$ and Y be an arbitrary Banach space. Consider the multidimensional Bernstein operators $(B_n^1)_Y : C(D_1, Y) \to C(D_1, Y)$,

$$(B_n^1)_Y(f, x) := \sum_{k_1=0}^{n} \cdots \sum_{k_p=0}^{n} p_{n,k_1}(x_1) \cdots p_{n,k_p}(x_p) f\left(\frac{k_1}{n}, \ldots, \frac{k_p}{n}\right), \tag{6.45}$$

$f \in C(D_1, Y)$, $x = (x_1, \ldots, x_p) \in D_1$, $n \in \mathbb{N}$.
We have for any $x = (x_1, \ldots, x_p) \in D_1$:

$$(B_n^1)_{\mathbb{R}}(e_0, x) = 1, \quad (B_n^1)_{\mathbb{R}}(\mathrm{id}, x) = x,$$

$$(B_n^1)_{\mathbb{R}}(\| \cdot - x \|_{\mathbb{R}^p}^2, x) = \sum_{j=1}^{p} \frac{x_j (1 - x_j)}{n}. \tag{6.46}$$

From relation (6.39) we obtain for any $f \in C(D_1, Y)$, $x = (x_1, \ldots, x_p) \in D_1$, $n \in \mathbb{N}$ and $h > 0$:

$$\|(B_n^1)_Y(f, x) - f(x)\|_Y \leq \left(p + \frac{1}{2} \cdot h^{-2} \sum_{j=1}^{p} \frac{x_j (1 - x_j)}{n} \right) \omega_2(f, h). \tag{6.47}$$

Consequently, we have

$$\|(B_n^1)_Y(f) - f\|_\infty \leq \frac{9}{8} p \cdot \omega_2\left(f, \frac{1}{\sqrt{n}}\right), \tag{6.48}$$

for any $f \in C(D_1, Y)$ and $n \in \mathbb{N}$.

A second type of multidimensional Bernstein operators can be defined on the simplex $D_2 := \{x = (x_1, \dots, x_p) \,|\, x_i \geq 0, \; x_1 + \dots + x_p \leq 1\}$ in the following mode:

$$(B_n^2)_Y(f, x) \tag{6.49}$$

$$:= \sum_{k_1 + \dots + k_p \leq n, \; k_i \geq 0} \binom{n}{k_1 \dots k_p} (x_1)^{k_1} \dots (x_p)^{k_p} (1 - x_1 - \dots - x_p)^{n-k_1-\dots-k_p}$$

$$\cdot f\left(\frac{k_1}{n}, \dots, \frac{k_p}{n}\right),$$

$$\binom{n}{k_1 \dots k_p} := \frac{n!}{k_1! \dots k_p!(n - k_1 - \dots - k_p)!}, \quad f \in C(D_2, Y), \tag{6.50}$$

$$x = (x_1, \dots, x_p) \in D_2, \; n \in \mathbb{N}.$$

Similar relations to (6.46) hold for B_n^2 and any $x = (x_1, \dots, x_p) \in D_2$. We obtain

$$\|(B_n^2)_Y(f, x) - f(x)\|_Y \leq \left(p + \frac{1}{2} \cdot h^{-2} \sum_{j=1}^{p} \frac{x_j(1 - x_j)}{n}\right) \omega_2(f, h), \tag{6.51}$$

for any $f \in C(D_2, Y)$, $x = (x_1, \dots, x_p) \in D_2$, $n \in \mathbb{N}$ and $h > 0$. Consequently, we have

$$\|(B_n^2)_Y(f) - f\|_\infty \leq \left(p + \frac{1}{2}\right) \cdot \omega_2\left(f, \frac{1}{\sqrt{n}}\right), \tag{6.52}$$

for any $f \in C(D_2, Y)$ and $n \in \mathbb{N}$.

6.2.4 Estimates with modulus $\tilde{\omega}_2$

In the previous subsection we considered second order moduli constructed using colinear points and the given estimates depend on the dimension of the space X and are not applicable to infinite dimensional spaces. In order to obtain estimates for general Banach spaces X, we consider now a "global" second order modulus.

In this subsection, without an express supposition, X will be a **Banach** space. When we specify, X will be a **Hilbert** space. Also, Y will be an arbitrary Banach space, $D \subset X$ will be compact and convex and μ will be a Borel positive measure on D.

Definition 6.2.6. *Set*

$$\tilde{\omega}_2(f, h) = \sup \left\{ \left\| \sum_{i=1}^{n} \lambda_i f(x_i) - f(x) \right\|_Y, \; x \in D, \; x_i \in D, \; \|x_i - x\|_X \leq h, \right.$$

$$x = \lambda_1 x_1 + \dots + \lambda_n x_n$$

$$\left. \lambda_i \in (0, 1), \; (1 \leq i \leq n), \; \lambda_1 + \dots + \lambda_n = 1 \right\}, \tag{6.53}$$

for $f \in C(D, Y)$, $h > 0$.

Remark 6.2.3. Obviously, we have

$$\omega_2^\star(f, h) \leq \tilde{\omega}_2(f, h), \quad \text{for all } f \in C(D, Y), \ h > 0. \tag{6.54}$$

Also, in the particular case $X = \mathbb{R}, Y = \mathbb{R}, D = I, I$ an interval, we have

$$\omega_2^\star(f, h) = \tilde{\omega}_2(f, h), \quad \text{for all } f \in C(I), \ h > 0. \tag{6.55}$$

Lemma 6.2.9. *Let $x, y \in D$ and the number $h > 0$, such that $\|x - y\|_X \geq h$. Define $v = \frac{y-x}{\|y-x\|_X}$. Let $f \in C(D, Y)$ be such that $\|f(x + hv) - f(x)\|_Y \leq \omega_2^\star(f, h)$. Then*

$$\|f(y) - f(x + h)\|_Y \leq h^{-2}\|y - x\|_X^2 \omega_2^\star(f, h). \tag{6.56}$$

Proof. Consider the interval $I = \{t \in \mathbb{R} \, | \, x + tv \in D\}$ and let the function $g \in C(I, Y)$, $g(t) = f(x + tv), \ t \in I$. Then we have $\|g(h) - g(0)\|_Y \leq \omega_2^\star(g, h)$. There is $t_y \in I, \ t_y \geq h$, such that $y = x + t_y v$. Then, similarly as in the proof of Lemma 2.2.4, by replacing the absolute value with the norm, we deduce

$$\|g(t_y) - g(h)\|_Y \leq h^{-2}(t_y)^2 \omega_2^\star(g, h).$$

From this we obtain relation (6.56). □

Theorem 6.2.9. *Suppose that $\mu(D) = 1$ and $x \in D$ is the barycentre of μ. Then the functional $F_Y : C(D, Y) \to Y$, induced by the measure μ, satisfies the estimate*

$$\|F_Y(f) - f(x)\|_Y \leq \left(F_{\mathbb{R}}(e_0) + h^{-2}F_{\mathbb{R}}(\|\cdot - x\|_X^2)\right) \tilde{\omega}_2(f, h), \ f \in C(D, Y), \ h > 0. \tag{6.57}$$

Proof. First we prove relation (6.57) in the particular case where F_Y is of the form

$$F_Y(f) = \sum_{j=1}^{r+1} v_j f(b_j), \ f \in C(D, Y), \tag{6.58}$$

where we have

a) $v_j > 0, 1 \leq j \leq r + 1$ and $v_1 + \cdots + v_{r+1} = 1$,

b) $b_j \in D \setminus \{x\}, 1 \leq j \leq r + 1$ and the vectors $b_1 - x, \ldots, b_r - x$ are linearly independent,

c) $x = v_1 b_1 + \cdots + v_{r+1} b_{r+1}$.

Note that conditions a) and c) are equivalent to the conditions given in the theorem in the case of discrete functionals.

Consider, in the beginning, the case $r \geq 1$. For any index $1 \leq j \leq r + 1$, define the point c_j in the following way:

$$c_j = \begin{cases} b_j, & \text{if } \|b_j - x\|_X \leq h, \\ x + h(b_j - x)/\|b_j - x\|_X & \text{if } \|b_j - x\|_X > h. \end{cases}$$

We have $x \in conv\{c_1, \ldots, c_{r+1}\}$. Hence, there are the numbers $q_1, \ldots, q_{r+1} \geq 0$, such that $q_1 + \cdots + q_{r+1} = 1$ and $q_1 c_1 + \cdots + q_{r+1} c_{r+1} = x$. Then, because $\|c_j - x\|_X \leq h$, for $1 \leq j \leq r + 1$, we obtain

$$\left\| \sum_{j=1}^{r+1} q_j f(c_j) - f(x) \right\|_Y \le \tilde{\omega}_2(f, h). \tag{6.59}$$

Define $X_1 := Sp\{b_1 - x, \dots, b_r - x\}$. Consider $\varphi_1 : X_1 \to Y$ to be the linear function given by the conditions

$$\varphi_1(c_j - x) = \sum_{j=1}^{r+1} q_j f(c_j) - f(c_j), \ 1 \le j \le r.$$

Then let $\varphi : X \to Y$ be an arbitrary linear function which extends the function φ_1 to the whole space X. Define $g \in C(D, Y)$, by

$$g = f + \varphi(\cdot - x).$$

Since the function $\varphi(\cdot - x)$ is affine, we obtain

$$\tilde{\omega}_2(f, h) = \tilde{\omega}_2(g, h) \quad \text{and} \quad \sum_{j=1}^{r+1} q_j f(c_j) - f(x) = \sum_{j=1}^{r+1} q_j g(c_j) - g(x). \tag{6.60}$$

Also we have $g(c_j) = g(c_{r+1})$, for $1 \le j \le r$. Hence, from (6.59) and (6.60) we obtain, for $1 \le j \le r + 1$

$$\|g(c_j) - g(x)\|_Y = \left\| \sum_{j=1}^{r+1} q_j g(c_j) - g(x) \right\|_Y = \left\| \sum_{j=1}^{r+1} q_j f(c_j) - f(x) \right\|_Y \le \tilde{\omega}_2(f, h).$$

Therefore it follows that

$$\|g(c_j) - g(x)\|_Y \le \tilde{\omega}_2(g, h). \tag{6.61}$$

We derive for any $1 \le j \le r + 1$ the inequality

$$\|g(b_j) - g(x)\|_Y \le (1 + h^{-2}\|b_j - x\|_X^2) \, \tilde{\omega}_2(g, h).$$

Indeed, in the case $b_j = c_j$ this follows directly from (6.61) and in the opposite case, it follows, by applying Lemma 6.2.9. Consequently, we obtain:

$$\|F_Y(f) - f(x)\|_Y = \left\| \sum_{j=1}^{r+1} v_j g(b_j) - g(x) \right\|_Y$$

$$\le \sum_{j=1}^{r+1} v_j \|g(b_j) - g(x)\|_Y$$

$$\le \sum_{j=1}^{r+1} v_j (1 + h^{-2}\|b_j - x\|_X^2) \, \tilde{\omega}_2(g, h)$$

$$= \left(F_{\mathbb{R}}(e_0) + h^{-2} F_{\mathbb{R}}(\| \cdot - x\|_X^2) \right) \tilde{\omega}_2(f, h). \tag{6.62}$$

If the functional F_Y is of the form (6.58), with $r = 0$, then $b_1 = x$, $\nu_1 = 1$, and the relation (6.62) is also true, obviously.

Now we extend the theorem in the case of the discrete functionals of the form

$$F_Y(f) = \sum_{k=1}^{n} \rho_k f(a_k), \ f \in C(D, Y),$$

where $\rho_1, \ldots, \rho_n > 0$, $\rho_1 + \cdots + \rho_n = 1$ and $a_k \in D$, for $1 \le k \le n$.

We apply Lemma 6.2.4 with these data and we obtain the points $b_{i,j} \in D$ and the numbers $\eta_{i,j} > 0$, $1 \le i \le s$, $1 \le j \le r_i + 1$, $r_i \ge 0$, which satisfy conditions 1), 2), 3) in this lemma. For any $1 \le i \le s$ and $1 \le j \le r_i + 1$, set $\rho_{i,j} = \eta_{i,j} / \left(\sum_{j=1}^{r_i+1} \eta_{i,j} \right)$.

Using the particular case of the theorem already proved, we obtain

$$\|F_Y(f) - f(x)\|_Y = \left\| \sum_{i=1}^{s} \left(\sum_{j=1}^{r_i+1} \eta_{i,j} f(b_{i,j}) - \left(\sum_{j=1}^{r_i+1} \eta_{i,j} \right) f(x) \right) \right\|_Y$$

$$\le \sum_{i=1}^{s} \left(\sum_{j=1}^{r_i+1} \eta_{i,j} \right) \left\| \sum_{j=1}^{r_i+1} \rho_{i,j} f(b_{i,j}) - f(x) \right\|_Y$$

$$\le \sum_{i=1}^{s} \left(\sum_{j=1}^{r_i+1} \eta_{i,j} \right) \sum_{j=1}^{r_i+1} \rho_{i,j} \left(1 + h^{-2} \|b_{i,j} - x\|_X^2 \right) \tilde{\omega}_2(f, h)$$

$$\le \sum_{i=1}^{s} \sum_{j=1}^{r_i+1} \eta_{i,j} \left(1 + h^{-2} \|b_{i,j} - x\|_X^2 \right) \tilde{\omega}_2(f, h)$$

$$= \sum_{k=1}^{n} \rho_k \left(1 + h^{-2} \|a_k - x\|_X^2 \right) \tilde{\omega}_2(f, h)$$

$$= \left(F_{\mathbb{R}}(e_0) + h^{-2} F_{\mathbb{R}}(\| \cdot - x\|_X^2) \right) \tilde{\omega}_2(f, h).$$

Now we prove the theorem in the general case. Let $F_Y : C(D, Y) \to Y$ be a linear positive functional induced by the measure μ and let $f \in C(D, Y)$. Let $\varepsilon > 0$, arbitrarily taken. From the uniform continuity, we can choose a number $\delta > 0$, such that for any $x, y \in D$, if $\|x - y\|_X < \delta$, we have $\|f(x) - f(y)\|_Y < \varepsilon$. Consider $\{D_1, \ldots, D_n\}$ a δ-compatible partition of D, with regard to the measure μ. Then choose the points $a_k \in D_k$ and denote $\rho_k = \mu(D_k)$, for $1 \le k \le n$. We eliminate from the partition the sets D_k, for which $\rho_k > 0$. If we consider the functional

$$F_Y^1(g) = \sum_{k=1}^{n} \rho_k g(a_k),$$

there follows from above the inequality

$$\|F_Y^1(f) - f(x)\|_Y \le \left(F_{\mathbb{R}}^1(e_0) + h^{-2} F_{\mathbb{R}}^1(\| \cdot - x\|_X^2) \right) \tilde{\omega}_2(f, h).$$

On the other hand, denoting $F_{D_k,Y}(g) = \int_{D_k} g \, d\mu$, we have

$$\|F_Y(f) - F_Y^1(f)\|_Y = \left\| \sum_{k=1}^{n} F_{D_k,Y}((f(\cdot) - f(a_k))|_{D_k}) \right\|_Y$$

$$\leq \sum_{k=1}^{n} \|F_{D_k,Y}((f(\cdot) - f(a_k))|_{D_k})\|_Y$$

$$\leq \sum_{k=1}^{n} F_{D_k,\mathbb{R}}(\|f|_{D_k}(\cdot) - f(a_k)\|_Y)$$

$$< \varepsilon \sum_{k=1}^{n} \mu(D_k) = \varepsilon.$$

In a similar mode we obtain

$$\left| \left(F_{\mathbb{R}}(e_0) + h^{-2} F_{\mathbb{R}}(\| \cdot - x\|_X^2) \right) \tilde{\omega}_2(f, h) - \left(F_{\mathbb{R}}^1(e_0) + h^{-2} F_{\mathbb{R}}^1(\| \cdot - x\|_X^2) \right) \tilde{\omega}_2(f, h) \right|$$
$$< \varepsilon^2 h^{-2} \tilde{\omega}_2(f, h).$$

Since $\varepsilon > 0$ was taken arbitrarily we obtain relation (6.57).   □

In the case where X is a Hilbert space and the norm $\| \cdot \|_X$ is generated by its inner product, we can derive a more general estimate.

Theorem 6.2.10. *Suppose that X is Hilbert space. The functional F_Y induced by the measure μ satisfies the estimate*

$$\|F_Y(f) - f(x)\|_Y \leq \|f(x)\|_Y \cdot |F_{\mathbb{R}}(e_0) - 1|$$
$$+ \left(F_{\mathbb{R}}(e_0) + h^{-1}\|F_X(\cdot - x)\|_X \right) \omega_1(f, h)$$
$$+ \left(F_{\mathbb{R}}(e_0) + h^{-2} F_{\mathbb{R}}(\| \cdot - x\|_X^2) \right) \tilde{\omega}_2(f, h) \qquad (6.63)$$

for $f \in C(D, Y)$, $h > 0$, $x \in D$.

Proof. First we have

$$\|F_Y(f) - f(x)\|_Y \leq \|f(x)\|_Y \cdot |F_{\mathbb{R}}(e_0) - 1| + \|F_Y(f - f(x))\|_Y.$$

Denote by z the barycentre of the measure μ. Also, let the functional

$$F_Y^1(g) = \frac{1}{F_{\mathbb{R}}(e_0)} \cdot F_Y(g), \quad f \in C(D, Y).$$

Using Theorem 6.2.9 and Lemma 6.2.6 we obtain

$$\|F_Y^1(f) - f(z)\|_Y \leq \left(F_{\mathbb{R}}^1(e_0) + h^{-2} F_{\mathbb{R}}^1(\| \cdot - z\|_X^2) \right) \tilde{\omega}_2(f, h)$$
$$\leq \left(F_{\mathbb{R}}^1(e_0) + h^{-2} F_{\mathbb{R}}^1(\| \cdot - x\|_X^2) \right) \tilde{\omega}_2(f, h).$$

Also we have $\|f(z) - f(x)\|_Y \leq \left(1 + h^{-1}\|z - x\|_X \right) \omega_1(f, h)$. Finally, using the relations above it follows that

$$\|F_Y(f - f(x))\|_Y = F_{\mathbb{R}}(e_0)\|F_Y^1(f - f(x))\|_Y$$
$$\leq F_{\mathbb{R}}(e_0)\|F_Y^1(f) - f(z)\|_Y + F_{\mathbb{R}}(e_0) \cdot \|f(z) - f(x)\|_Y$$
$$\leq F_{\mathbb{R}}(e_0) \cdot \left(F_{\mathbb{R}}^1(e_0) + h^{-2}F_{\mathbb{R}}^1(\| \cdot -x\|_X^2)\right)\tilde{\omega}_2(f, h)$$
$$+ F_{\mathbb{R}}(e_0) \cdot \left(F_{\mathbb{R}}^1(e_0) + h^{-1}\|F_X^1(\cdot - x)\|_X\right)\omega_1(f, h).$$

Then from the definition of the functional F^1 we obtain (6.63).   □

Remark 6.2.4. The constant 1 which appears in the estimates (6.57) and (6.63) in front of the terms $\| f(x)\|_Y \cdot |F_{\mathbb{R}}(e_0) - 1|$, $h^{-1}\|F_X(\cdot - x)\|_X \omega_1(f, h)$, $F_{\mathbb{R}}(e_0)\tilde{\omega}_2(f, h)$ and $h^{-2}F_{\mathbb{R}}(\| \cdot -x\|_X^2)\tilde{\omega}_2(f, h)$ are optimal. This fact can be seen by considering the particular case $X = \mathbb{R}$, $Y = \mathbb{R}$, $D = [a, b]$, see relation (2.51).

Now we derive results for operators. Let $L_Y : C(D, Y) \to \mathcal{F}(D, Y)$ be the linear operator induced by the family of positive Borel measures $\{\mu_x\}_{x\in D}$. From Theorems 6.2.9 and 6.2.10, see also Lemma 6.2.2, we obtain

Corollary 6.2.5. *Suppose that the conditions*
i) $L_{\mathbb{R}}(e_0) = e_0$, *and*
ii) $L_{\mathbb{R}}(\varphi) = \varphi$, *for any* $\varphi \in X^\star$
are satisfied. Then,

$$\|L_Y(f, x) - f(x)\|_Y \leq \left(L_{\mathbb{R}}(e_0, x) + h^{-2}L_{\mathbb{R}}(\| \cdot -x\|_X^2, x)\right)\tilde{\omega}_2(f, h), \quad (6.64)$$

for $f \in C(D, Y)$, $h > 0$, $x \in D$.

Corollary 6.2.6. *Suppose additionally that X is a Hilbert space. Then*

$$\|L_Y(f, x) - f(x)\|_Y \leq \|f(x)\|_Y \cdot |L_{\mathbb{R}}(e_0, x) - 1|$$
$$+ \left(L_{\mathbb{R}}(e_0, x) + h^{-1}\|L_X((\cdot - x), x)\|_X\right)\omega_1(f, h)$$
$$+ \left(L_{\mathbb{R}}(e_0, x) + h^{-2}L_{\mathbb{R}}(\| \cdot -x\|_X^2, x)\right)\tilde{\omega}_2(f, h) \quad (6.65)$$

for $f \in C(D, Y)$, $h > 0$, $x \in D$.

For the sequences of linear positive operators $((L_n)_Y)_n$ induced by families of Borel positive measures, we obtain the following therems of convergence, which generalize Theorems 6.2.7 and 6.2.8. The proofs are similar.

Theorem 6.2.11. *Let X be an arbitrary Banach space. Suppose that the conditions*
i) $(L_n)_{\mathbb{R}}(e_0) = e_0$,
ii) $(L_n)_{\mathbb{R}}(\varphi) = \varphi$, *for all* $\varphi \in X^\star$,
iii) $\lim_{n\to\infty} \|\Psi_n\|_\infty = 0$
hold. Then, relation (6.42) holds.

Theorem 6.2.12. *Suppose that X is a Hilbert space. Also, suppose that we have*
i) $\lim_{n\to\infty} \|(L_n)_{\mathbb{R}}(e_0) - e_0\|_\infty = 0$,
ii) $\lim_{n\to\infty} \|\Psi_n\|_\infty = 0$.
Then relation (6.42) holds.

Examples 6.2.2. Let the multidimensional Bernstein operators $(B_n^1)_Y : C(D_1, Y) \rightarrow C(D_1, Y)$, be as defined in Examples 6.2.1. From relation (6.57) we obtain

$$\|(B_n^1)_Y(f, x) - f(x)\|_Y \leq \left(1 + h^{-2} \sum_{j=1}^{p} \frac{x_j(1 - x_j)}{n} \right) \tilde{\omega}_2(f, h), \qquad (6.66)$$

for any $f \in C(D_1, Y)$, $x = (x_1, \ldots, x_p) \in D_1, n \in \mathbb{N}$ and $h > 0$. Consequently, we have

$$\|(B_n^1)_Y(f) - f\|_\infty \leq \left(1 + \frac{p}{4} \right) \tilde{\omega}_2 \left(f, \frac{1}{\sqrt{n}} \right), \quad f \in C(D_1, Y), \ n \in \mathbb{N}. \qquad (6.67)$$

Also, for the operators $(B_n^2)_Y : C(D_2, Y) \rightarrow C(D_2, Y)$, we obtain

$$.\|(B_n^2)_Y(f, x) - f(x)\|_Y \leq \left(1 + h^{-2} \sum_{j=1}^{p} \frac{x_j(1 - x_j)}{n} \right) \tilde{\omega}_2(f, h), \qquad (6.68)$$

for any $f \in C(D_2, Y)$, $x = (x_1, \ldots, x_p) \in D_2, n \in \mathbb{N}$ and $h > 0$. Consequently,

$$\|(B_n^2)_Y(f) - f\|_\infty \leq 2\, \tilde{\omega}_2 \left(f, \frac{1}{\sqrt{n}} \right), \quad f \in C(D_2, Y), \ n \in \mathbb{N}. \qquad (6.69)$$

References

1. U. Abel and M. Ivan, Durrmeyer variants of Bleimann, Butzer and Hahn operators, In: *Mathematical Analysis and Approximation Theory*, (Proceeding of RoGer seminar 2002; ed. by A. Lupaş), Burg Verlag, Sibiu (2002), pp. 1–8.
2. J.A. Adell and J. de la Cal, Preservation of moduli of continuity by Bernstein-type operators, In: *Approximation, Probability and Related Fields*, (ed. by G. A. Anastassiou and S. T. Rachev), Plenum Press, New York, (1994), 1–18.
3. J. Adell and A. Pérez-Palomares, Global smothness preservation for the Bernstein polynomials, In: *Approximation and optimization*, (Proc. Int. Conf. Approximation and Optimization, Cluj-Napoca 1996; ed. by D.D. Stancu et al.), vol I, Transilvania Press, Cluj-Napoca, (1997).
4. O. Agratini, *Approximation by Linear Operators* (in Romanian), Cluj Univ. Press, (2000).
5. F. Altomare and M. Campiti, *Korovkin-type Approximation Theory and its Applications*, Walter de Gruyter, Berlin, New York, (1994).
6. G.A. Anastassiou and S. Gal, *Approximation Theory: Moduli of Continuity and Global Smoothness Preservation*, Birkhäuser, Boston, Basel, Berlin, (2000).
7. O. Aramă, Properties concerning the monotony of the sequence of polynomials of S. Bernstein, (in Romanian), *Studii şi cercetări (Cluj)* **8** no. 3-4, (1957), 195–210.
8. C. Badea, On a Korovkin type theorem for simultaneous approximation, *J. Approx. Theory* **62**, (1990), 223–234.
9. C. Badea, I. Badea and H. H. Gonska, Improved estimates on simultaneous approximation by Bernstein operators, *Rev. Anal. Numér. Théor. Approx.* **22**, (1993), 1–22.
10. B. Bajsanski and R. Bojanic, A note on approximation by Bernstein polynomials, *BAMS* **70**, (1964), 675–677.
11. H. Bavinck , Approximation processes for Fourier-Jacobi expansions, *Applicable Analysis* **5**, (1976), 293–312.
12. H. Berens and G. G. Lorentz, Inverse theorems for Bernstein polynomials, *Indiana University Math. J* **21**, (1972), 693–798.
13. H. Berens, and T. Xu, On Bernstein–Durrmeyer polynomials with Jacobi weights, In: *Approximation Theory and Functional Analysis*, (ed. by C.K. Chui), (1991), pp. 25–43.
14. S.N. Bernstein, Démonstration du théorème de Weierstrass, fondée sur le calcul des probabilités, *Commun. Soc. Math. Kharkow*, (2) **13**, (1912-13), 1–2.
15. G. Bleimann, P.L. Butzer and L. Hahn, A Bernstein-type operator approximating continuous functions on the semi-axis, *Indag. Math.* **42**, (1980), 255–262.
16. H. Bohman, On approximation of continuous and analytic functions, *Ark. Mat.* **2**, (1952), 43–56.
17. H. Brass, Eine Verallgemeinrung der Bernsteinschen Operatoren, *Abhandlugen aus den Math. Seminar der Univ. Hamburg* **36**, (1971), 111–122.
18. Yu.A. Brudnyi, On a certain method for approximation of bounded functions, given on an segment (Russian), In: *Studies in Contemporary Problems in Constructive Theory of Functions*, (Proc. Second All-Union Conf. Baku 1962; ed. by I.I. Ibragimov), Baku: *Izdat. Akad. Nauk Azerbaidžan*, (1965), 40–45.

19. P.L. Butzer and H. Berens, *Semi-groups of Operators and Approximation*, Springer, Berlin, Heidelberg, New York, (1967).

20. Jia-ding Cao, On linear approximation methods (in Chinese), *Acta Sci. Natur. Univ. Fudan* **9** (1964), 43–52.

21. E. Censor, Quantitative results for positive linear approximation operators, *J. Approx. Theory* **4** (1971), 442–450.

22. C. Cişmaşiu, A probabilistic interpretation of Voronovskaja's theorem, *Bul. Univ. Braşov, Ser. C* **27**, (1985), 7–12.

23. C. Cottin and H.H. Gonska, Simultaneous approximation and global smoothness preservation, *Rend. Circ. Mat. Palermo* (2) Suppl. **33**, (1993), 259–279.

24. C.G. Esseen, Über die asymptotisch beste Approximation stetiger Funktionen mit Hilfe von Bernstein-Polynomen, *Numer. Math.* **2** (1960), 206–213.

25. M.M. Derriennic, Sur l'approximation de fonctions intégrables sur [0, 1] par des polynômes de Bernstein modifiés, *J. Approx. Theory* **31**, no. 4, (1981), 325-343.

26. M.M. Derriennic, On multivariate approximation by Bernstein-type, *J. Approx. Theory*, **45**, no. 2, (1985), 155–166.

27. R. De Vore, *The Approximation of Continuous Functions by Positive Linear Operators*, Springer, Berlin, Heidelberg, New York, (1972).

28. R. De Vore and G.G. Lorentz, *Constructive Approximation*, vol. I, Springer, Berlin, (1993).

29. Z. Ditzian, Multidimensional Jacobi-type Bernstein-Durrmeyer operators, *Acta Sci. (Szeged)* **60**, (1995), 225–243.

30. Z. Ditzian and V. Totik, *Moduli of Smoothness*, Springer, New York, (1987).

31. J.L. Durrmeyer, Une formule d'inversion de la transformée de Laplace: applications à la théorie des moments, Thèse de 3ème cycle, Faculté des Sciences Univ. Paris, (1967).

32. M. Felten, Direct and inverse estimates for Bernstein polynomials, *Constructive Approximation*, **14**, (1998), 459–468.

33. G. Freud, On approximation by positive linear methods II, *Studia Sci. Math. Hungar.* **3**, (1968), 365–370.

34. I. Gavrea, The approximation of the continuous functions by means of some positive operators, *Resultate Math.* **30** no. 1-2, (1996), 55–66.

35. I. Gavrea, H. Gonska, R. Păltănea and G. Tachev, General estimates for the Ditzian–Totik modulus, *East Journal of Approximation*, **9**, no. 2, (2003), 175–194.

36. H.H. Gonska, Quantitative Aussagen zur Approximation durch positive lineare Operatoren, Dissertation, Universität Duisburg, (1979).

37. H.H. Gonska, On approximation of continuously differentiable functions by positive linear operators, *Bull. Austral. Math. Soc.* **27**, (1983), 73–81.

38. H.H. Gonska, On approximation in spaces of continuous functions, *Bull. Austral. Math. Soc.* **28**, (1983), 411–432.

39. H.H. Gonska, Two problems on best constants in direct estimates, (Problem Section of Proc. Edmonton Conf. Approximation Theory; ed. by Z. Ditzian et al.), 194 Providence, RI. Amer. Math. Soc. (1983).

40. H.H. Gonska, Quantitative Korovkin type theorems on simultaneous approximation, *Math. Z.* **186**, (1984), 419–433.

41. H.H. Gonska, On approximation by linear operators: Improved estimates, *Anal. Numér. Théor. Approx.* **14** (1985), 7–32.

42. H.H. Gonska, Quantitative Approximation in C(X), Habilitationsschrift, *Schriftenreihe des Fachbereichs Mathematik*, Universität Duisburg, (1985).

43. H.H. Gonska and J. Meier-Gonska, A Bibliography on Approximation of Functions by Bernstein Type Operators, (1955-1982), In: Approximation Theory IV (Proc. Int. Symp. College Station 1983; ed. by C.K. Chui, L.L. Schumaker, and J.D. Ward), New York: Acad Press (1983), 739-785; Supplement 1986, In: Approximation Theory V , New York: Academic Press (1986), 621–653.

44. H.H. Gonska and R. K. Kovacheva: The second order modulus revisited: remarks, applications, problems, *Conf. Sem. Mat. Univ. Bari*, **257** (1994), 1–32.

45. H.H. Gonska and D-x Zhou: On an extremal problem concerning Bernstein operators, *Serdica Math. J.*, (1995), 137–150.

46. H.H. Gonska and G. Tachev, On the constants in ω_2^φ-inequalities, *Rend. Circ. Mat. Palermo* Serie II, Suppl. **68**, (2002), 467–477.

47. T.N.T. Goodman and A. Sharma, A modified Bernstein-Schomberg operator, In: *Constructive Theory of Functions "87"* (ed. by B. Sendov et al.) Bulgarian Academy of Science, Sofia, (1987), pp. 166–173.

48. M. Ivan, Interpolation methods and their applications, (in Romanian), Ph. D. Thesis, Babeş-Bolyai University, Cluj-Napoca (1982).

49. K.G. Ivanov, On Bernstein Polynomials, *C.R. Acad. Bulg.* **35**, (1982), 893–896.

50. M.A. Jiménez Pozo, Quantitative theorems of Korovkin type in bounded function spaces, In: *Constructive Function Theory 1981.* Varna, Bulgar, Acad. Sci. Sofia, (1983), pp. 488–494.

51. H. Johnen, Inequalities connected with moduli of smoothness, *Mat. Vesnik* **3**, (1972), 389–303.

52. D.K. Kacsó, Approximation by means of piecewise linear functions, *Result. Math.* **35**, (1999), 89–102.

53. L.V. Kantorovich, Sur certains développements suivant les polynômes de la forme de S. Bernstein, I, II, *C. R. Acad. SSSR*, (1930), 563–568, 595–600.

54. S. Karlin and W.J. Studden, *Tchebycheff Systems with Applications in Analysis and Statistics*, Interscience, New York, (1966).

55. H.-B. Knoop and P. Pottinger, Ein Satz vom Korovkin-Typ für C^k-Räume, *Math. Z.* **148**, (1976), 23–32.

56. P.P. Korovkin, On convergence of linear positive operators in the space of continuous functions, *Dokl. Akad. Nauk SSSR* **90**, (1953), 961–964.

57. P.P. Korovkin, *Linear Operators and the Theory of Approximation* (Russian), Fizmatgiz, Moscow, (1959).

58. T. Lindvall, Bernstein polynomials and the law of large numbers, *Math. Scientist* **7**, (1982), 127–139.

59. G.G. Lorentz, Bernstein Polynomials, Univ. Press. Toronto, (1953), 2nd ed. New York: Chelsea (1986).

60. G.G. Lorentz, Inequalities and the saturation classes of Bernstein polynomials, In: *On Approximation Theory*, (ed. by P.L. Butzer et al.) Birkhäuser, Basel, (1964), 200–207.

61. G. G. Lorentz and L. L. Schumaker, Saturation of positive operators, *J. Approx. Theory* **5**, (1972), 413–424.

62. A. Lupaş, Die Folge der Betaoperatoren, Dissertation, Universität Stuttgart, (1972).

63. A. Lupaş, Contributions to the theory of approximation by linear operators (in Romanian), Ph. D. Thesis, Babeş-Bolyai University, Cluj-Napoca, (1975).

64. A. Lupaş, The approximation by means of some linear positive operators. In: Approximation Theory (Proc. Int. Dortmund Meeting on Approximation Theory 1995, ed. by M.W. Müller et al.), Berlin: Akad Verlag, (1995), 251–275.

65. R.G. Mamedov, On the order of the approximation of differentiable functions by linear positive operators, (in Russian), *Doklady S.S.S.R.* **128**, (1959), 674–676.

66. R.G. Mamedov, On the asymptotic value of the approximation of repeatedly differentiable functions by linear positive operators (in Russian), *Doklady SSSR* **146**, (1962), 1013–1016.

67. D.H. Mache. A link between Bernstein polynomials and Durrmeyer polynomials with Jacobi weights, In: *Approximation and Interpolation*, (Proc. Approximation Theory VIII; ed. by Ch. Chui and L.I. Schumaker), World Scientific Publishing Co., (1995), pp. 403–410.

68. G. Moldovan, On approximation of continuous functions by Bernstein polynomials, (in Romanian), *Studia Univ. Babeş-Bolyai* Ser. Math.-Phys. **11**, (1966), 63–71.

69. B.Mond, Note: On the degree of approximation by linear positive operators, J. Approx. Theory **18**, (1976), 304–306.

70. B.Mond and R. Vasudevan, On approximation by linear positive operators. *J. Approx. Theory* **30**, (1980), 334–336.

71. G. Mühlbach, Operatoren vom Bernsteinchen Typ, *J. Approx. Th.* **3**, (1970), 274–292.

72. M. W. Müller, Die Folge der Gammaoperatoren, Dissertation, Stuttgart, (1967).

73. A.B. Németh, Korovkin theorem for nonlinear 3-parameter families, *Mathematica (Cluj)* **11(34)**, I, (1969), 135–136.

74. S.M. Nikolski, On the best approximation by polynomials of functions which satisfy a Lipschitz condition (Russian), *Izv. Akad. Nauk SSSR* **10**, (1946), 295–318.

75. T. Nishishiraho, Korovkin type approximation closures for vector-valued functions, *Ryukyu Math. J.* **9**, (1996), 53–69.

76. P.E. Parvanov and B. D. Popov, The limit case of Bernstein's operators with Jacobi-weights, *Mathematica Balcanica* **8**, Fasc. 2-3, (1994), 165–177.

77. R. Păltănea - Sur un operateur polynomial défini sur lénsemble des fonctions intégrables, *Babeş Bolyai Univ., Fac. Math., Res. Semin.* **2**, (1983), 101–106.

78. R. Păltănea, L'éstimation de l'approximation des fonctions continues par les opérateurs de Brass, *Research Semin. Fac. Math. "Babeş-Bolyai" Univ.* **6**, (1984), 261–263.

79. R. Păltănea, L'estimation de l'approximation des dérivées d'ordre r par les polynômes de Brass, *Research Semin. Fac. Math. "Babeş-Bolyai" Univ.* **7**, (1986), 207–210.

80. R. Păltănea, Une propriété d'extrémalité des valeurs propres des opérateurs polynômiaux de Durrmeyer généralisés. *L'Analyse Numér. et la Th. de l'Approx.* **15** (1)(1986), 57–64.

81. R. Păltănea, Une classe générale d'opérateurs polynômiaux. *L'Analyse Numér. et la Th. de l'Approx.* **17** (1) (1988), 49–52.

82. R. Păltănea, Improved constant in approximation with Bernstein operators, *Research Semin. Fac. Math. "Babeş-Bolyai" Univ.* **6**, Univ. Babeş-Bolyai, Cluj-Napoca (1988), 261–268.

83. R. Păltănea, General estimates for linear positive operators that preserve linear functions, *Anal. Numér. Théor. Approx.* **18(2)**, (1989), 147–159.

84. R. Păltănea: On the estimate of the pointwise approximation of functions by linear positive functionals, *Studia Univ. Babeş-Bolyai* **53**, no. 1, (1990), 11–24.

85. R. Păltănea, Une généralization de la notion de convexité, *Research Semin. Fac. Math. "Babeş-Bolyai" Univ.* **6**, (1990), 193–196.

86. R. Păltănea, Approximation operators and their connections with some particular allures (in Romanian), Ph. D. Thesis, Babeş-Bolyai University, Cluj-Napoca, (1992).

87. R. Păltănea, Best constants in estimates with second order moduli of continuity, In: *Approximation Theory*, (Proc. Int. Dortmund Meeting on Approximation Theory 1995, ed. by M.W. Müller, D. Mache, M. Felten), Berlin: Akad. Verlag (1995), pp. 251–275.

88. R. Păltănea, New second order moduli of continuity, In: Approximation and optimization (Proc. Int. Conf. Approximation and Optimization, Cluj-Napoca 1996; ed. by D.D. Stancu et al.), vol I, Transilvania Press, Cluj-Napoca, (1997), pp. 327–334.

89. R. Păltănea, Convexity of higher order that is invariant under symmetries, (Proc. Internat. Conf. on "Symmetry and antisymmetry in mathematics"), Univ. Transilvania Braşov, (1996), 93-96.

90. R. Păltănea, The preservation of the property of the quasiconvexity of higher order by Bernstein operators, *Anal. Numér. Théor. Approx.* **25(1-2)**, (1996), 195–201.

91. R. Păltănea, Optimal estimates with moduli of continuity, *Result. Math.* **32** (1997), 318–331.

92. R. Păltănea, On an optimal constant in approximation by Bernstein operators, *Rend. Circ. Mat. Palermo* **52**, suppl. (1998), 663–686.

93. R. Păltănea, On the invariant subspace of some type of linear operators, *Studii în metode de analiză numerică şi optimizare (Chişinău)* **1** (1), (1998), 44–46.

94. R. Păltănea, On the transformation of the second order modulus by Bernstein operators, *L'Analyse Numér. et la Th. de l'Approx.* **27**, nr. 2, (1998), 309–313.

95. R. Păltănea, Estimates of the degree of approximation with second order moduli of continuity, In: Proceedings of the annual meeting of the Romanian Society of Mathematical Sciences, Cluj-Napoca, 1998, Digital Data Publishing, Cluj (1999), 131–134.

96. R. Păltănea, Saturation theorem for certain sequences of positive linear operators, In: *Analysis, Functional Equations, Approximation and Convexity*, (Proc. of the conference held in honour of professor Elena Popoviciu), Carpatica Press, Cluj-Napoca, (1999), pp. 227–231.

97. R. Păltănea, An improved estimate with the second order modulus of continuity, In: *Proc. of the "Tiberiu Popoviciu" Itinerant Seminar of Functional Equations, Approximation and Convexity*, (ed. by E. Popoviciu), Srima Press, Cluj-Napoca, (2000), 167–171.

98. R. Păltănea R, New type of estimates with moduli of continuity, In: *RoGer Seminar 2000 - Braşov*, (Proc. of the 4th Romanian- German Semin. on Approx. Theory and its Appl.; ed. by H.Gonska and al.), *Schriftenreihe des Fachbereichs Mathematik*, Gerhard Mercator Univ. Duisburg, SM-DU-485, (2000), 110-114.

99. R. Păltanea, On a limit operator, In: *Proc. of the "Tiberiu Popoviciu" Itinerant Seminar of Functional Equations, Approximation and Convexity*, (ed. by E. Popoviciu), Srima Press, Cluj-Napoca (2001), pp. 169–180.

100. R. Păltănea, Two papers on Durrmeyer-type operators, *Schriftenreihe des Fachbereichs Mathematik*, Gerhard Mercator Univ. Duisburg, SM-DU-502, (2001).

101. R. Păltănea, On the estimates with second order modulus with parameter, temporary reference: *Schriftenreihe des Fachbereichs Mathematik*, Gerhard Mercator Univ. Duisburg, SM-DU-507, (2001).

102. R. Păltănea, Generalized Brass operators, temporary reference *Schriftenreihe des Fachbereichs Mathematik*, Gerhard Mercator Univ. Duisburg, SM-DU-518, (2001).

103. R. Păltănea, Approximation of continuous functions by a sequence of generalized Durrmeyer type operators, *Proc. Anual Sess. Romanian Math. Soc.*, Univ. Transilvania Braşov, (2001), 198–201.

104. R. Păltanea, Estimates with generalized second order moduli, In: *Proc. of the "Tiberiu Popoviciu" Itinerant Seminar of Functional Equations, Approximation and Convexity*, (ed. by E. Popoviciu), Srima Press, Cluj-Napoca, (2002), pp. 197–210.

105. R. Păltănea, Estimates with second order moduli , *Rend. Circ. Mat. Palermo* **68** Suppl. (2002), 727–738.

106. R. Păltănea, Approximation by Durrmeyer operators with general weights, In: *Proc. Internat. Symp. on Numerical Analysis and Approximation Theory dedicated to the 75-th Anniversary of Professor Dr. D.D. Stancu*, (ed. R. Trînbiţas), Cluj University Press, (2002), pp. 396–404.

107. R. Păltănea, Estimates of approximation by linear operators in the multidimentional case, In: *Mathematical Analysis and Approximation Theory*, (Proceeding of RoGer seminar 2002; ed. by A. Lupaş, L. Lupaş and H. Gonska), Burg Verlag, Sibiu (2002), 207–220.

108. R. Păltănea, Vector variants of some approximation theorems of Korovkin and of Sendov and Popov, In: *Constructive Theory of Functions Varna 2002*, (ed. by. B.D. Bojanov), Darba Publ. House, Sofia, (2003), 366–373.

109. R. Păltănea, Optimal constant in approximation by Bernstein operators, *J. Comput. Analysis Appl.* **5**, no. 2, Kluwer Academic, (2003), 195–235.

110. R. Păltănea, Estimates for positive linear operators using an arbitrary Chebychev systyem. In: *Proc. of the "Tiberiu Popoviciu" Itinerant Seminar of Functional Equations, Approximation and Convexity*, (ed. by E. Popoviciu), Srima Press, Cluj-Napoca, (2004), 75–84.

111. R. Păltănea, Approximation of functions in Banach spaces using positive linear operators, In: *Mathematical Analysis and Approximation Theory*, (Proceedings of RoGer seminar 2004; ed. by M. Ivan), Mediamira Science Publisher, Cluj-Napoca, (2004), 5–20.

112. J. Peetre, Exact interpolation theorems for Lipschitz continous functions, *Ricerche Mat.* **18**, (1969), 239–259.

113. E. Popoviciu, *Mean Value Theorems in Mathematical Analysis and their Connections with the Theory of Interpolation*, (in Romanian) Dacia Press, Cluj, (1972).

114. E. Popoviciu, Sur une allure de quasi-convexité d'ordre supérieur, *Anal. Numér. Théor. Approx.* **11**, (1982), 129–137.

115. T. Popoviciu: Sur l'approximation des fonctions convexes d'ordre supérieur, *Mathematica* (Cluj) **10**, (1935), 49–54.

116. T. Popoviciu, On the best approximation of continuous functions by polynomials (in Romanian), Inst. Arte Grafice Ardealul, Cluj, (1937).

117. T. Popoviciu, Les fonctions convexes, Herman & Cie, Paris, (1945).

118. T. Popoviciu, On the proof of Weierstrass' theorem using interpolation polynomials (in Romanian), *Lucrările Ses. Gen. Şt. Acad. Române din 1950* **1-4**, (1950), translated in english by D. Kasćo, *East J. Approx.* **4**, no. 1, (1998), 107–110.

119. I. Raşa and T. Vladislav, Numerical Analysis, Splines, Bernstein Operators. Casteljau's Algorithm (in Romania), Technical Press Bucureşti, (1998).

120. A. Sahai and G. Prasad, Sharp estimates of approximation by some positive linear operators, *Bull. Austral. Math. Soc.* **29**, (1984), 13–18.

121. I. J. Schoenberg, On variation diminishing approximation methods, In: *On Numerical Approximation*, (ed. by R.E. Langer), Univ. of Wisconsin Press, Madison, (1959), 249–271.

122. F. Schurer and F.W. Steutel, On the degree of approximation of functions in $C^1[0, 1]$ by Bernstein polynomials, TH-Report 75-WSK-07 (Onderafdeling der Wiskunde, Technische Hogeschool Eindhoven) (1975).

123. F. Schurer and F.W. Steutel, The degree of local approximation of function in $C^1[0, 1]$ by Bernstein polynomials, *J. Approx. Theory* **19** (1977), 69–82.

124. B. Sendov and V. Popov, The convergence of the derivatives of the positive linear operators, (Russian), *C.R.Acad. Bulgare Sci.* **22**, (1969), 507–509.

125. B. Sendov and V. Popov, *The Averaged Moduli of Smoothness*, John Wiley & Sons, New York, (1988).

126. O. Shisha and B. Mond, The degree of convergence of linear positive operators, *Proc. Nat. Acad. Sci. USA*, **60** (1968), 1196–1200.

127. P.C. Sikkema: Über den Grad der Approximation mit Bernstein-Polynomen, *Numerische Math.* **1**, (1959), 221–239.

128. P. C. Sikkema: Der Wert einiger Konstanten in der Theorie der Approximation mit Bernstein-Polynomen, *Numerische Math.* **3** (1961), 107–116.

129. S.P. Sing and O.P. Varshney, On positive linear operators. *J. Orissa Math. Soc.* **1**, (1982), no. 2, 51–56.

130. D.D. Stancu, Sur l'approximation des dérivées des fonctions par les dérivées correspondantes de certains polynômes de type Bernstein, *Mathematica* (Cluj), **2**(25) (1960), 335–348.

131. D.D. Stancu, Approximation of functions by a new class of linear polynomial operators, *Rev. Roumaine Math. Pures Appl.* **13**, no. 8, (1968), 1173–1194.

132. D.D. Stancu, Approximation of functions by means of some new classes of positive linear operators, In: *Numerical Integration* (Proc. Conf. Math. Res. Inst. Oberwolfach 1981; ed. by G. Hämmerlin), Basel: Birkhäuser, (1972), 241–251.

133. E.L. Stark, Bernstein-polynome, 1912-1955. In: *Functional Analysis and Approximation* (Proc. Conf. Math. Res. Inst. Oberwolfach 1980; ed. by P. Butzer, B. Sz.-Nagy, E. Görlich), Basel:Birkhäuser, (1981), 443–461.

134. V. Totik, An interpolation theorem and application to positive operators, *Pacific J. Math.* **111**, 447–481.

135. V. Totik, Approximation by Bernstein polynomials, *American J. Math.* **116**, (1994), 995–1018.

136. O.P. Varshney and S. P. Sing, On degree of approximation by positive linear operators, *Rend. Mat.* **2**(7), no. 1, (1982), 219–225.

137. T. Vladislav and I. Raşa, Numerical analysis, abstract Chauchy's problem, Altomare's projectors, (in Romanian), Technical Press, Bucureşti, (1999).

138. E. Voronovskaja, Determination de la forme asymptotique d'approximation des fonctions par les pôlynomes de S. N. Bernstein, *Dokl. Akad. Nauk SSSR Ser. A*, (1932), 79–85.

139. S. Waldron, A generalized beta integral and the limit of the Bernstein-Durrmeyer operator with Jacobi weights, *J. Approx. Theory*, **122**, no. 1, (2003), 141–150.

140. S. Wigert, Sur l'approximation par polynômes des fonctions continues, *Ark. Math. Astr. Fys.* **22** B (1932), 1–4.

141. D-X. Zhou, On smoothness characterization by Bernstein type operators, *J. Approx. Theory* **81**, (1995), 303–315.

142. D-X. Zhou On a problem of Gonska, *Results Math.* **28**, no. 1-2, (1995), 169–183.

143. Xin-long Zhou, On Bernstein polynomials, (Chinese). *Acta Math. Sinica*, **28** (1985), no. 6, 848–855.

144. V.V. Zhuk, Functions of the Lip 1 class and S.N. Bernstein's polynomials (Russian), *Vestnik Leningrad Univ. Mat. Mekh. Astronom.* (1989), vyp. 1, 25–30.

Index